Sorghum Molecular Breeding

R. Madhusudhana • P. Rajendrakumar
J.V. Patil
Editors

Sorghum Molecular Breeding

Springer

Editors
R. Madhusudhana
Plant Breeding
ICAR-Indian Institute of Millets Research
Hyderabad, Telangana, India

P. Rajendrakumar
Biotechnology
ICAR-Indian Institute of Millets Research
Hyderabad, Telangana, India

J.V. Patil
Plant Breeding
ICAR-Indian Institute of Millets Research
Hyderabad, Telangana, India

ISBN 978-81-322-2421-1 ISBN 978-81-322-2422-8 (eBook)
DOI 10.1007/978-81-322-2422-8

Library of Congress Control Number: 2015946619

Springer New Delhi Heidelberg New York Dordrecht London

Printed on acid-free paper

Springer (India) Pvt. Ltd. is part of Springer Science+Business Media (www.springer.com)

Preface

Sorghum is an important staple food crop in the semi-arid tropics of Asia and Africa. Sorghum is more nutritious than fine cereals and is the principal source of energy, protein, vitamins, and minerals for millions of the poorest people in these regions. It grows well in harsh environments with minimum inputs, where other crops yield poorly. The area of sorghum declined globally over the past few decades, primarily due to susceptibility to biotic and abiotic factors and marginal economics. In this scenario, genetic enhancement of grain and fodder yield is a major challenge to the sorghum breeders. The application of modern breeding approaches such as DNA marker-assisted breeding and transgenics will help in addressing this challenge more effectively. The use of DNA marker technology, in improving the breeding efficiency and designing superior cultivars with greater speed, precision, and value, has been well demonstrated in major crops.

DNA marker technology has undergone a major change during the last three decades. Significant advances in the sequencing technologies and bioinformatics resulted in quicker genome sequencing of a crop species at a relatively cheaper cost. This has brought about a revolution in marker discovery, high-throughput genotyping for QTL mapping, and gene discovery of economically important traits, making the application of marker-assisted selection (MAS) in breeding programmes more affordable. Pyramiding of desirable genes through MAS has now become a practice in several crop breeding programmes. The application of this technology for the mapping and dissection of complex traits will now permit wholly new approaches to the improvement of sorghum. Combining molecular marker and transgenic approaches with conventional breeding schemes can increase the overall selection gain and, therefore, the efficiency of a breeding programme.

Due to the advances in sequencing, genotyping, phenotyping, QTL mapping, genetic transformation, and tissue culture technologies, we are beginning to visualize the practical solutions for the genetic enhancement of sorghum through DNA marker-assisted breeding and transgenics. At this point, it is essential to look back and critically review the advancements made till now so that we can formulate suitable strategies for the future sorghum improvement programmes. To support this task, we have requested the contributors to present the information in a simple way so that it appeals to audiences from diverse backgrounds.

As a reference book, we are sure that this book on "Sorghum Molecular Breeding", a one stop information source for sorghum molecular breeding research, will be of great help to students, teachers, managers, breeders, and biotechnologists, especially in sorghum, for planning future breeding strategies for the genetic improvement of sorghum, assisted by molecular breeding tools for achieving greater selection response and breeding efficiency. To accomplish this, we have attempted to cover important aspects of sorghum molecular breeding and transgenic research with emphasis on the marker development, application of DNA markers in genetic diversity, QTL mapping and heterosis breeding, bioinformatics resources, and transgenics.

This book is an account of comprehensive and critical review of up-to-date information featuring the latest approaches, technologies, resources, and practical progress in the area of marker-assisted breeding and transgenics in sorghum in the domain of genetic improvement. This book consists of 10 chapters accommodated in four sections, viz. Sorghum Introduction, Advances in DNA Marker Research, Advances in Genomics Research, and Advances in Transgenic Research.

Presenting an overview of sorghum molecular research, this book aims to expose the insights gained by several studies in sorghum molecular breeding involving genetics and breeding principles, molecular biology, bioinformatics, computational biology, and biotechnology. As a result, this book is intended to serve as a resource material to inter-disciplinary research groups comprising geneticists, breeders, biotechnologists, bioinformaticians, and students.

We are extremely grateful to all the learned contributors and sincerely thank them for their cooperation in compiling useful and updated information on different aspects of sorghum molecular breeding. We place on record our sincere thanks to Dr. S. Ayyappan, DG, ICAR, and Dr. SK Datta, DDG (Crop Science), ICAR, for their encouragements. We record our sincere gratitude to all the staff of IIMR who extended full support in several ways during the preparation of the book. We are confident that the book will be widely accepted by students, teachers, and researchers in the field of sorghum breeding in particular and plant breeding and life sciences in general.

Hyderabad, Telangana, India

R. Madhusudhana
P. Rajendrakumar
J.V. Patil

Contents

Part I Sorghum: Introduction

1 **Sorghum: Origin, Classification, Biology and Improvement** ... 3
K. Hariprasanna and J.V. Patil

Part II Advances in DNA Marker Research

2 **DNA Markers in Diversity Analysis** ... 23
Sujay Rakshit and M. Swapna

3 **Linkage Mapping** ... 47
R. Madhusudhana

4 **Application of DNA Markers for Genetic Improvement** ... 71
R. Madhusudhana

5 **Heterosis Prediction Using DNA Markers** ... 101
P. Rajendrakumar

Part III Advances in Genomics Research

6 **Genomics and Bioinformatics Resources** ... 117
P. Rajendrakumar and Sujay Rakshit

7 **Post-genome Sequencing Developments** ... 155
K.N. Ganapathy and P. Rajendrakumar

8 **Molecular Marker Development Using Bioinformatic Tools** ... 179
P. Rajendrakumar

Part IV Advances in Transgenic Research

9 **Advances in Genetic Transformation** ... 199
K.B.R.S. Visarada and N. Sai Kishore

10 **Genetic Engineering for Novel Traits** ... 217
B. Venkatesh Bhat, D. Balakrishna, P. Madhu, and J.V. Patil

Contributors

D. Balakrishna Biotechnology, ICAR-Indian Institute of Millets Research, Hyderabad, Telangana, India

B. Venkatesh Bhat Genetics & Cytogenetics, ICAR-Indian Institute of Millets Research, Hyderabad, Telangana, India

K.N. Ganapathy Plant Breeding, ICAR-Indian Institute of Millets Research, Hyderabad, Telangana, India

K. Hariprasanna Plant Breeding, ICAR-Indian Institute of Millets Research, Hyderabad, Telangana, India

N. Sai Kishore GITAM Institute of Science, GITAM University, Visakhapatnam, Andhra Pradesh, India

P. Madhu ICAR-Indian Institute of Millets Research, Hyderabad, Telangana, India

R. Madhusudhana Plant Breeding, ICAR-Indian Institute of Millets Research, Hyderabad, Telangana, India

J.V. Patil Plant Breeding, ICAR-Indian Institute of Millets Research, Hyderabad, Telangana, India

P. Rajendrakumar Biotechnology, ICAR-Indian Institute of Millets Research, Hyderabad, Telangana, India

Sujay Rakshit Plant Breeding, ICAR-Indian Institute of Millets Research, Hyderabad, Telangana, India

M. Swapna ICAR-Indian Institute of Millets Research, Hyderabad, Telangana, India

K.B.R.S. Visarada Genetics & Cytogenetics, ICAR-Indian Institute of Millets Research, Hyderabad, Telangana, India

About the Editors

Dr. R. Madhusudhana, Principal Scientist (Plant Breeding), ICAR-IIMR

Dr. R. Madhusudhana is presently working as the principal scientist (Plant Breeding) at the ICAR-Indian Institute of Millets Research (formerly Directorate of Sorghum Research), Hyderabad, India. He graduated with a bachelor's degree in agricultural sciences and completed his master's in genetics and plant breeding from the University of Agricultural Sciences, Dharwad, India. He received his Ph.D. (Genetics) from the Indian Agricultural Research Institute (IARI), a premier institute of higher studies in agricultural sciences located in New Delhi, India. He was awarded with several gold medals for his academic achievements during his bachelor's, master's, and doctoral studies and was awarded I rank in Plant Breeding in Agricultural Research Services (ARS), ICAR, New Delhi. He was the recipient of DST-BOYSCAST fellowship and did his post-doc in the Institute of Grassland and Environmental Research, Aberystwyth, UK.

He joined ICAR-Indian Institute of Millets Research as a scientist in 1996 and then on successfully handled several national and international projects. He was involved in the development of two sorghum high yielding cultivars, CSH23 and CSV20, which were released for all India cultivation. Presently, he is working on molecular breeding of sorghum employing the tools of DNA markers for QTL (Quantitative Trait Loci) mapping of important traits and marker-assisted selection. He has trained several graduate students and guided six Ph.D. students and one M.Sc. student. He was also involved in the successful conduct of several training programs at IIMR (Indian Institute of Millets Research). He has more than 40 peer-reviewed foreign and national journal papers, four books and book chapters, and 15 other technical publications to his credit. He is a Life Member of the Indian Society of Genetics and Plant Breeding, New Delhi, India, and Society of Millets Research, Hyderabad, India.

Dr. P. Rajendrakumar, Senior Scientist (Biotechnology), ICAR-IIMR

Dr. P. Rajendrakumar, born in Puducherry, India, pursued his bachelor's degree in agriculture (1990–1994), master's (1995–1997), and doctoral degree in plant breeding and genetics (1997–2001) from Tamil Nadu Agricultural University, Coimbatore, India. He was a recipient of Dr. R. Appadurai Award, Dr. K. Ramaiah Award, and Dr. M.S. Swaminathan Award during his Ph.D. years. After completing his Ph.D. degree, he worked as a research associate at the Centre for Cellular and Molecular Biology (2001–2004) and Directorate of Rice Research (2004–2008) in Hyderabad, India.

He joined the National Research Centre for Sorghum, Hyderabad, which was upgraded to Directorate of Sorghum Research and recently to ICAR-Indian Institute of Millets Research, as a senior scientist (Biotechnology) in 2008. He was the recipient of a three-year research project under the special grant, Rapid Grant for Young Investigators from the Department of Biotechnology, Government of India, in 2010. His research areas of interest include bioinformatics, marker-assisted breeding, and genomics. Current areas of research include the development of molecular markers through in silico analysis and application of genomics tools for understanding the molecular basis of grain yield heterosis and improvement of grain protein quality and iron and zinc content in sorghum. He has also organized a 21-day training on "Molecular Breeding Approaches for Genetic Enhancement of Millet Crops" as one of the course co-directors in 2014. He was awarded Junior Research Fellowship (JRF) by the Indian Council of Agricultural Research (ICAR), New Delhi, India, for pursuing his master's degree and Senior Research Fellowship (SRF) by the ASPEE Agricultural Research and Development Foundation, Mumbai, India, for pursuing his doctoral studies. He has guided one M.Sc. and one Ph.D. student. He is a Life Member of Indian Society of Plant Breeders, Coimbatore, India, and Society of Millets Research, Hyderabad, India. He has published 15 research papers in various peer-reviewed international journals and 10 research papers in national journals.

Dr. J.V. Patil, Director, ICAR-IIMR

Dr. J.V. Patil received his Ph.D. (Plant Breeding) from Mahatma Phule Krishi Vidyapeeth (MPKV), Rahuri, Maharashtra, India. He rendered his teaching services and research activities at MPKV, Rahuri, as an associate professor (Plant Breeding) and professor (Genetics and Plant Breeding). He developed several high yielding cultivars of safflower (1), pulses (*Mungbean-1, French bean-1, Pigeonpea-2, Chickpea-5*), and sorghum (8). Dr. Patil joined as the director ICAR-Indian Institute of Millets Research, Hyderabad, in August 2010 and continues to render his services.

He advocated several innovative and economical farming techniques such as planting methods and in situ rainwater conservation in chickpea and post-rainy sorghum. Under his leadership, a successful model on value chain on sorghum is established, which has brought reputation not only to IIMR but also to ICAR in commercializing value-added sorghum technologies and reviving sorghum economy. He has 165 peer-reviewed foreign and national journal papers, 31 books and book chapters, and 90 other technical publications to his credit. He has guided as many as 16 Ph.D. and 16 M.Sc. students.

Dr. Patil is a recipient of "NAAS Fellowship Award," "Baliraja' Late Annasaheb Shinde Smruti Krishi Sanshodhan Gaurav Puraskar," and Maharashtra state's "Vasantrao Naik Krishi Puraskar" for outstanding research in the field of crop improvement and Maharashtra state's "Vasantrao Naik Award for Best Marathi Literature 2010 and 2012" and "Bharat Krishak Samaj Award 2012." He also received the team awards such as "CGIAR's Baudouin Award 2002" and "ICRISAT's Doreen Mashier Award 2002."

Part I

Sorghum: Introduction

1 Sorghum: Origin, Classification, Biology and Improvement

K. Hariprasanna and J.V. Patil

Contents

1.1 **Introduction** 4

1.2 **Origin and Taxonomy** 4
1.2.1 Classification 5
1.2.2 Cultivated Sorghum 5

1.3 **Morphology** 8
1.3.1 Root 8
1.3.2 Stem 8
1.3.3 Leaves 8
1.3.4 Inflorescence 8
1.3.5 Seed 10

1.4 **Growth Stages** 10

1.5 **Reproductive Biology** 10
1.5.1 Panicle Initiation 10
1.5.2 Panicle Emergence 11
1.5.3 Anthesis and Pollination 11
1.5.4 Seed Development 12

1.6 **Male Sterility Systems** 12

1.7 **Crop Improvement** 13

1.8 **Genetic Gain over Years** 18

1.9 **Future Scenario of Sorghum Improvement** 18

References 18

Abstract

Sorghum is a staple food crop for millions of poor people in the semi-arid tropics of Africa and Asia. It is one of the important dryland crops grown in marginal soils and a source of feed, fodder and biofuel apart from food. It is a short-day C_4 plant, and its easy adaptability to hot and dry agroecologies makes it a climate change-compliant crop. There are five basic races and ten intermediate races under cultivated taxa based on fundamental spikelet types. Sorghum is considered as an often cross-pollinated species, with outcrossing up to 6 % depending on the genotype and growing conditions. Extensive efforts in crop improvement have resulted in the development of a number of high-yielding cultivars with substantial yield increment over the years. The discovery and utilisation of the male sterility system have led to the successful commercial exploitation of heterosis. A number of biotic and abiotic yield-limiting factors, and changes in consumption pattern and demand have resulted in a steady decline in cultivated area over the years. Much progress has been achieved in the field of sorghum biotechnology, including genomics over the last two decades. Adoption of genomic tools and molecular breeding strategies can help in tailoring sorghum cultivars with desired traits to enhance the productivity under various limiting factors in the years to come.

K. Hariprasanna (✉) • J.V. Patil
Plant Breeding, ICAR-Indian Institute of Millets Research, Rajendranagar, Hyderabad, Telangana 500 030, India
e-mail: hari@millets.res.in

R. Madhusudhana et al. (eds.), *Sorghum Molecular Breeding*, DOI 10.1007/978-81-322-2422-8_1

Keywords
Sorghum • Morphology • Growth stages • Genetic improvement • Genetic gain

1.1 Introduction

Sorghum (*Sorghum bicolor* [L.] Moench) is the world's fifth major cereal in terms of production and area harvested. It is a staple food crop for millions of the poorest and most food-insecure people in the semi-arid tropics of Africa and Asia. It is one of the important dryland food crops grown on marginal lands in more than 100 countries. It is a genus with many species and subspecies, and there are several types of sorghum, including grain sorghums, forage sorghums (for pasture and hay), sweet sorghums (for syrups) and broomcorn. The crop is agronomically suited to hot and dry agroecologies where other food grains do not grow easily. Sorghum is a dual-purpose crop; both grain and stalks or stover are highly valued for human and animal consumption, respectively. In developed countries like the USA and Australia, it is predominantly used for feed purposes (ICRISAT 2004). In large parts of the developing world, stover represents an important output of sorghum cultivation. More than 80 % of global sorghum area of 42.12 m ha (FAO 2014) lies in developing countries, mainly in the African and Asian continents, where sorghum grain is grown primarily for food. The remaining area of 16–20 % is predominantly in the developed world, especially cultivated by large-scale commercial farms, which produce sorghum mainly for animal feed. The normal area, production and yield levels in top 20 sorghum-producing countries along with that in the beginning of the last decade are given in Table 1.1.

1.2 Origin and Taxonomy

It is difficult to determine when and where the domestication of sorghum occurred (de Wet et al. 1970). Murdock (1959) has suggested that the Mande people around the headwaters of the Niger River may have domesticated sorghum. The origin and early domestication of sorghum is hypothesised to have taken place in northeastern Africa or at the Egyptian-Sudanese border around 5,000–8,000 years ago (Mann et al. 1983). The largest diversity of cultivated and wild sorghum is also found in this part of Africa. The secondary centre of origin of sorghum is the Indian subcontinent, with evidence for early cereal cultivation discovered at an archaeological site in western parts of Rojdi (Saurashtra) dating back to about 4,500 years (Vavilov 1992; Damania 2002). It is supposed that African slaves brought sorghum seeds with them to the USA, and thus the crop got introduced to the USA, which is the largest sorghum-growing country at present.

Sorghum Taxonomy

Kingdom – *Plantae*
Sub-kingdom – *Tracheobionta*
Superdivision – *Spermatophyta*
Division – *Magnoliophyta*
Class – *Liliopsida*
Sub-class – *Commelinidae*
Order – *Cyperales*
Family – *Poaceae* (grass)
Tribe – *Andropogoneae*
Sub-tribe – *Sorghinae*
Genus – *Sorghum*
Species – *Sorghum bicolor*
- Sub-species – *Sorghum bicolor ssp. arundinaceum* – common wild sorghum
- Sub-species – *Sorghum bicolor ssp.* bicolor – grain sorghum
- *Sub-species* – *Sorghum bicolor ssp. drummondii* – Sudan grass

Species – *Sorghum almum* – Columbus grass
Species – *Sorghum halepense* – Johnson grass
Species – *Sorghum propinquum*

Sorghum was first described by Linnaeus in 1753 under the name *Holcus*. In 1794, Moench separated the genus *Sorghum* from the genus *Holcus*. In 1805, Person suggested the name

Table 1.1 Sorghum area, production and yield in major producing countries

	Area (lakh ha)		Production (lakh tonnes)		Yield (kg/ha)	
Country	2001–2003	2011–2013	2001–2003	2011–2013	2001–2003	2011–2013
USA	31.70	20.79	108.89	72.00	3,416	3,431
Nigeria	67.40	52.97	75.44	68.32	1,119	1,294
Mexico	18.86	17.46	60.78	65.69	3,214	3,762
India	94.65	66.04	72.13	60.88	762	920
Argentina	5.62	9.39	28.14	41.16	5,014	4,380
Ethiopia	12.76	18.27	16.26	39.65	1,280	2,169
Sudan	59.42	61.65	41.36	36.71	688	576
Australia	7.49	6.29	18.07	21.34	2,402	3,400
China	7.83	5.32	29.62	20.22	3,785	3,809
Brazil	5.55	7.39	11.69	20.07	2,044	2,721
Burkina Faso	15.46	17.57	14.52	17.90	938	1,016
Niger	23.72	30.30	6.83	11.57	290	379
Cameroon	4.11	7.78	5.40	11.33	1,317	1,456
Mali	8.16	11.61	6.29	10.74	773	998
Chad	7.62	8.60	5.14	8.55	675	986
Tanzania	5.99	7.87	5.09	8.26	804	1,054
Egypt	1.57	1.47	9.08	7.82	5,769	5,318
Yemen	3.35	4.88	2.95	4.37	868	895
Venezuela	2.64	2.22	5.59	3.87	2,128	1,745
Bolivia	0.53	1.42	1.36	3.75	2,536	2,740
World	**432.03**	**408.63**	**574.61**	**588.86**	**1,329**	**1,443**

Sorghum vulgare for *Holcus sorghum* (L.). In 1961, Clayton proposed the name *Sorghum bicolor* (L.) Moench as the correct name for cultivated sorghum and brought all the sorghums together under this name (Clayton 1961), which is currently being used. Sorghum is classified under the family Poaceae (see box). There seems no argument that sorghum plants are African in origin, but the domestication event(s) may have taken place elsewhere and more than once (Kimber 2000). De Wet and his colleagues suggested that sorghum had a diverse origin and probably arose from *Sorghum verticilliflorum*, which is usually found in areas where sorghum is cultivated. There is tremendous variation in *S. verticilliflorum*, and it as well as other wild species readily crosses with cultivated sorghum.

1.2.1 Classification

The most detailed classification of *Sorghum* was made by Snowden (1936, 1955). He described 31 cultivated species and 17 related wild species. At present, his species are more appropriately considered to be races of one species. Snowden (1936) later sub-divided the sorghums into the following sections, sub-sections and series.

Section – *Eu-sorghum*
- Sub-section – *Arundinacea*
  - Series – *Spontanea* (10 wild species)
  - Series – *Sativa* (31 cultivated species)
- Sub-section – *Halepensia* (4 wild grasses)

Section – *Para-sorghum* (8–10 annual and perennial grasses)

Snowden's classification (1936) was later refined by Garber (1950) and by Doggett (1970) (Fig. 1.1).

1.2.2 Cultivated Sorghum

The cultivated races are placed in *Sorghum bicolor* subsp. *bicolor*. Harlan and de Wet (1972) have developed a simplified, informal

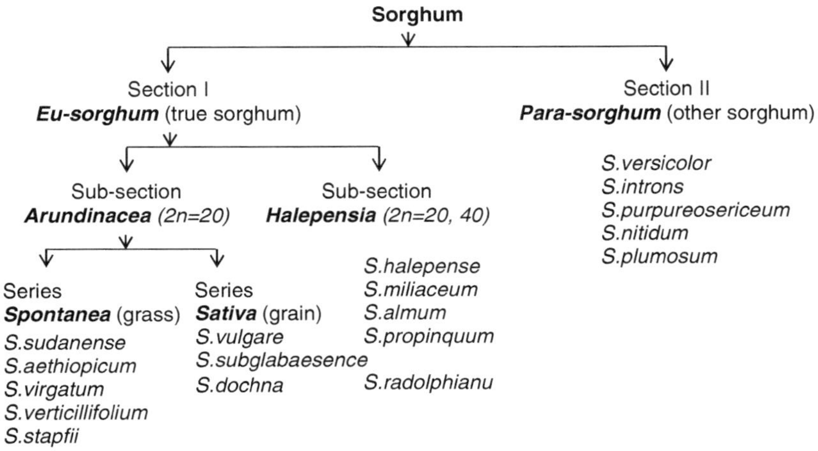

Fig. 1.1 Classification of sorghum

classification useful to plant breeders for the cultivated sorghums and their closest wild relatives. The cultivated taxa, covering 28 (out of 31) species of Snowden's series *Sativa* belonging to a primary gene pool, are partitioned into five basic races and 10 intermediate races under *S. bicolor* subsp. *bicolor* based on fundamental spikelet types. All the 15 races of cultivated sorghum can be identified by mature spikelets alone, although head type is sometimes helpful. The five basic races are briefly described below:

1.2.2.1 *Bicolor*

It is the primitive type. Panicles are open and medium in size. The rachis is typically long, with long, slightly stiff branches terminating with the long clasping glumes. The glumes are thick and coriaceous with obscure nerves, and the tips of the lower glumes are depressed and hairy. The pedicellate spikelets are persistent and the pedicels are short. Grains are elongated, sometimes slightly obovate, nearly symmetrical dorsoventrally and generally small in size; glumes clasp the grain, which may be completely covered or ¼ exposed at the tip (Fig. 1.2a). Seed, glumes and plant parts generally are highly pigmented. Bicolor plants are fairly low yielding and medium in height and tend to tiller profusely (Mann et al. 1983).

1.2.2.2 *Guinea*

Panicles are long, loose, glabrous and pendulous. The rachis is long with short branches. Sessile spikelets open when mature, thus exposing the grain. The glumes are involuted, open widely and are hairy with conspicuous awns. Pedicellate spikelets are both persistent and deciduous. Grains are small to medium, flattened dorsoventrally, sub-lenticular in outline and twisting at maturity 90°, between gaping involute glumes that are nearly as long to longer than the grain (Fig. 1.2b). The grain tends to be light coloured or slightly pigmented. Plants are medium to tall and tend to be low yielding (Mann et al. 1983).

1.2.2.3 *Caudatum*

One of the most important races agronomically and provides genes for high yield and seed quality. Panicles are dense to slightly open, medium to large, oblong, with a stout peduncle. The rachis and primary branches are rigid. Sessile spikelets are obovate to elliptical, while pedicellate spikelets are deciduous. Glumes are coriaceous, pubescent and shorter than the large grain. Grains are markedly asymmetrical, and the side next to the lower glume is flat or even somewhat conclave and the opposite side is rounded and bulging; the persistent style often at the tip of a beak points towards the lower glume; glumes are ½ of the length of the grain or less (Fig. 1.2c). Grains

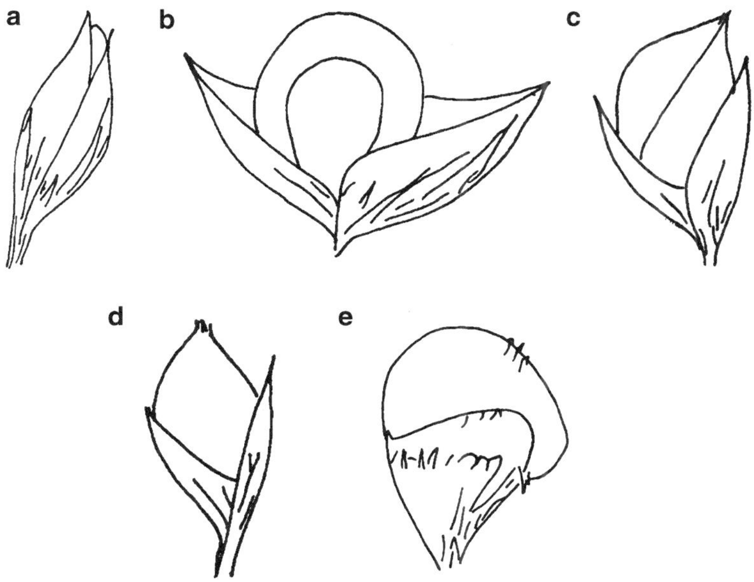

Fig. 1.2 Diagrammatic representation of spikelets of five basic races. (**a**) *Bicolor.* (**b**) *Guinea.* (**c**) *Caudatum.* (**d**) *Kafir.* (**e**) *Durra*

are usually chalky white or pigmented. Plants are generally medium to tall and are usually high yielding (Mann et al. 1983).

1.2.2.4 *Kafir*

Panicles are erect, elongated, mostly semi-compact and cylindrical. The rachis is long and branches and sessile spikelets tend to be hairy, but the glumes are almost glossy at maturity. Glumes are much shorter than the grain. Grains are approximately symmetrical, more or less spherical, not twisting, glumes clasping and variable in length (Fig. 1.2d). Plants are of medium height and generally high yielding (Mann et al. 1983).

1.2.2.5 *Durra*

Panicles tend to be stiff, dense, compact and pubescent and ovate to oblong in shape. The rachis is stout, hidden and sometimes recurved with short, semi-erect and often hairy branches. The pedicellate spikelets are large and persistent, while the sessile spikelets are obovate elliptic. Grains are rounded obovate, wedge-shaped at the base and broadest slightly above the middle. The glumes are very wide and the tip is of a different texture from the base, and often these glumes are with a transverse crease across the middle and coriaceous on the lower half (Fig. 1.2e). Glumes tend to be lightly pigmented. Plants are medium to tall and of good quality.

In Africa, according to Harlan (1972), *guinea* is primarily West African with a secondary centre in Malawi-Tanzania. *Caudatum* is most abundant from east Nigeria to eastern Sudan and southward into Uganda. *Kafir* is primarily a race of East Africa, south of the equator and Southern Africa. *Durra* is dominant in Ethiopia and westward across the continent in the driest zones near the Sahara. The hybrid races are expected in overlapping areas. The *bicolor* race occurs on a minor scale almost everywhere in Africa. The sweet types used for chewing are usually *bicolors*, and some are used for beer. Indian sorghums are mostly *durra*, *guineas* and *guinea-kafirs*, with some *bicolors* grown on a minor scale. The American grain sorghums are almost entirely *kafir-caudatums*. Broomcorns, sorgos and Sudan grass fall under the *bicolor* race. The intermediates that are caused by hybridisation of races exhibit characters of both parents. The intermediate races are:

1.	*Guinea-bicolor*	6.	*Guinea-kafir*
2.	*Caudatum-bicolor*	7.	*Guinea-durra*
3.	*Kafir-bicolor*	8.	*Kafir-caudatum*
4.	*Durra-bicolor*	9.	*Durra-caudatum*
5.	*Guinea-caudatum*	10.	*Kafir-durra*

1.3 Morphology

1.3.1 Root

The roots are adventitious and fibrous. The roots can be divided into a primary and secondary system. An embryonic or primary root first appears upon germination. Several such roots develop; these are not branched or are sparsely branched. Primary roots have a limited growth and their functions are soon taken over by the secondary roots. Secondary roots, which develop from the first node, develop into the extensive root system. The primary roots subsequently die. Brace roots may appear later on the lower most nodes and may be numerous if the plant is unadapted, but these are not effective in water and nutrient uptake. The cultivated sorghums are either non-rhizomatous or very weakly rhizomatous. The root system survives to support the tillers and ratoon crop from adventitious buds at the base of the parent stem. Well-developed rhizomes are found in the sub-species *halepense* (adopted from House 1985).

1.3.2 Stem

The stem or culm is erect and made up of a series of alternating nodes and internodes. It is solid with a hard cortex or rind and a softer pith. Vascular bundles are scattered throughout the stem, but they are more near the peripheral area. The vascular bundles in the central portion of the stem are larger than those at the periphery, and these central bundles branch into leaf midribs, while the peripheral bundles branch to form the smaller veins in the leaf blade. The pith may be sweet or insipid, juicy or dry. The leaf is attached to the stem at the node, which appears as a ring at the base of the leaf sheath. Buds form at each node except at the node to which the flag leaf is attached. These buds may develop to form axillary tillers. The basal tillers form at the first node (adopted from House 1985).

1.3.3 Leaves

Leaves may be concentrated near the base or uniformly distributed and arranged alternating to the opposite side with parallel venation. Leaf consists of a sheath and a blade. The sheath is attached to the node and surrounds the internode. The leaf sheath is often covered with a waxy bloom. The angle of attachment of leaves to the stem varies. The leaf blade is long, narrow and pointed. The leaf blade may be straight or bend like an arc. The tip of the leaf may even drop down. The length and width of leaf blade vary widely. The midrib is prominent, greenish or white and flattened or slightly conclave on the upper surface and convex on the lower one. There is a short membranous ligule at the junction of the leaf blade with the sheath. The number of leaves vary depending on the genotype; in well-adapted plants, there will be usually 14–17 leaves. Stomata occur on both surfaces of the leaf (adopted from House 1985).

1.3.4 Inflorescence

1.3.4.1 Panicle

It may be short and compact or loose and open. The rachis, the central axis of the panicle, may be completely hidden by dense panicle branches or exposed and differs greatly in shape and length. The rachis may be striated, hairy or glabrous. Several primary branches are borne at each node and these branches vary in length and strength. Each primary branch bears secondary branches, which in turn bear spikelets. The panicle usually grows erect at the apex of the culm but may be recurved depending on the genotype. The wild- and forage-type sorghums have a rather loose pyramidal panicle with spreading branches. The shape and density of the panicle are important

DUS (distinctiveness, uniformity and stability) characters often claimed by the breeders.

1.3.4.2 Raceme

It always consists of one or several spikelets. One spikelet is always sessile (fertile) and the other pedicellate (sterile), except the terminal sessile spikelet, which is accompanied by two pediceled spikelets. The racemes vary in length according to the number of nodes and length of internodes. On the pediceled spikelet, the pedicels vary in length from 0.5 to 3.0 mm and usually are very similar to the internodes (adopted from House 1985).

1.3.4.3 Sessile Spikelets

The sessile spikelet varies in shape from lanceolate to almost round and ovate and is sometimes depressed in the middle. It is green coloured at flowering, changing to shades of straw, cream, yellow, red, brown, purple or almost black at grain maturity. The intensity and extent of colouring on the glumes is variable. Glumes vary from hairy to almost hairless. The glumes are hard and tough in most species. Some species have thin and brittle glumes, while others have thin and papery glumes. The lower glume is somewhat flattened and conforms more or less to the shape of the spikelet, while the upper one is more convex or boat shaped. The seed may be enclosed by the glume or may protrude from it, just visible to almost completely exposed. There are two lemmas, each a delicate white tissue. The lower lemma is elliptic or oblong about equal in length to the glume. The upper lemma is shorter and more ovate and may be awned. A small palea is also present. Two lodicules are placed on either side of the ovary at its base. Androecium consists of one whorl of three stamens. The anthers are attached at the base of the ovule by long thread-like filaments that are versatile and yellowish. Gynoecium is centrally placed and consists of two pistils with one ovule from which two feathery stigmas protrude (Figs. 1.3 and 1.4).

1.3.4.4 Pedicellate Spikelets

These are much narrower than sessile spikelets, usually lanceolate in shape. They may be smaller or longer than the sessile spikelets or of the same size. They possess only anthers but (very rarely)

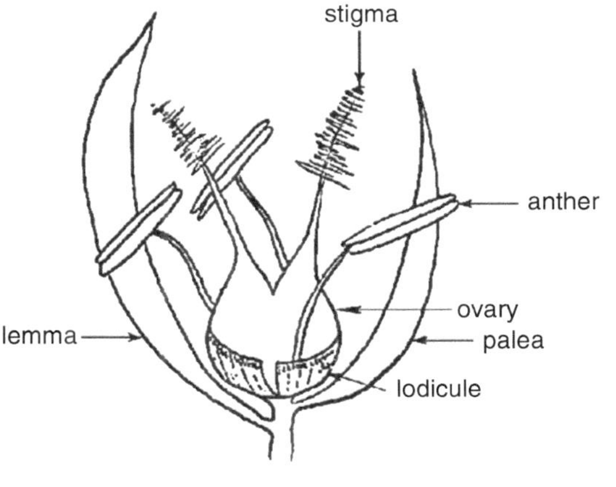

Fig. 1.3 Diagrammatic representation of sorghum floret

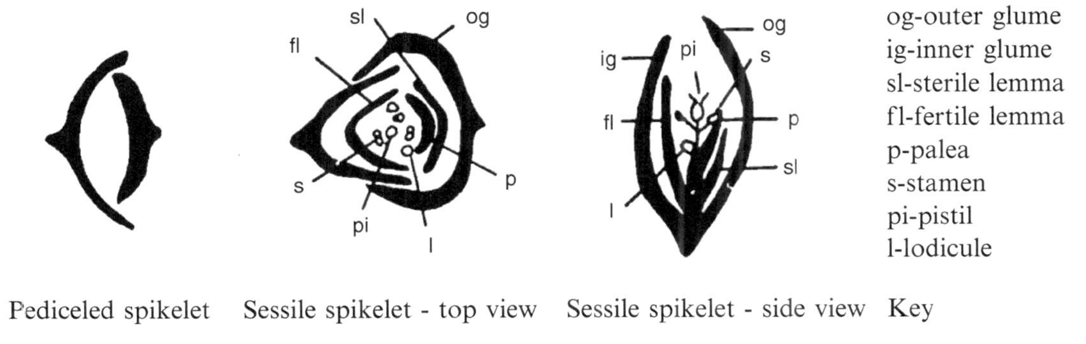

Fig. 1.4 Components of the pediceled and sessile spikelets (Adopted from House 1985)

may have a rudimentary ovary. The lemmas are much reduced in size and only rarely an awn will be present on the upper lemma.

The floral characters like lemma-awn formation, stigma colouration, stigma length, length of pedicel, anther length and colour and glume colour are important DUS test traits in sorghum.

1.3.5 Seed

The sorghum grain is a caryopsis. Grain is usually partially enclosed by glumes, which are removed during threshing. The shape of the seed is oval to round, from 4 to 8 mm in diameter and varying in size, shape and colour depending on the cultivar. The seed coat consists of the pericarp and testa. The pericarp is the outermost layer of the seed and consists of the epicarp, hypodermis, mesocarp and endocarp. The testa is situated directly below the endocarp and encloses the endosperm. If only the pericarp is coloured, the seed is usually yellow or red. Pigment in both the pericarp and testa results in a dark-brown or red-brown colour. Apart from the role of the testa in the colouring of the seed, it contains tannins with a bitter taste (Plessis 2008). Brown-seeded types are high in tannins, which lower palatability. The embryo consists of plumule (foliage leaves), coleoptile (shoot sheath) and radical coleorhiza (root sheath) referred to as scutellum (Fig. 1.5). The endosperm is starchy and consists of hard and soft endosperm. In general, seed composition is endosperm, 82 %; embryo, 12 %; and seed coat, 5–6 % (Carter et al. 1989).

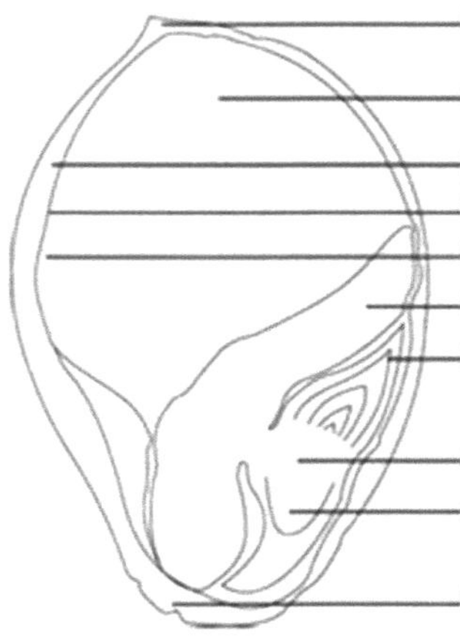

Fig. 1.5 Schematic representation of a longitudinal cross section of a sorghum seed

1.4 Growth Stages

The growth and development of sorghum can be divided into three stages: GS1, planting to panicle initiation; GS2, panicle initiation to flowering; and GS3, flowering to physiological maturity. The duration of these stages varies depending on the cultivar, adaptation, climatic conditions, date of planting and temperature. Vanderlip and Reeves (1972) described the temperate sorghum growth stages on a 0–9 scale. The growth stage and duration of each stage in a rainy-season sorghum hybrid (CSH 16) with a total duration of about 110 days have been described in Table 1.2.

1.5 Reproductive Biology

A proper understanding of the floral biology, pollination control and seed development is essential for designing effective breeding strategies. The breeding methods and procedures to be adopted for genetic enhancement are largely determined by the mode of reproduction. Sorghum can be considered as an often cross-pollinated species, with outcrossing normally ranging between 0.6 and 6 %, depending on the genotype, but even up to 30 % has also been reported (Miller 1982). The reproductive biology of sorghum is covered briefly hereunder:

1.5.1 Panicle Initiation

Sorghum is a short-day plant, and blooming is hastened by short days and long nights, though varieties differ in their photoperiod sensitivity (Quinby and Karper 1947). Floral initiation takes place 30–40 days after germination. Usually, the floral initial is 15–30 cm above the ground when the plants are about 50–75 cm tall (House 1985).

Table 1.2 Growth stage characteristics in CSH 16 during rainy season (Rao et al. 2004)

Growth stage number	Days from emergence[a]	Duration (days)	Characteristics for identification of stage
0	0	0	Emergence: coleoptiles visible at the soil surface (first leaf is seen with a round tip)
1	6	6	3-leaf stage: collar of third leaf visible
2	16	10	5-leaf stage: collar of fifth leaf visible
3	32	16	Growing point differentiation (panicle initiation): approximately 8–9 fully expanded leaves
4	50	18	Final flag leaf visible (tip of flag leaf visible in the whorl)
5	60	10	Boot: panicle extends into the flag leaf sheath
6	68	8	50 % flowering (50 % of the plants in a row completed 50 % anthesis)
7	80	12	Soft dough: squeezing grains between fingers results in a little or no milk
8	96	16	Hard dough: grain is hard and firm when pressed between fingers
9	106	10	Physiological maturity: black layer (spot) appears on the hilar end (point of attachment of grain on floret) at the base of the seed

[a]Planting to emergence takes about 4 days

Floral initiation marks the end of the vegetative phase. The time required for transformation from the vegetative primordial to reproductive primordial is largely influenced by the genotype and environment. The grand growth period in sorghum follows the formation of a floral bud and consists largely of cell enlargement. Hybrids usually take less time to reach panicle initiation, more days to expand the panicle and a longer grain filling period than the parental lines (Maiti 1996).

1.5.2 Panicle Emergence

During the period of rapid cell elongation, floral initial develops into an inflorescence. About 6–10 days before flowering, developing inflorescence inside the leaf sheath of flag leaf will appear as a boot-shaped structure. This will occur in about 55–60 days from germination in a variety that takes 60–65 days to flower and is referred to as the booting stage. Flowering will complete in 8–10 days depending upon the weather parameters prevailing during the crop season.

1.5.3 Anthesis and Pollination

Anthesis starts with the exertion of the complete panicle. Flowers begin to open 2–3 days after the complete emergence of the panicle. A floret opening or anthesis is achieved by swelling of the lodicules and is followed by the exertion of anthers and stigmas between the lemma and palea. The anthesis starts at the tip of panicle and progresses downwards over a period of 4–5 days or even longer depending on the environmental conditions. Anthesis takes place first in the sessile spikelets from top to bottom of the inflorescence. It takes about 6 days for the completion of anthesis in the panicle with maximum flowering at 3 or 4 days after anthesis begins. Flowering proceeds downwards to the base in a horizontal plane on the panicle. When flowering of the sessile spikelets is halfway down the panicle, pedicellate spikelets start opening at the top of the panicle and proceed downwards. The flowering phase of pedicellate spikelets overtakes that of sessile spikelets before they reach the base of the inflorescence (Maiti 1996). Anthesis takes place during the morning hours and frequently occurs just before or just after sunrise but may be delayed on cloudy and wet mornings. It normally starts around midnight and proceeds up to 10:00 h depending on the cultivar, location and weather. Maximum flowering is observed between 6:00 and 9:00 h. As all panicles in a field do not flower at the same time, pollen is usually available for a period of 10–15 days. At the time of flowering, the glumes open and all the three anthers fall free, while the two stigmas protrude, each on a stiff style. The anthers dehisce when they are dry and pollen is blown into the air. The pollen in the anthers remains alive several hours after pollen shedding. Flowers remain open for 30–90 min. Dehiscence of the anthers for pollen diffusion takes place through the apical pore. The pollen drifts to the stigma, where it germinates; the pollen tube, with two nuclei, grows down the style to fertilise the egg and form a 2n nucleus. Glumes close shortly after pollination, though usually the empty anthers and stigmas still protrude. The florets of some of the very long glumed types do not open for fertilisation resulting in cleistogamy. Sorghum is predominantly a self-pollinating crop, but cross-pollination also takes place to some extent depending on the genotype, panicle type and wind direction and velocity. Stigmas exposed before the anthers dehisce are subjected to cross-pollination (Aruna and Audilakshmi 2008).

1.5.4 Seed Development

Seeds are borne on raceme branches on the panicles. After fertilisation, the ovule begins to develop as a light-green, almost cream-coloured sphere. After about 10 days, it begins to take size and becomes darker green. The development of grains follows a sequence of stages comprising milky dough, soft dough and hard dough to the final physiological maturity, when a black layer is formed at the hilar region due to the formation of the callus tissue. It takes about 30 days for the seeds to reach maximum dry weight (physiological maturity). The seeds began to turn from green to the different colours depending on the genotype at the time of maturity. The seeds contain about 30 % moisture at physiological maturity, and the level reduces to 10–15 % at 20–25 days after attaining physiological maturity (House 1985). During this period, the seeds lose up to 10 % of dry weight. The seed can be harvested at any time from physiological maturity to seed dryness. After harvest, the seed has to be dried to a uniform moisture level of 11–12 % for storage.

1.6 Male Sterility Systems

The phenomenon of pollen sterility in sorghum was first reported independently by Ayyangar and Ponnaiya (1937) in India and by Stephens (1937) in America. In each case, anthers were devoid of pollen, but stigmas were receptive and monogenic inheritance was observed with male sterility recessive (Stephens and Quinby 1945). Later on, Stephens and Holland (1954) indicated that male sterility in some cross-combinations was caused by the intercalation between the milo cytoplasm and *kafir* nuclear factors, thus describing the cytoplasmic-genetic male sterility (CMS).

A number of male sterile cytoplasms have been identified and studied in sorghum (Rao 1962; Pi and Wuu 1963; Hussaini and Rao 1964; Webster and Singh 1964; Nagur and Menon 1974; Schertz and Ritchey 1978; Pring et al. 1982; Lee et al. 1989; Xu et al. 1995). These cytoplasms are known to have originated in regions as geographically diverse as India, America and Africa (Sane et al. 1996). They are broadly classified into groups designated as A_1, A_2 (Schertz and Ritchey 1978), A_3 (Quinby 1980), A_4 (Worstell et al. 1984), Indian A_4 (A_{4M}, A_{4VZM}, A_{4G}) (Rao et al. 1984), A_5, A_6, 9E (Webster and Singh 1964), KS cytoplasms (Ross and Heckerott 1972), etc., depending on their maintainer and restorer crosses, of which mainly the A_1 cytoplasm (milo) has been commercially exploited in sorghum. Fertility restoration of A_1 or milo cytoplasm is most simple and complete among all CMS systems available. Thus, till now most of the released sorghum hybrids across the world are based on the A_1 cytoplasm (Reddy and Stenhouse 1994).

1.7 Crop Improvement

Sorghum improvement in India can be illustrated as a brilliant example, which has significantly contributed towards the green revolution in dry lands. Efforts were made to improve the sorghum cultivars in India since the 1930s. Most of the varieties till the 1960s were the result of pure line selection practised in local landraces. The locals were tall, photosensitive, late maturing, flowering after the rainfall ceased and characterised by localised adoption and low harvest index. However, the hybridisation between the local cultivars reflected about 5 % yield improvement at experimental level, which could not make any impact on production till the 1960s. The hybridisation and selection up to a limited extent among improved landraces could not bring genetic improvement to a perceptible level, but basic traits for local adaptation were preserved (Aruna 2014).

Since most of the local cultivars were tall, photosensitive and late maturing with low yields, development of relatively short-duration, photoperiod-insensitive sorghums had become the primary objective of almost all the sorghum improvement programmes. In rainy season, genetic improvement of varieties was achieved during the 1960s by the production of relatively short-duration photoperiod-insensitive sorghums with short height by manipulating the gene for height and maturity by introducing an American germplasm and adopting temperate × temperate and temperate × tropical crosses. The major genotypic changes brought about during the 1960s triggered cultivar-input-management interaction and resulted in quantum jumps in productivity imparting stability to production. This resulted in a quantum jump in the productivity from 560 kg/ha in 1970 to 1,000 kg/ha in 2000.

Apart from the development of short-statured high-yielding cultivars, emphasis was also laid on the exploitation of hybrid vigour as evidenced in the USA. The real development of hybrids in sorghum became feasible with the discovery of genetic male sterility (ms_2) by Stephens (1937) and the subsequent discovery of cytoplasmic male sterility ms_{c1} (Stephens and Holland 1954; Doggett 1988). An early hybrid in the USA was RS 610, a cross of Combine kafir 60A (CMS) with Combine 7078 which offered a substantial yield increase over the varieties. In 1962, the Indian Council of Agricultural Research launched the Accelerated Sorghum and Millet Improvement Project (ASMIP) with an objective to initiate the hybrid breeding in sorghum and millets. Later on, the All India Coordinated Sorghum Improvement Project (AICSIP) was launched in 1969 with 18 main centres spread over in 13 state agricultural universities in 11 major sorghum-growing states. Through the efforts of Rockefeller Foundation, a wide range of germplasm was made available in India which involved male steriles, several converted lines and tropical varieties collected from the Indian sub-continent and several African countries. The first commercial sorghum hybrid CSH1, an early duration and dwarf hybrid, was released in 1964 for all India cultivation using the parental lines bred in the USA and supplied by the Rockefeller Foundation. With the release of CSH 1, sorghum became the second crop after maize in developing high-yielding hybrids using the CMS system. This hybrid became most popular with the farmers as it had a high yield potential and is suited to light-soil and low-rainfall areas. During the next four decades, remarkable progress has been achieved by diversifying the parental lines for yield, maturity, height, disease and insect tolerance and quality by utilising indigenous and exotic germplasm. The early efforts made to identify heterotic combinations among male sterile and converted dwarf lines resulted in the development of other hybrids, CSH 2 in 1965 and CSH 3 in 1970. In spite of their better yield potential and suitability of growing in both rainy and post-rainy seasons, these hybrids could not become popular because of the seed production problem in both of these hybrids due to their common pollen parent IS 3691 being shorter than respective male sterile lines.

Though the use of dwarfing, earliness and photoinsensitivity traits were helpful in realising higher grain yields in the hybrids, the fodder yield was sacrificed to a major extent. Hence, further genetic improvement was oriented to rectify the problems through genetic enhancement and use of improved derivatives. Further hybrid breeding was based on these genetically enhanced indigenously bred parental lines. Efforts were augmented to develop new dual-purpose hybrids. Keeping in view the importance of fodder, the development of fodder sorghum hybrids was also emphasised, albeit the progress was lower. Till date, a total of 31 hybrids (CSH 1 to CSH 31, Tables 1.3a and 1.3b) have been released at the national level, which is a standing testimony of success of Indian sorghum breeding not only in terms of yield enhancement (grain or fodder) but also in terms of diversification of parental lines and progressive advances in the incorporation of acceptable levels of resistance against major pests and diseases. The hybrids played a major role in pushing up productivity and production, particularly in the case of rainy-season sorghum.

Table 1.3a Performance of rainy-season-adapted sorghum hybrids

Hybrid	Year of release	Parentage	Grain yield (t/ha)	Stover yield (t/ha)	Duration (days)	Area for which it is recommended
CSH 1	1964	CK 60A × IS 84	2.5–3.0	5.5–6.5	95–100	Low rainfall and light soils of MH, KA, MP, GJ, UP, RJ and TN
CSH 2	1965	CK 60A × IS 3691	3.0–3.5	7.0–7.5	120–125	Assured rainfall areas of MH, KA, UP, RJ and TN
CSH 3	1970	2219A × IS 3691	3.5–3.8	7.0–7.5	115–120	Assured rainfall tracts of MH, Telangana region of AP, TN, Malwa plateau of MP and Bundelkhand region of UP
CSH 4	1973	1036A × Swarna	3.5–3.8	7.0–7.5	110–115	-do-
CSH 5	1975	2077A × CS3541	3.8–4.0	8.5–9.0	110–115	All rainy areas and late-kharif tracts of AP and summer irrigated areas in TN and KA. Ideally suited for inter-cropping and ratooning
CSH 6	1977	2219A × CS3541	3.0–3.5	7.5–8.0	95–100	Low rainfall and light soils, suitable for inter-cropping and ratooning
CSH 9	1981	296A × CS3541	3.8–4.0	8.5–9.5	105–110	All rainy sorghum-growing areas except humid areas of KA and TN
CSH 10	1984	296A × SB 1085	3.5–3.8	12–14	105–110	MH, KA, AP, MP, RJ and UP
CSH 11	1986	296A × MR 750	3.8–4.0	10–12	105–110	All rainy sorghum-growing areas
CSH 13	1991	296A × RS 29	3.8–4.0	14–15	110–115	All sorghum-growing areas
CSH 14	1992	AKMS 14A × AKR 150	3.8–4.0	8.0–9.0	100–105	All rainy sorghum-growing areas, medium to heavy soils, low-rainfall areas
CSH 16	1997	27A × C43	4.0–4.5	9.0–10.0	110–115	All rainy sorghum-growing areas

(continued)

Table 1.3a (continued)

Hybrid	Year of release	Parentage	Grain yield (t/ha)	Stover yield (t/ha)	Duration (days)	Area for which it is recommended
CSH 17	1998	AKMS 14A × RS 673	4.0–4.2	9–10	105–110	Rainy sorghum-growing areas of KA, GJ, MP and TN
CSH 18	1999	IMS 9A × Indore 12	4.0–4.5	12.0–14.0	110–115	All rainy sorghum-growing areas
CSH 20MF	2005	2219 A × UPMC 503	–	24–25	105–110	Multi-cut fodder sorghum for all India cultivation
CSH 22SS	2005	ICSA 38 × SSV 84	1.7–1.8	42–45 (Cane)	115–125	Sweet sorghum hybrid for all India cultivation
CSH 23	2005	MS 7A × RS 627	4.0–4.2	8.5–9.0	105–110	All rainy sorghum-growing areas
CSH 24MF	2008	ICSA 467 × Pantchari 6	–	23–24	105–110	Multi-cut fodder sorghum for all India cultivation
CSH 25	2008	PMS 28 A × C 43	4.0–4.5	12.5–13.0	110–115	Rainy sorghum areas under normal time of sowing, especially for MH and KA
CSH 26	2011	MLSA 848 × R 4000	4.3	14.6	109	For all sorghum-growing states of India
CSH 27	2011	279A × CB11	3.9–4.0	13.0–14.0	105–108	All rainy sorghum-growing areas
CSH 28	2012	IRAT 204 × SPV 1134	4.7	15	111	MH, KA, MP, South GJ and North AP
CSH 29	2012	501A × 606 R	4.9	15.2	110	MH, KA, MP, South GJ and North AP
CSH 30	2012	415A × CB 33	4.3–4.4	12–14	105–110	Rainy sorghum-growing areas of MH, KA, MP, South GJ and North AP

AP Andhra Pradesh, *MH* Maharashtra, *KA* Karnataka, *MP* Madhya Pradesh, *GJ* Gujarat, *UP* Uttar Pradesh, *RJ* Rajasthan, *TN* Tamil Nadu

Table 1.3b Performance of post-rainy-season-adapted sorghum hybrids

Hybrid	Year of release	Parentage	Grain yield (t/ha)	Stover yield (t/ha)	Duration (days)	Area for which it is recommended
CSH 7R	1977	36A × 168	2.5–3.0	5.0–6.0	110–115	Entire Deccan post-rainy tracts of MH, KA, AP and also suitable for GJ
CSH 8R	1977	36A × PD3-1-11	2.5–3.0	3.5–3.7	110–115	Entire Deccan post-rainy tracts of MH, KA and AP
CSH 12R	1986	296A × M148-138	2.5–2.8	4.0–5.0	115–120	-do-
CSH 13	1991	296A × RS29	3.2–3.5	5.0–5.5	110–115	All sorghum-growing areas
CSH 15R	1995	104A × RS 585	3.0–3.5	5.5–6.0	110–115	MH, South KA and Northwest AP
CSH 19R	2000	104A × R 354	2.5–3.0	5.5–6.0	115–120	All post-rainy sorghum-growing areas of the country
CSH 31R	2014	MLSA 1426 × 6644R	2.5	7.5	115	For all sorghum-growing states of India

The varietal improvement programme could also identify high-yielding open-pollinated varieties (OPVs) simultaneously. Till date, CSV 1 to CSV 35 (Tables 1.4a and 1.4b) have been released at the national level, and many more varieties have been released in various states. Some of these varieties are dual-purpose types. However, varieties were less preferred by the farmers during rainy season. Better preference was received by dual-purpose varieties such as CSV 10, CSV 13, SPV 462 and CSV 15 in some restricted pockets. A major advantage of varieties over

Table 1.4a Performance of rainy-season-adapted sorghum varieties

Variety	Year of release	Grain yield (t/ha)	Stover yield (t/ha)	Duration (days)	Area for which it is recommended
CSV 1 (Swarna)	1968	3.0–3.5	8.0–9.0	95–100	Sorghum-growing areas of MH, GJ, KA and AP
CSV 2	1973	3.0–3.5	8.0–9.0	105–110	MH (Vidarbha and Marathwada), MP and adjoining areas of RJ, Bundelkhand and North Telangana of AP
CSV 3	1973	3.5–4.0	8.5–9.5	105–110	All rainy sorghum-growing areas
CSV 4	1974	3.0–3.58	8.0–9.0	105–110	All rainy sorghum-growing areas and humid areas due to ability to tolerate grain mold
CSV 5	1975	3.0–3.5	8.0–9.0	110–115	All rainy, early *rabi* and summer seasons. Suited for humid areas of TN, KA and MH
CSV 6	1975	3.2–3.5	8.0–9.0	115–120	All rainy sorghum-growing areas
CSV 9	1983	3.0–3.5	8.5–9.0	110–115	All rainy sorghum-growing areas
CSV 10	1985	3.0–3.5	8.5–9.2	110–115	All rainy sorghum-growing areas. Most suitable for Maharashtra, Karnataka, Rajasthan and AP
CSV 11	1985	3.0–3.6	9.5–10.0	110–115	All rainy sorghum-growing areas
SPV 462	1985	3.0–3.3	9.7–10	105–115	Dual-purpose variety, also released as CO 26 in TN and as DSV 2 in KA
CSV 13	1986	3.0–3.5	9.5–10.0	110–115	All rainy sorghum-growing areas
CSV 15	1996	3.5–3.8	11.5–12.5	110–115	All rainy sorghum-growing areas
CSV 17	2002	2.5–3.2	6.5–7.0	95–100	Rainy, low-rainfall and drought-prone areas of the country
CSV 19SS	2004	0.8–1.0	40–45	115–120	Sweet sorghum variety for MH, KA, AP, MP and GJ
CSV 20	2006	3.1–3.2	13.0–13.5	105–110	All rainy sorghum-growing areas
CSV 21F	2006	–	11.3–11.5	110–115	Forage sorghum variety
CSV 23	2007	2.5–3.0	14.0–15.0	110–115	All rainy sorghum-growing areas
CSV 24SS	2011	1.2–1.3	39–40	119	Sweet sorghum variety
CSV 27	2011	2.8–3.0	16.0–17.0	115–120	All rainy sorghum-growing areas, dual purpose
CSV 28	2012	2.8	17	110	Dual-purpose variety
CSV 30F	2012	4.3	13–14	116	Forage sorghum variety
CSV 32	2013	5.0	14.2	107	All rainy sorghum-growing areas
CSV 33	2014	4.5	11.8	96–100	All rainy sorghum-growing areas
CSV 34	2014	4.5	12.9	113	All rainy sorghum-growing areas
CSV 35	2014	4.1	12.6	109	All rainy sorghum-growing areas

Table 1.4b Performance of post-rainy-season-adapted sorghum varieties

Variety	Year of release	Grain yield (t/ha)	Stover yield (t/ha)	Duration (days)	Area for which it is recommended
CSV 7R	1974	2.0–2.5	6.5–7.5	120–125	All post-rainy sorghum-growing areas of MH, KA and AP. Suitable for early planting
CSV 8R	1979	2.5–3.0	7.0–7.5	115–120	All post-rainy sorghum-growing areas of MH, KA and AP. Suitable for early planting, medium to deep soils
CSV 14R	1992	2.2–2.5	5.0–5.5	110–115	All post-rainy sorghum-growing areas
CSV 18	2005	3.5–3.8	8.5–9.0	120–125	Post-rainy sorghum-growing areas of MH, KA and AP
CSV 22R	2007	2.2–2.3	7.0–7.1	116–120	All post-rainy sorghum-growing areas
CSV 26R	2011	1.0–1.1	4.0–4.5	110–115	Post-rainy tract of Deccan, MH, KA and AP
CSV 29R	2012	2.5–2.8	6.5–7.0	118–120	Post-rainy tract of Deccan, MH, KA and AP
M 35-1	1969	2.0–2.5	6.0–6.5	115–120	Post-rainy tract of Deccan, MH, KA and AP
Swati	1985	2.0–2.5	5.0–5.5	120–125	All post-rainy sorghum-growing areas
CSV 216R (Phule Yashoda)	2000	2.0–2.5	7.5–8.0	120–125	All post-rainy sorghum-growing areas, suitable for deep soils
Phule Maulee	1999	1.5–2.0	4.5–5.0	110–115	Suitable for post-rainy season under shallow to medium soils
Phule Anuradha	2008	1.5–1.8	3.0–3.5	100–112	Suitable for shallow soils in rainfed areas
Phule Revati	2010	2.0–2.5	7.5–8.2	118–120	Suitable for deep soils with irrigation
Phule Uttara	2005	1.0–1.2	4.0–4.5	110–115	Suitable for *Hurda* (green tender grains)
Phule Vasudha	2007	3.0–3.5	7.0–7.5	115–120	Suitable for deep soils
Phule Chitra	2006	2.0–2.5	5.5–6.5	110–115	Suitable for medium soils
Selection-3	1994	0.5–0.6	1.5–1.8	90–100	Suitable for shallow soils
Parbhani Moti	2004	3.2–3.5	6.0–6.5	120–125	Suitable for medium to deep soils
PKV Kranti	2006	3.0–3.5	5.0–7.5	115–120	Medium to deep soils and irrigated conditions
Phule Suchitra	2012	2.4–2.8	6.0–6.5	120–125	Suitable for medium soils

hybrids was their relative better grain quality and multiple resistance or tolerance against major pests and diseases. The dual-purpose varieties CSV 15, CSV 20, CSV 23 and CSV 27 could establish higher grain and fodder yield potential than the potential hybrids released earlier (Patil and Mishra 2014).

Though significant improvement in case of rainy-season sorghum could be achieved, post-rainy sorghum did not record as much progress. As most of the post-rainy sorghums are grown under receding moisture situation without any supplementary irrigation, the exploitation of heterosis has been limited. Although several hybrids have been developed and released for post-rainy-season cultivation, the area covered with hybrids is almost negligible. The lack of appropriate hybrids with acceptable grain quality adapted to different agro-ecological situations of post-rainy season characterised by terminal drought, low temperatures and biotic stresses like shoot fly infestation is a major constraint for higher productivity (Rao et al. 1986; Rana et al. 1997). Six hybrids and five varieties were hitherto centrally released for post-rainy season. The first post-rainy-season sorghum hybrid CSH 7R and the latest hybrid CSH 19R were released keeping in view the importance of fodder. The post-rainy-season varieties CSV 8R, CSV 14R, CSV 18 and Swathi were better received than the hybrids CSH 7R and CSH 8R. However, the hybrids CSH 15R and CSH 19R are more productive, but the acceptability among farmers is not high as farmers are hesitant to invest on hybrid seeds during post-rainy-season or dry season without irrigation (Patil and Mishra 2014). Besides nationally released cultivars, a number of state-released varieties are very popular among the farmers in different post-rainy sorghum-growing belts.

1.8 Genetic Gain over Years

Genetic gain for grain yield across years in the Indian sorghum improvement programme was worked out for the period of 1970–2009. The yield gain was prominent in rainy-season hybrid trials (18.5 kg/ha/year), whereas both in post-rainy-season hybrid and varietal trials, it was insignificant. The annual yield gain in rainy-season sorghum is nearly 1.5 times that of post-rainy sorghum. In variety trials, a very high gain in grain yield at 90 kg/ha/year was recorded through the 1980s, after which non-significant changes in grain yield gain were observed. Much progress in rainy-season variety trials after 1985 was not observed. Across the years in India, the gap between potential and farm yield declined 0.32 % per year among rainy-season cultivars and 0.46 % per year among post-rainy-season cultivars. From the study, it was evident that though substantial progress has been made towards yield gain, this was not represented by increased production because of the extensive loss of the sorghum area to other remunerative crops (Rakshit et al. 2014).

1.9 Future Scenario of Sorghum Improvement

The genetic gains in breeding trials coupled with better crop management have impacted the yield gains in India over the years. However, the area under sorghum is rapidly declining over the years, more so in the recent past, because of a number of biotic and abiotic stress factors that limit the yield at farmers' fields. Change in the consumer food habits and easy availability of fine cereals also have impacted the demand for sorghum resulting in lesser people taking up the sorghum farming. Hence, the development of more heterotic hybrids and open-pollinated varieties endowed with tolerance of various biotic stress factors like shoot fly, stem borer and grain mold through deployment of biotechnological tools is the need of the hour. Over the last two decades, much progress has been made in the field of sorghum biotechnology both towards transgenic research and DNA marker studies including genomics (Rakshit and Patil 2014). Molecular breeding strategies can also help in developing cultivars with enhanced productivity under moisture stress with stay-green traits and delayed senescence. With the availability of the whole-genome sequence, it should be possible to target specific genes and employ genomic tools to tailor the sorghum for various future needs.

References

Aruna C (2014) High yielding sorghum cultivars for kharif. In: Chapke RR, Vinayagam SS, Patil JV (eds) Improved sorghum cultivation and value-addition perspectives. Directorate of Sorghum Research, Hyderabad, pp 89–93. ISBN 81-89335-50-2

Aruna C, Audilakshmi S (2008) Reproductive biology and breeding behaviour of sorghum. In: Reddy Belum VS, Ramesh S, Ashok Kumar A, Gowda CLL (eds) Sorghum improvement in the new millennium. International Crops Research Institute for the Semi-Arid Tropics, Patancheru, pp 7–12

Ayyangar GNR, Ponnaiya BWX (1937) The occurrence and inheritance of earheads with empty anther sacs in sorghum. Curr Sci 5:309

Carter PR, Hicks DR, Oplinger ES, Doll JD, Bundy LG, Schuler RT, Holmes BJ (1989) Grain Sorghum (Milo). Alternative field crops manual. University of

Wisconsin & University of Minnesota. http://www.hort.purdue.edu/newcrop/afcm/index.html
Clayton WD (1961) Proposal to conserve the generic name *Sorghum* Moench (Gramineae) versus *Sorghum* Adans (Gramineae). Taxon 10:242
Damania AB (2002) The Hindustan centre of origin of important plants. Asian Agri-Hist 6(4):333–341
de Wet JMJ, Harlan JR, Price EG (1970) Origin of variability in the Spontanea complex of Sorghum bicolor. Am J Bot 57(6):704–707
Doggett H (1970) Sorghum, Tropical agriculture series. Longman Scientific & Technical, London, 403 p
Doggett H (1988) Utilization of grain sorghum. In: Sorghum (ed) Longman Scientific & Technical. Wiley/CBRS, New York
FAO (2014) Statistical database, Food and Agriculture Organization of the United Nations, Rome. http://faostat.fao.org/. Accessed 1 Aug 2014
Garber ED (1950) Cytotaxonomic studies in the genus Sorghum. Univ Calif Publ Bot 23:283–361
Harlan JR (1972) Genetic resources in sorghum. In: Rao NGP, House LR (eds) Sorghum in seventies. Oxford & IBH, New Delhi, pp 1–13
Harlan JR, de Wet JMJ (1972) A simplified classification of cultivated sorghum. Crop Sci 12(2):172–176
House LR (1985) A guide to sorghum breeding, 2nd edn. ICRISAT, Patancheru, 206 p
Hussaini SH, Rao PV (1964) A note on the spontaneous occurrence of cytoplasmic male sterility in Indian sorghums. Sorghum Newsl 27:27–28
ICRISAT (2004) Sorghum, a crop of substance. International Crops Research Institute for the Semi-Arid Tropics, Patancheru, 97 p
Kimber CT (2000) Origins of domesticated sorghum and its early diffusion to India and China. In: Smith CW, Frederiksen RA (eds) Sorghum: origin, history, technology and production. Wiley, New York, pp 3–98
Lee SH, Muthukrishnan S, Sorensen EL, Liang GH (1989) Restriction endonuclease analysis of mitochondrial DNA from sorghum with fertile and male sterile cytoplasms. Theor Appl Genet 77:379–382
Maiti R (1996) Sorghum science. Oxford & IBH Publishing Company, New Delhi, 352 p
Mann JA, Kimber CT, Miller FR (1983) The origin and early cultivation of sorghums in Africa, Bulletin (Texas agricultural experiment station), 1454. Texas Agricultural Experiment Station, College Station
Miller FR (1982) Genetic and environmental response characteristics of sorghum. ICRISAT, Patancheru, pp 393–402
Murdock GP (1959) Staple subsistence crops of Africa. Geogr Rev 50:521–540
Nagur T, Menon P (1974) Characterization of different male sterility-inducing cytoplasms in sorghum. Sorghum Newsl 17:18
Patil JV, Mishra JS (2014) Handbook of sorghum. Astral International Pvt. Ltd., New Delhi, 241 p. ISBN 9789351243250
Pi CP, Wuu KD (1963) The inheritance of cytoplasmic genetic male sterility in sorghum. Bot Bull Acad Sin 4:15–22
Plessis Jean du (2008) Sorghum production. Department of Agriculture, Republic of South Africa. available on the web www.nda.agric.za/publications
Pring DR, Conde MF, Schertz KF (1982) Organelle genome diversity in sorghum: male-sterile cytoplasm. Crop Sci 22:414–421
Quinby JR (1980) Interaction of genes and cytoplasms in male sterility in sorghum. In: Proceedings of the 35th corn and sorghum research conference, American Seed Trade Association, Chicago, Illinois, USA
Quinby JR, Karper RE (1947) The effect of short photoperiod on sorghum varieties and first generation hybrids. J Agric Res 75:295–300
Rakshit S, Patil JV (2014) Sorghum. In: Chopra VL (ed) Breeding field crops II: advances. Studium Press (LLC), Houston, pp 79–96
Rakshit S, Hariprasanna K, Gomashe S, Ganapathy KN, Das IK, Ramana OV, Dhandapani A, Patil JV (2014) Changes in area, yield gains, and yield stability of sorghum in major sorghum-producing countries, 1970 to 2009. Crop Sci 54:1571–1584
Rana BS, Swarnlata Kaul, Rao MH (1997) Impact of genetic improvement on sorghum productivity in India. In: Proceedings of the international conference on genetic improvement of sorghum and pearl millet, 22–27 Sept 1996, Lubbock, Texas, pp 142–165
Rao NGP (1962) Occurrence of cytoplasmic-genetic male sterility in some Indian sorghums. Indian J Genet Plant Breed 22:257
Rao NGP, Tripathi DP, Rana BS (1984) Genetic analysis of cytoplasmic systems in sorghum. Indian J Genet 44:48–49
Rao NGP, Jaya Mohan Rao V, Reddy BB (1986) Progress in genetic improvement of rabi sorghums in India. Indian J Genet Plant Breed 46(2):348–354
Rao SS, Seetharama N, Kiran Kumar KA, Vanderlip RL, Rao SS, Seetharama N, Kiran Kumar KA, Vanderlip RL (2004) Characterisation of sorghum growth stages, NRCS bulletin series no. 14. National Research Centre for Sorghum, Rajendranagar, 20 p
Reddy BVS, Stenhouse JW (1994) Sorghum improvement for the semi-arid tropic region: past, current and future research thrust in Asia. Punjabrao Krishi Vidyapeeth Research J 18:155–170
Ross WM, Heckerott HL (1972) Registration of seven isocytoplasmic sorghum germplasm lines. Crop Sci 12:720
Sane AP, Nath P, Sane PV (1996) Cytoplasmic male sterility in sorghum: organization and expression of mitochondrial genes in Indian CMS cytoplasms. J Genet 75(2):151–159
Schertz KF, Ritchey JM (1978) Cytoplasmic-genic male sterility systems in sorghum. Crop Sci 18:890–893
Snowden JD (1936) Cultivated races of sorghum. Allard and Sons, London, p 274

Snowden JD (1955) The wild fodder sorghums of the section Eu-sorghum. J Linn Soc Bot (London) 55:191–260

Stephens JC (1937) Male sterility in sorghum: its possible utilization in production of hybrid seed. J Am Soc Agron 29:690

Stephens JC, Quinby JR (1945) The Ms_2AV_{10} linkage group in sorghum. J Agric Res 70:209–218

Stephens JC, Holland RF (1954) Cytoplasmic male-sterility for hybrid sorghum seed production. Agron J 46:20

Vanderlip RL, Reeves HE (1972) Growth stages of sorghum. Agron J 64:13–16

Vavilov NI (1992) In: Dorofeyev VF (ed) Origin and geography of cultivated plants. Cambridge University Press, Cambridge, 498 p

Webster OJ, Singh SP (1964) Breeding behavior and histological structure of non-dehiscent anther character in *Sorghum vulgare* pers. Crop Sci 4:656–658

Worstell JV, Kidd HJ, Schertz KC (1984) Relationships among male sterility inducing cytoplasms of sorghum. Crop Sci 24:186–189

Xu GW, Cui YX, Schertz KF, Hart GE (1995) Isolation of mitochondrial DNA sequences that distinguish male-sterility inducing cytoplasms in *Sorghum bicolor* (L.) Moench. Theor Appl Genet 90:1180–1187

Part II

Advances in DNA Marker Research

DNA Markers in Diversity Analysis

2

Sujay Rakshit and M. Swapna

Contents

2.1 **Introduction** 24

2.2 **Assaying Genetic Diversity Using Different Marker Systems** 24

2.3 **Approaches for Diversity Analysis** 25
2.3.1 Quantifying Genetic Diversity 26
2.3.2 Quantifying Genetic Relationships 27
2.3.3 Displaying Relationships 29
2.3.4 The Use of Diverse Data Sets in Diversity Analysis 32

2.4 **Molecular Diversity Among Cultivated and Wild Sorghums** 34

2.5 **Functional Diversity Analysis Using Gene and EST-Based Markers** 37

2.6 **Establishment of Core and Mini-Core Collections Using Molecular Markers** 38

2.7 **Applications of Diversity Analysis for Genetic Improvement in Sorghum** 40

2.8 **Conclusions** 42

References 42

Abstract

The study on genetic diversity is critical to success in plant breeding, as it provides information on the quantum of genetic divergence, which serves as a platform for specific breeding objectives. Parental combinations likely to create segregating progenies with maximum genetic potential for further selection, designing introgression program, and selection of parental combinations toward maximization of heterosis are dependent on diversity analysis. Genetic diversity may be assessed using different marker systems, which encompasses morphological, biochemical, and molecular (DNA) markers. With recent advances in genomics research, DNA markers assume much more significance. Using different marker systems, genetic diversity in crop plants may be accessed at species level, at the population level, among germplasm accessions, at an individual genotype level like among pure lines or clones, etc. Intra- and inter-population diversity is measured by various statistical measures, which depend on the type of data set and objective of the study. The relationship between individuals may be displayed using cluster analysis, principal component analysis, principal coordinate analysis, and multi-dimensional scaling. Large array of statistical packages are available to conduct diversity analysis, and a considerable progress has been made with sorghum in analyzing diversity using morphological and molecular

S. Rakshit (✉)
Plant Breeding, ICAR-Indian Institute of Millets Research, Rajendranagar, Hyderabad, Telangana 500 030, India
e-mail: sujay@millets.res.in

M. Swapna
ICAR-Indian Institute of Millets Research, Rajendranagar, Hyderabad, Telangana 500 030, India

R. Madhusudhana et al. (eds.), *Sorghum Molecular Breeding*, DOI 10.1007/978-81-322-2422-8_2

markers alone or in combination. In recent past, functional diversity is being assayed using gene and EST-based markers. Using various strategies, core and mini-core collections have been established in sorghum, which are a very important resource for genomic studies in the crop.

Keywords
Sorghum • Genetic distance • Molecular markers • Core collections • Functional markers

2.1 Introduction

Crop improvement has played a very important role in addressing the food needs of the growing human population since time immemorial. The basic principle of plant breeding is selection on existing variation. The variation may be genetic (fixable) or environmental (non-fixable), which may be natural or created. Without genetic variation for a trait, no progress in plant breeding is possible. Thus, the study on genetic diversity is critical to success in plant breeding, as it provides information on the quantum of genetic divergence, which serves as a platform for specific breeding objectives. In a broader sense, classification or grouping of an individual or population compared to other individuals or populations is referred to as genetic diversity. It is always a relative measure, as the distance between any pair of entries in the study may be greater or lesser depending on all pair-wise comparisons that can be made in the study. Contrary to genetic diversity, through genetic fingerprinting, an individual or a population may be identified unambiguously based on the presence or absence of specific alleles at different marker loci among individuals or based on frequencies of alleles of the markers in a population. Fingerprinting is an absolute measure and does not change depending on other individuals or populations under study (Abdel-Mawgood 2012). Through diversity analysis, parental combinations can be identified which is likely to create segregating progenies with maximum genetic potential for further selection (Aremu 2011). Genetic variability in diverse populations can help us in designing the introgression program. Knowledge on genetic relationship among parents increases the possible heterosis in hybrids and reduces the chance of reselection within related germplasm. Breeders mostly work with a narrow genetic base; hence, there is a need to create variability within breeding populations using genetically diverse materials. This chapter deals with different marker systems and approaches to assess genetic diversity in crop plants, with particular reference to sorghum.

2.2 Assaying Genetic Diversity Using Different Marker Systems

Several researchers of diverse crops have employed different data sources and data types to study genetic diversity. Conventionally, morphological traits have been extensively used in deciphering genetic diversity in crop plants (Rakshit et al. 2012). The use of morphological traits to characterize germplasm at the species level is easier, but the identification of genotypes within a species based on morphological traits alone is relatively difficult (Dhaliwal et al. 2009). Moreover, it is now well established that morphological characters alone do not reliably portray the genetic relationships among the genotypes (Kumar et al. 2008). This is mainly because the morphological expression is stage or organ specific, often influenced by the environment; genetic control of these traits is often complex; they are limited in number, often intra- and intergenic interactions; genotype × environment interactions make scoring of such traits difficult; and only a fraction of genome is accessed with such traits (Rakshit et al. 2002). Biochemical markers have played a limited role to address the limitations of morphological markers. DNA markers could address these limitations effectively and have become the markers of choice across plant species including sorghum (Deu et al. 2006; Medraoui et al. 2007; Ritter et al. 2007; Wang et al. 2009; Rakshit et al. 2012; Ganapathy et al. 2012; Billot et al. 2013; Ramu et al. 2013).

DNA markers, also referred to more commonly as molecular markers, are based on DNA sequence polymorphisms. DNA sequence variations determine the genetic diversity of organisms. Hence, by evaluating polymorphisms directly at the DNA level, the genetic diversity can be measured. Using DNA markers, which follow Mendelian inheritance, it is possible to trace the evolutionary history of the species by phylogenetic analysis, to study genetic relationship and population structures, and to map such markers (Hoshino et al. 2012). Based on technical principles, molecular markers may be grouped into four classes: (1) nucleic acid hybridization-based markers, e.g., restriction fragment length polymorphisms (RFLPs) (Botstein et al. 1980); (2) polymerase chain reaction (PCR)-based markers, e.g., random amplification of polymorphic DNAs (RAPD) (Williams et al. 1990), amplified fragment length polymorphisms (AFLP) (Vos et al. 1995), microsatellites, or simple sequence repeats (SSRs) (Zietkiewicz et al. 1994); (3) single nucleotide polymorphisms (SNPs) (Chen and Sullivan 2003); and (4) array-based platforms like Diversity Arrays Technology (DArT) (Kilian et al. 2005), restriction site-associated DNA (RAD) (Miller et al. 2007a, b), single feature polymorphism (SFP) (Borevitz et al. 2003), etc. Using the technical features of mentioned marker systems, some additional molecular markers are also used for diversity analysis. For example, internal transcribed spacer (ITS) of eukaryotic systems is being used for diversity analysis either by direct sequence comparison of the region or by restriction digestion of PCR-amplified product of ITS regions. Another is the use of chloroplast DNA in inter-specific diversity analysis. With rapid progress in next-generation sequencing (NGS) tools, now sequencing has become much faster and cheaper. Genotype by sequencing (GBS) is a recent method to detect large sequence polymorphism in a very short period of time at an affordable cost (Elshire et al. 2011).

Among the techniques mentioned, RFLPs (codominant marker) though highly reproducible are sparingly used due to cost and technical difficulty. With the advent of PCR technology in the mid-1980s (Saiki et al. 1985; Mullis and Faloona 1987), the approach of marker analysis has dramatically changed (Hoshino et al. 2012). Though very easy to handle, RAPD's (dominant marker) use has reduced drastically due to its low reproducibility. Among the PCR-based markers, microsatellites are most popularly used as a DNA marker for all practical purposes across plant species because they are hypervariable, codominant, robust, chromosome specific, and multi-allelic in nature (Kumar et al. 2008). These markers not only appear in chromosomal DNA but on eukaryotic organellar genomes as well (Abdel-Mawgood 2012). AFLP markers, though hypervariable, are not under extensive application due to its technical complications and dominant nature. Array-based and NGS-based markers are very potent, but their large-scale application in the crop diversity analysis is yet to be demonstrated. Each marker system differs in terms of their cost, technical complications, consistency and reproducibility of results, and the ability to detect polymorphism (Schlotterer 2004; Schulman 2007; Bernardo 2008). This chapter is beyond the scope to go into the details of these tools. Technical details and application of various marker systems in molecular diversity analysis have been reviewed by various authors (Rakshit et al. 2002; Arif et al. 2011; Abdel-Mawgood 2012).

2.3 Approaches for Diversity Analysis

Genetic diversity in crop plants is analyzed at different levels: at species level, at population level, among germplasm accessions, at an individual genotype level like among pure lines or clones, etc. As the nature of each genetic material is different, sampling strategies also vary from case to case (Mohammadi and Prasanna 2003). Generally, the number of individuals sampled per population, the number of loci sampled, genotypic and allelic compositions of population, reproductive system, and effective population size influence the sampling variances of diversity measures (Nei and Chesser 1983; Namkoong

1988; Weir 1990). Three basic steps are involved in the process of diversity analysis: (1) quantifying genetic diversity, measure of intra- and inter-population genetic diversity; (2) quantifying genetic relationships, diversity and differentiation at the nucleotide level and calculation of the distances (geometric or genetic) among all pair of subjects; and (3) displaying the relationships, classification or clustering and ordination. The variables used for this may be qualitative (binary or categorical – ordinal and nominal) or quantitative (continuous or discrete). Categorical as well as quantitative variables can be converted to binary variables by making groups and defining each group as present (1) or absent (0). However, such grouping has its own limitations because weight will be given to the category of a character while calculating similarity coefficients, and the more categories a variable will have, the more weight it will receive. While the variables with higher category will be combined with other binary or categorical variables with few categories, the variable will receive more weight, and the result will be biased.

2.3.1 Quantifying Genetic Diversity

2.3.1.1 Intrapopulation Genetic Diversity

This may be quantified either by the number of variants (rate of polymorphism, *pj*; proportion of polymorphic loci; the richness of allelic variants, *A*; and average number of alleles per locus, *n*) or by the frequency of variants (effective number of alleles, A_e; average expected heterozygosity, H_e; or Nei's genetic diversity, *D*). A locus is considered polymorphic when the most common allele has a frequency of less than 0.95, and a rare allele has frequencies less than 0.005. The proportion of polymorphic loci *P* is defined by the formula, n_{pj}/n_{total}, where n_{pj} is the number of polymorphic loci and n_{total} is the total number of loci. The percentage of variable loci in a population is expressed by this parameter. In codominant marker system, we can differentiate homozygotes and heterozygotes, and this expression can be employed more effectively in such marker systems than dominant markers. Richness of allelic variants (*A*) indicates the number of variants in a sample. It is sensitive to the sample size and can only be applied with codominant markers. Average number of alleles per locus *n* is $(1/K\sum_{i=1}^{k} n_i)$, where *K* is the total number of loci and n_i is the number of alleles detected in the *i*th locus. It is also best applied with codominant markers. An effective number of alleles (A_e) are calculated by the formula $(1/\sum_{i}^{n} p_i^2)$, where p_i is the frequency of the *i*th allele in a locus. The measure tells about the number of alleles that would be expected in a locus in each population. Average expected heterozygosity (H_e) or Nei's genetic diversity (*D*) is the probability that, at a single locus, any two alleles, chosen at random from the population, are different to each other. Three calculations are possible in this regard: (1) a locus with two alleles ($h_j = 1 - p^2 - q^2$), (2) a locus *j* with *i* alleles ($h_j = 1 - \sum p_i^2$), and (3) average for several loci ($H = \sum_{j}^{l} h_j / l$), where h_j is heterozygosity per locus, *p* and *q* are allele frequencies, and *l* is the total number of loci. H_e estimates the extent of genetic variability in the population, and it ranges from 0 to 1 and maximizes when there are many alleles with equal frequencies.

2.3.1.2 Inter-population Genetic Diversity

This may be quantified by inter-population differentiation for one locus (g_{ST}), inter-population differentiation for several loci (G_{ST}), population's contribution to total genetic diversity, F statistics (Wright), and analysis of molecular variance (AMOVA). g_{ST} is calculated as $1-(h_S/h_T)$, where $h_s = \tilde{n}/(\tilde{n}-1)[1-(1/s)\sum\sum x_{ij}^2 - h_0/2\tilde{n}]$ and $h_T = 1 - \sum[(1/s)\sum x_{ij}]^2 + (h_S/\tilde{n}s) - (h_0/2\tilde{n}s)]$; $\tilde{n}$ is the harmonic average of population size, *s* is the number of population, h_0 is the average observed heterozygosity, and X_{ij} is the estimated frequency of the *i*th allele in the *j*th population. Codominant markers are more suited to such estimation as compared to dominant markers. $G_{ST} = D_{ST}/H_T$, where H_T is total genetic diversity ($H_S + D_{ST}$), H_S is intrapopulation genetic diversity, and D_{ST} is inter-population diversity. The next parameter for inter-population genetic diversity, population's contribution to total gene diversity, is calculated by removing a population so that its contribution to the total gene diversity may be evaluated as $C_{T(K)} = (H_T - H_{T/K})/H_T$, $C_{S(K)} = (H_S - H_{S/K})/H_T$, and $C_{ST(K)} = (D_{ST} - D_{ST/K})/H_T$. $C_{T(K)}$ is the contribution of K to total diversity, $C_{S(K)}$ is the contribution of K to intrapopulation diversity, $C_{ST(K)}$ is the contribution of K to inter-population diversity, H_T is the total gene diversity, H_S is the intrapopulation genetic diversity, D_{ST} is the inter-population diversity, $H_{T/K}$ is the total gene diversity after removing population K, $H_{S/K}$ is the intrapopulation gene diversity after removing population K, and $D_{ST/K}$ is the inter-population gene diversity after removing population K. Only codominant markers can be used for such analysis. Wright's F statistic refers to genetic structure of populations, and the equation for its calculation is $(1 - F_{IT}) = (1 - F_{IS})(1 - F_{ST})$, where $F_{IT} = 1 - (H_I/H_T)$, $F_{IS} = 1 - (H_I/H_S)$, and $F_{ST} = 1 - (H_S/H_T)$; H_T is the total gene diversity or expected heterozygosity in the total population as estimated from the pooled allele frequencies, H_I is the intrapopulation gene diversity or average observed heterozygosity in a group of populations, and H_S is the average expected heterozygosity estimated from each sub-population. F statistics may be used to analyze structures of sub-divided populations and to measure the genetic distance among these. The underlying assumption of this analysis is that those sub-populations do not intermate and have different allele frequencies to those of the total population. F_{ST} from 0 to 0.05 signifies small, 0.05 to 0.15 signifies moderate, 0.15 to 0.25 signifies large, and >0.25 signifies very large genetic differentiation between two populations. Variation within a species can also be measured by the analysis of molecular variance (AMOVA). A hierarchical or nested model is used for this analysis. In AMOVA, the hierarchical levels of gene diversity may include (1) continents, which may contain lesser hierarchical levels; (2) geographic regions within a continent; (3) areas within a region in a continent; (4) populations within an area of a region in a continent; or (5) individuals within a population in an area of a region in a continent.

2.3.2 Quantifying Genetic Relationships

This deals with quantifying diversity and differentiation at nucleotide levels using sequencing data or restriction data and quantifying genetic distances. To understand diversity and differentiation at the nucleotide level, each nucleotide is assumed as locus.

2.3.2.1 Nucleotide Diversity

Intrapopulation nucleotide diversity is denoted by π, and $\pi = n/(n-1)\sum X_i X_j \pi_{ij}$, where n is the number of sequences being analyzed among the individuals of the population, X_i is the estimated frequency of the ith sequence in the population, X_j is the estimated frequency of the jth sequence in the population, and π_{ij} is the proportion of different nucleotides between sequences i and j. The extent of nucleotide diversity among several sequences in a given region of the genome is measured by this parameter, which ranges from 0 to 1, and is equivalent to the measure of allelic diversity within a locus. This parameter informs about nucleotide sequences, and the model assumes haplotypes (haploid genotypes). On the other hand, inter-population nucleotide diversity is measured by four parameters, viz., V_{XY} (measures population divergence based on the degree of sequence variation in 1 sequence and 2 populations), V_W (measures average diversity in a population based on several sequences), V_b (measures the total differentiation in several populations), and N_{ST} (measures the relative differentiation) according to following formulae:

$$V_{XY} = d_{XY} - (\pi_X + \pi_y)/2$$

$$V_b = [1/\{s(s-1)\}]\sum_X\sum_Y V_{XY}$$

$$V_W = 1/s\sum\pi_X$$

$$N_{ST} = V_b/(V_b + V_W),$$

where V_{XY} is divergence among populations X and Y, π_X is the nucleotide diversity in population X, d_{XY} is the probability that two random nucleotides in populations X and Y be different, and s is the number of populations. The level of differentiation among nucleotide sequences in populations is known from this analysis. Sequence data from a sample of individuals in each population is needed for this analysis. It requires specific computer software like CLUSTAL W, MALIGN, and PAUP, having sequence alignment features for calculation. Using restriction data, the intrapopulation nucleotide diversity (π) can be measured by the formula $\pi = -1/r(In\ G)$, where r is the number of nucleotides recognized at the restriction site and $ln\ G$ is the natural logarithm of the probability of no substitution in the restriction site.

Inter-population diversity is measured as $V_{XY} = d_{XY} - (\pi_X + \pi_Y)/2$, where V_{XY} is divergence between populations X and Y, π_X and π_Y are restriction diversity in the populations, and d_{XY} is fragment diversity among two populations. d_{XY} is equal to $-(2/r)\ ln\ (G_{XY})$, where $G_{XY} = F_{XY}(3 - 2G°_{XY})1/4$. $G°$ is $F_{XY}{}^{1/4}$, where F_{XY} is the proportion of shared alleles among populations X and Y and calculated as $\{2\sum X_{iX} X_{iY})/(\sum(X_{iX} + X_{iY})\}$, where X_{iX} is estimated frequency of the i fragment in population X. Diversity in the restriction sites of a sample of two or more populations is estimated with this parameter. Software like BIOSYS and GENEPOP can be used for this calculation. RAPD data can also be used in this analysis, where the value of "r" is replaced by the primer length ($r = 10$).

2.3.2.2 Genetic Distance

Genetic distance (GD) is "any quantitative measure of genetic difference, be it at the sequence level or the allele frequency level, that is calculated between individuals, populations or species" (Beaumont et al. 1998). The GD may be calculated following two models: equilibrium model and disequilibrium model. Equilibrium model assumes that the distance remains constant over time, i.e., equilibrium always exists between migration and genetic drift, while in disequilibrium model, distance changes with time due to migration and genetic drift. For all practical purposes, the second model is used. Distances between operational taxonomic units (OTUs – individuals, accessions, or populations) may be of two types, geometric distance or genetic distance. In diversity studies where comparisons are made using morphological or marker data collected from the OTUs, geometric distances are used. In this calculation, evolutionary aspects are not considered. Hence, the dendrograms obtained cannot be interpreted as phylogenetic trees giving information about evolution or divergence among groups. In contrast, the genetic distance of any given OTU can be incorporated into phylogeny studies. It can be used with both codominant and dominant markers. However, in case of dominant markers, two generations of the same population need to be studied to measure the segregation of loci (Lynch and Milligan 1994).

Various statistical measures are used to measure the GD, which depends on the type of data set and objective of the study. Metroglyph and score index are often used to study morphological variation. The metroglyph analysis is a semi-graphical analysis of complex problem using the index score method as described by Anderson (1957). In this analysis, a scatter diagram is first plotted taking two most variable characters. Subsequently, other morphological characters are represented as rays at different positions on the glyph. Each germplasm line bears a serial number and is represented as a glyph which is the intersection point of mean values of X and Y coordinates. The sum of index values with regard to all the characters allotted to an individual is the indication of the individual worth. The performance of a genotype is adjudged by the value of the index score of that genotype. Euclidian or straight-line measure of distance is a frequently used statistic for estimating GD between OTUs and is calculated as $d_{(i,j)} = [(x_1 - y_1)^2 + (x_2 - y_2)^2 + \ldots + (x_p - y_p)^2]^{1/2}$, where i and j are two individuals having morphological characters (p) as $x_1, x_2, \ldots, x_p$ and $y_1, y_2, \ldots, y_p$, respectively (Mohammadi and Prasanna 2003). Smith et al. (1991) denoted Euclidian distance, $d_{ij} = \sum\left[T_{1(i)} - T_{2(i)}\right)^2 / \mathrm{var}T_{(i)}]^{1/2}$, where T_1 and T_2 are the values of ith trait in lines 1 and 2, respectively, and $varT_{(i)}$ is the variance of ith trait. Gower's (1971) measure of distance accommodates various types of characters like dichotomous, qualitative, and quantitative, and it is calculated between individual i and j as $DG_{ij} = 1/p\sum W_k d_{ijk}$, where p is the number of character and d_{ijk} is the contribution of kth character to the total distance between two individuals; $d_{jk} = |x_{ik} - x_{jk}|$, where x_{ik} and x_{jk} are the values of the kth character for i and j individuals, respectively; and $w_k = 1/R_k$, where R_k is the range of the kth character in the sample (Franco et al. 1997).

Using molecular data, common measure of distance is that of Rogers (1972), which is calculated as $RD_{jk} = 1/2\left[\sum\left(X_{ai} - X_{aj}\right)^2\right]^{1/2}$, where X_{ai} and X_{aj} are the frequencies of allele a in individuals i and j. With binary variables, similarity or dissimilarity matrices are formed between all the possible pairs of OTUs. Similarity is estimated by the number of coincidences and following commonly used estimates:

Nei and Li's (1979) coefficient, $GD_{NL} = 1 - [2N_{11}/(2N_{11} + N_{10} + N_{01})]$

Jaccard's (1908) coefficient, $GD_J = 1 - [N_{11}/(N_{11} + N_{10} + N_{01})]$

Simple matching coefficient (Sokal and Michener 1958), $GD_{SM} = 1 - [(N_{11} + N_{00})/(N_{11} + N_{10} + N_{01} + N_{00})]$

Modified Rogers' distance, $GD_{MR} = [(N_{10} + N_{01})/2N]^{1/2}$,

where N_{11} is the number of alleles present in both OTUs, N_{00} is the number of alleles absent in both OTUs, N_{10} is the number of alleles present in OTU i, and N_{01} is the number of alleles present in OTU j (Mohammadi and Prasanna 2003). The GD_{SM} considers that absence corresponds to homozygous loci. It can be used with dominant marker data (RAPD and AFLP) since absences correspond to homozygous recessives. The GD_J counts alleles present for either individual (i or j), and double absences are considered as missing data. In case of false-positive or false-negative data, the measure turns to be biased. The GD_{NL} counts the percentage of shared alleles among two individuals and gives more weight to those alleles that are present in both. In this, the absence of an allele is given less significance. GD_J and GD_{NL} can be applied with codominant marker data (RFLP, SSR). GD_J and GD_{NL} estimates may differ in the presence of heterozygous loci (Link et al. 1995). GD_{SM} and GD_{MR} are a kind of Euclidean measures of distance. Euclidean metric properties of GD_{SM} allow its use in hierarchical clustering strategies (Mohammadi and Prasanna 2003).

Several models are used to estimate genetic distance. Common ones are *mutation of infinite alleles* (e.g., Nei's genetic distance) and *stepwise mutation model* (e.g., distance using microsatellites). In the first model, each mutation event is presumed to give rise to a new allele (e.g., isozymes). It is assumed that if two genes are the same then, no mutation has occurred and in case of two genes being different, an unknown number of mutations have occurred. Since time t when two genes diverged from an ancestor, the average number of mutations is $2t\mu$, where μ is the rate of mutation. μ is multiplied by 2 because two independent genes are being considered. The probability that two genes are common by descent after time t is $P = e^{-2t\mu}$. In the second model, the assumption is that mutation is a progressive change and the fragments migrating similar distances have fewer mutations. This model is applicable for the SSRs, where mutation is assumed to change the number of repeats, increasing or decreasing step by step. It is observed that the square of the difference in the number of repeats between two microsatellites is proportional to the time of divergence from a common ancestor.

2.3.3 Displaying Relationships

With an increase in sample size in diversity analysis, methods to classify and order the variability assume much importance. Displaying relationships (classification or clustering) is the process of grouping (or clustering) objects in categories or classes based on their common attributes or relationships. Multivariate analysis tools can simultaneously analyze multiple measurements on an individual and are being widely employed to present diversity based on wide data set. Many algorithms are used for this purpose. Some commonly used algorithms are cluster analysis, principal component analysis (PCA), principal coordinate analysis (PCoA), and multi-dimensional scaling (MDS).

2.3.3.1 Cluster Analysis

Cluster analysis is "a group of multivariate techniques whose primary purpose is to group individuals or objects based on the characteristics they possess, so that individuals with similar descriptions are mathematically gathered into same cluster" (Hair et al. 1995). Broadly, two clustering methods are in use: first, distance-based methods, where a pair-wise distance matrix is analyzed by the clustering algorithm resulting in graphical representations (dendrogram or tree) where clusters are visually identified, and, second, model-based methods, in which statistical methods like maximum-likelihood or Bayesian methods are used to infer about parameters corresponding to each cluster and cluster members (Mohammadi and Prasanna 2003). Distance-based clustering methods are most commonly used in diversity analysis in crop plants, and it may be hierarchical, non-hierarchical, or overlapping. Hierarchical grouping proceeds most commonly by a series of successive mergers of a group of individuals. This grouping starts with single individuals, where most similar individuals are first grouped together and progressively similar groups are merged to form the full dendrogram for the individuals. These methods are referred to as agglomerative hierarchical methods. Three methods, viz., simple linkage, complete linkage, and average linkage methods, are used in the process. The main difference in these three methods is the way the proximity of groups is defined. Simple linkage method minimizes the inter-group distance by taking the distance to the neighbor with highest similarity. Because of this reason, simple linkage method is also referred to as "nearest neighbor" method. Complete linkage method minimizes the inter-group distance by taking the distance to the individual with minimal similarity, i.e., farthest neighbor. Both these methods work well in regular and compact groups but become difficult to perform when different groups are not well distributed in space. Average linkage minimizes the inter-group distance by taking the average pair-wise distances among all individuals in the sample. This is often referred to as UPGMA (unweighted paired group method using arithmetic averages) (Sneath and Sokal 1973; Panchen 1992). This is the most commonly used algorithm and it follows Ward's minimum variance method (Ward Jr 1963). In non-hierarchical grouping, each individual is assigned to a unique group by comparing it with the initial classes so that its positioning is the most appropriate (Everitt 1980). In this, tree or dendrogram is not created and it is rarely used to study intraspecific diversity in crops. In overlapping groups, individuals may belong to more than one group.

While choosing clustering method, the first step is to gather knowledge of the species under study in terms of its diversity, reproduction system, ploidy number, and levels of heterozygosity. This is followed by careful selection of the genetic characters to be analyzed. Finally, it is most essential to employ different clustering methodologies and assess the level of agreement obtained with each of them. "Whatever algorithm is used for generating the dendrogram, it is useful to carry out bootstrapping of the allele frequencies (followed by calculation of genetic distances, etc.) to assess the reliability of the nodes" (Mohammadi and Prasanna 2003).

Among the various clustering methods, a single method may not be always effective in explaining genetic associations. Most of the comparative studies between different clustering algorithms, UPGMA followed by Ward's method in combination with GD_J or GD_{SM} coefficients, provided more consistent results (Milligan and Cooper 1985; Peeters and Martinelli 1989; Kantety et al. 1995; Rincon et al. 1996; Lombard et al. 2000). The efficiency of different clustering algorithms may be compared by estimating "co-phenetic correlation coefficient." It is a product–moment correlation coefficient, which measures the agreement between the similarity/dissimilarity matrices as input of cluster analysis and the dissimilarity–similarity indicated by the phenogram as output of analysis. Co-phenetic correlation coefficient $r \geq 0.9$ indicates a very good fit; $0.8 \leq r \leq 0.9$ indicates good fit; $0.7 \leq r \leq 0.8$

indicates poor fit; and $r<0.7$ is indicative of a very poor fit (Rohlf 1992). It may be noted that this interpretation is subjective as statistical test for this is not possible as an individual coefficient in the dissimilarity matrices is not independent (Rincon et al. 1996). The low r values are only indicative of the distortion that might have occurred while constructing the phenogram, not the utility of the output. The congruence of the cluster information derived from the phenogram to prior idea about the structure of groups according to passport information is another simple way to compare different clustering methods. The basic assumption of most of clustering algorithm is that data are ultrametric, which mostly constraints all mentioned algorithms that can be addressed using alternative methods like neighbor joining or Fitch–Margoliash method (Swofford et al. 1996).

2.3.3.2 Principal Component Analysis (PCA) and Principal Coordinate Analysis (PCoA)

Both PCA and PCoA can be used to obtain two- and three-dimensional scatter plot of individuals, in such a way that the genetic distances among the individuals are reflected in the geometrical distances in the plot. PCA is a data reduction method in which total variation in the original characters is reduced into a limited number of uncorrelated new variables, called principal components (PCs) (Wiley 1981). In this analysis, eigenvalues, which define the amount of total variation displayed on the PC axes, are calculated first. PC1 summarizes maximum variability present in the original data, followed by PC2 and so on, which were not explained by the previous PCs (Jolliffe 1986). PCA is applied on two types of data matrices: a variance–covariance matrix and a correlation matrix. When data of two characters are taken on the same scale, variance–covariance matrix is used; otherwise correlation matrix is used. PCA also helps in determining the optimum cluster numbers in a study, where the objective is to maximize the variation explained by the first PC of each cluster. PCoA starts with similarities or dissimilarities between individuals and produces a low-dimensional plot of the data such that the distances between the points in the plot are close to original dissimilarities. PCoA is recommended over PCA when there are lots of missing data, and/or a number of individuals are lesser than characters (Rohlf 1972). PCA and PCoA are useful when the first two or three PCs explain most of the variation. Often PCA or PCoA is used as a pattern-finding method to complement cluster analysis. This is dependent on how much variability is explained by the first 2–3 PCs. Melchinger (1993) demonstrated that if the first two or three PCs explain <25 % of variability, cluster analysis is more sensitive and reliably depicts pedigree relationship, than PCA or PCoA. However, the major advantage of ordination methods over cluster analysis is that these methods help in the identification of individuals intermediate between two groups (Lessa 1990).

2.3.3.3 Multi-dimensional Scaling (MDS)

MDS also uses similarity/dissimilarity matrix between a set of individuals (n) to put them in a few dimensions (m) in such a way that inter-individual proximities approximate the original similarities/dissimilarities (Johnson and Wichern 1992). Differences between close individuals are reflected better by MDS as compared to PCA of PCoA. In MDS, smaller or greater distances between individuals are not represented by the same scale.

Whatever grouping is obtained through any of the algorithm used, the grouping needs to be validated through external, internal, relative, or resampling techniques. In external validation, the distance matrix is compared with other information like genealogy. Internal validation quantifies the distortion due to grouping methods through co-phenetic matrices (Sokal and Rohlf 1994), which has been discussed earlier. In relative validation, similarity between different methods is compared. Common resampling techniques are bootstrapping and jackknife. The techniques resample the actual data to find out its subtler patterns. Bootstrapping is a resampling method by replacement with the same data matrix, which allows calculation of standard errors, confidence intervals, and other measures of statistical accuracy.

This procedure is particularly effective to estimate the statistical support for the internal branches in a tree (Felsenstein 1985). In general, it may be said that internal tree branches with >70 % bootstrap value are true at 95 % confidence level (Hillis and Bull 1993). Parametric bootstrapping is preferable than non-parametric bootstrapping to understand genetic relationships. Contrary to bootstrapping, resampling is performed without replacement in jackknife technique (Efron 1979). Though it is a simple resampling technique, its inherent limitations are that the number of resampling units is limited and much information on the distribution of estimates cannot be obtained. Numerous software are available for assessing genetic diversity, which are mostly available freely in the public domain. Many perform similar tasks, with the main differences being in the user interface, type of data input and output, and platform (Table 2.1).

2.3.4 The Use of Diverse Data Sets in Diversity Analysis

Input data in diversity analysis may be morphological (qualitative and quantitative), biochemical, and molecular. These may be used individually or together in grouping of individuals. When characters with different scales are used as input data for cluster analysis, data may be standardized by dividing each variable with means of either its standard deviation or its range. Thus, equal weightage is provided to all the characters. Milligan and Cooper (1985) suggested standardization by range to be a better option than the standard deviation. In case of qualitative and molecular marker data, which are normally binary in nature, such standardization is not needed. Morphological trait data may be transformed to binary data as described by Sneath and Sokal (1973). While using different data sets, two main questions arise: firstly, whether analysis should be carried out based on individual data set or combined data set and, secondly, how to combine the data effectively. In this process, the most important point of consideration is congruence among the results derived from individual data sets. Co-phenetic correlation among the results obtained from different marker system gives an idea about the congruence of different results. However, consensus on the utility of combining data from different marker systems is lacking. Combining data from different marker systems should be carried out cautiously because many of the morphological traits may be highly correlated and may lead to bias in the results. Assigning different weight to different traits may address the inaccuracy of traits in terms of their genetic nature and contributions to genetic diversity. However, this is also not foolproof as there is no consensus on weight to be given to different traits. In this regard, combining parentage and genetic marker information may give a better estimate of genetic relationship (Souza and Sorrells 1991).

Information on the combination of morphological and molecular data in studying sorghum diversity is scanty. Geleta et al. (2006) combined qualitative data of 10 traits and quantitative data of 16 traits into AFLP and SSR data in studying diversity of 45 sorghum accessions by converting the morphological traits to binary data. Their study showed that molecular markers were slightly more efficient than morphological traits in estimating genetic diversity. Co-phenetic correlation study of the diversity indices obtained by different marker systems showed that estimated values of genetic relationship given for AFLP and SSR markers, and SSR and morphological markers were significantly related, though AFLP showed non-significant correlation with morphological traits. Dendrogram obtained by different marker systems provided largely similar information, though molecular markers were more efficient in differentiating all the lines, than morphological traits. In a similar study, Rakshit et al. (2012) also demonstrated that SSR markers in combination with morphological (both quantitative and qualitative) traits can successfully differentiate closely related accessions of *Maldandi* sorghum. They reported 38 % congruence between the groupings obtained using quantitative, qualitative, and SSR data. Though some information in this regard has been generated, still consensus on the ways to merge different

Table 2.1 Software used for diversity analysis

Software	Author	Availability
Arlequin	Excoffier L and Lischer HEL	http://cmpg.unibe.ch/software/arlequin35/
DnaSP	Rojas J and Rozas R	http://www.ub.es/dnasp
PowerMarker	Liu K and Muse SV	http://statgen.ncsu.edu/powermarker/index.html
MEGA2	Kumar S and others	http://www.megasoftware.net
PAUP	David Swofford	http://paup.csit.fsu.edu/
TFPGA	Miller MP	http://www.marksgeneticsoftware.net/
GDA	Lewis PO and Zaykin D	http://tomato.biol.trinity.edu/programs/index.php/Genetic_Data_Analysis
GENEPOP	Raymond M and Rousset F	http://wbiomed.curtin.edu.au/genepop
NTSYSpc	Rohlf FJ	http://exetersoftware.com/cat/ntsyspc/ntsyspc.html
STRUCTURE	Pritchard J	http://pritch.bsd.uchicago.edu/structure.html
GeneStrut	Constantine CC and others	http://www.vet.murdoch.edu.au/vetsch/imgad/GenStrut.htm
POPGENE	Yeh FC, Yang RC, and Boyle T	ftp://ftp.microsoft.com/Softlib/MSLFILES/HPGL.EXE
MacClade	David DR and Maddison WP	http://macclade.org/macclade.html
PHYLIP	Felsentein J	http://evolution.genetics.washington.edu/phylip.html
SITES	Hey J	http://lifesci.rutgers.edu/~heylab/ProgramsandData/Programs/SITES/SITES_documentation.htm#Contents
CLUSTAL W	Thompson and others	http://www.ebi.ac.uk/Tools/msa/clustalw2/
MALIGN	Gladstein D and Wheeler W	http://research.amnh.org/scicomp/book/export/html/75

Modified from http://www.bioversityinternational.org/fileadmin/bioversityDocs/Training/molecular_markers_volume_2/english/MolMarkers%20Vol2%20IV_Software_programs.pdf sets of data and their practical utility is lacking. In depth analysis in this regard is needed.

2.4 Molecular Diversity Among Cultivated and Wild Sorghums

Among the earliest efforts on characterization and cataloguing of the sorghum world collection, the works of Snowden (1936), Murty and Govil (1967), Miller (1968), Harlan and de Wet (1972), and others are significant. These pioneering works have laid the foundation of classification of sorghum germplasm. Harlan and de Wet (1972) developed a simplified classification of traditional sorghum cultivars based on spikelet and grain morphology into five basic races, *bicolor*, *caudatum*, *durra*, *guinea*, and *kafir*, and ten intermediate races (in all pair-wise combinations of the basic races). Ollitrault (1987) reported application of biochemical markers to assess neutral genetic variation and group sorghum accessions by race and origin. With the advent of molecular markers, different DNA marker systems have been employed to study the patterns of genetic diversity among sorghum accessions from the ex situ germplasm collection. RFLPs have been employed effectively to study genetic diversity in sorghum (Aldrich and Doebley 1992; Deu et al. 1994, 2006; Cui et al. 1995). Subsequently, other marker systems like RAPD (Menkir et al. 1997; Ayana et al. 2000a, b; Jeya Prakash et al. 2006), AFLP (Menz et al. 2004; Wu et al. 2006; Arya et al. 2008), and SSR (Brown et al. 1996; Taramino et al. 1997; Ghebru et al. 2002; Casa et al. 2005; Deu et al. 2008; Sagnard et al. 2011; Rakshit et al. 2012; Ganapathy et al. 2012; Billot et al. 2013; Ramu et al. 2013) have been deployed to study molecular diversity in sorghum. In many studies, different marker systems have been used in combination to study the diversity (Tao et al. 1993; de Oliveira et al. 1996; Smith et al. 2000; Agrama and Tuinstra 2003; Uptmoor et al. 2003; Geleta et al. 2006). In the recent past, DArT markers have been developed and deployed in diversity and population structure analysis in sorghum (Mace et al. 2008; Bouchet et al. 2012). SNP resource is also now available for such analysis (Nelson et al. 2011; Zheng et al. 2011). Table 2.2 summarizes the major publications on sorghum molecular diversity. It may be observed that with the advent of DNA technology, currently SSR markers are relied more in diversity analysis than any of the other marker systems. Reported studies on sorghum diversity using any of the marker systems deal mostly with a limited number of genotypes. However, with the availability of automated marker analysis and sophisticated analysis software programs, Billot et al. (2013) have studied the diversity in global composite germplasm collection (GCGC) of sorghum comprising of 3,367 accessions using 41 of the 48 reference markers of microsatellite kit reported by Billot et al. (2012) (http://sorghum.cirad.fr/SSR_kit), since these markers were reliable. These 41 markers gave rise to a total of 783 alleles with an average of 19.2 alleles per locus. Landraces (87 % of total accessions studied) contributed 94 %, breeding lines (8 % of total accessions studied) captured 57 %, and wild and weedy accessions contributed 65 % of detected SSR alleles. Across the accessions, 54.2 % alleles were of less than 1 % frequency, while 78.7 % were below 5 % frequency. Three times more private (or rare), alleles were detected in wild and weedy samples as compared to landraces. *Kafir* race presented smallest numbers of alleles per marker and private alleles and a lower genetic diversity (H_e=0.41 versus H_e of 0.60–0.76 for other four basic races). Highest numbers of alleles (86.8 %) were detected among the accessions of African origin. In Africa, Eastern Africa exhibited the largest gene diversity, followed by Central Africa, while Southern Africa was the poorest. Among Asian countries, the Middle East represented higher genetic diversity compared to India and East Asia. The study showed that cultivated sorghum cultivars were structured according to geographic region and race within the region, which was also earlier suggested (de Oliveira et al. 1996; Ghebru et al. 2002). A total of 13 groups of variable size was observed among the 3,367 accessions studied. The peripheral groups in Western Africa, Southern Africa, and

Table 2.2 Important publications on molecular diversity analysis

Marker system	Number of genotypes	Major observation	Reference
RFLP	56	Nuclear and chloroplast diversity revealed that cultivated sorghums were derived from wild species *arundinaceum*	Aldrich and Doebley (1992)
RAPD and RFLP	36	Phenetic analysis of band sharing was consistent with the current sub-species grouping of accessions	Tao et al. (1993)
RFLP	94	Multivariate analysis put the accessions in six clusters, which broadly corresponded to major sub-races	Deu et al. (1994)
RFLP	53	Parsimony analysis put the accessions in two major clusters	Cui et al. (1995)
SSR	17	Average diversity observed was 0.56. Identified SSR primers from maize to be used in sorghum	Brown et al. (1996)
RFLP, RAPD, and SSR	84	Different races from the same location were more closely related than from different locations	de Oliveira et al. (1996)
RAPD	190	86 % of total variation among accessions, 14 % among races, and 13 % among regions. PCA failed to put genotypes into racial groups	Menkir et al. (1997)
SSR	9	Observed high degree of polymorphism was detected with SSR markers and all lines could be differentiated with 13 SSR markers	Taramino et al. (1997)
SSR and RFLP	50	Observed moderate correlation with molecular and pedigree distances	Smith et al. (2000)
RAPD	93	Low level of differentiation among wild sorghum population, both at population and regional basis	Ayana et al. (2000a, b)
SSR	28	Cluster analysis detected 7–10 major sub-groups, where landraces were grouped in distinct clusters	Ghebru et al. (2002)
SSR and RAPD	22	SSR-based distances correlated more with distance based on geographic origin and racial diversity	Agrama and Tuinstra (2003)
RAPD, AFLP, and RFLP	46	UPGMA showed good fit to similarity estimates more with AFLP	Uptmoor et al. (2003)
SSR and AFLP	50	Cluster analysis failed to give distinct grouping of B and R lines suggesting absence of well-defined heterotic grouping. Marker at 1–2 cm apart gave better agreement dendrogram with pedigree information rather solely depending on randomly distributed AFLP and SSRs	Menz et al. (2004)
SSR	104	Landraces retained 86 % if diversity. Neighbor joining put wild species in distinct group	Casa et al. (2005)

(continued)

Table 2.2 (continued)

Marker system	Number of genotypes	Major observation	Reference
AFLP, SSR, and morphological markers	45	Shannon diversity index was highest with morphological markers followed by AFLP and SSR. Genetic relationship based on AFLP–SSR was highly related to morphological SSR but that using morphological and AFLP markers alone did not correlate well	Geleta et al. (2006)
RFLP	210	Neighbor joining analysis put all Asian accessions in a distinct group. Morphological races influence the diversity much than other components	Deu et al. (2006)
RAPD	32	Identified two clusters with two distinct out groups	Jeya Prakash et al. (2006)
AFLP	26	Genotypes were clustered into two major clusters	Wu et al. (2006)
AFLP	24	Diversity pattern matched well with pedigree information	Arya et al. (2008)
SSR	484	In Niger, genetic differentiation between botanical races was the first structuring factor, but the geographical distribution and the ethnic group to which farmers belonged were also significantly associated with genetic diversity partitioning	Deu et al. (2008)
SSR	377	Developed the community resource for association mapping in sorghum	Casa et al. (2008)
DArT	90	Cluster analysis discriminated well between all 90 genotypes. The study showed the potentiality of DArT makers in diversity analysis	Mace et al. (2008)
SSR	446	Analyses highlighted a strong racial structure of genetic diversity within cultivated sorghum. Introgressed nature of most of the wild and weedy sorghum was observed. A strong spatial genetic structure within each pool, due to spatially limited dispersal, and consequent gene flow between the wild and the crop pools were recorded	Sagnard et al. (2011)
SSR	82	SSR markers in combination with morphological traits can effectively differentiate closely related Maldandi accessions	Rakshit et al. (2012)
SSR	82	Very low variability was recorded within *rabi* accessions. Parental lines of hybrids could be differentiated based on fertility groups	Ganapathy et al. (2012)
DArT	177	Model-based and neighbor-joining diversity analyses identified six groups and confirmed previous evolutionary hypotheses	Bouchet et al. (2012)
SSR	31	Clustering analysis based on the genetic dissimilarity identified eight clusters, and grouping was in good agreement with pedigree, race, and geographic origin	Madhusudhana et al. (2012)

(continued)

Table 2.2 (continued)

Marker system	Number of genotypes	Major observation	Reference
EST–SSR	384	Grouping of accessions is identical in distance-based and model-based clustering methods. Genotypes were grouped primarily based on race within the geographic origins	Ramu et al. (2013)
SSR	3,367	The largest number of alleles was concentrated in Central and Eastern Africa. Cultivated sorghum was found to be structured according to geographic regions and race within the region. A total of 13 groups of variable size was recorded	Billot et al. (2013)

East Asia found to be more homogeneous and distinctly differentiated. Little correspondence between races and marker-based grouping was detected in the majority of races except *kafir*, suggesting races to be referred to as morphotypes. Wild and weedy accessions were highly diverse and scattered among cultivated accessions. This suggested that large genes flow between different types of sorghum. Based on the diversity level, they developed a core reference set of 384 accessions capturing 78.3 % of diversity. This reference set would facilitate the study of functional diversity in sorghum in the days to come as well as in association mapping.

2.5 Functional Diversity Analysis Using Gene and EST-Based Markers

SSR markers are the most commonly used markers in diversity analysis as evident from the above discussion. However, the major disadvantage of the anonymous SSR in studying diversity is that these are random DNA markers and access the diversity which often may not have any relevance to the functional diversity among the accessions. With rapid progress in genome sequencing projects, large-sequence information is available in public databases, which include both genetic and non-genetic sequences. Genetic sequences (referred to as expressed sequence tags or ESTs) derived from cDNAs are an important resource to access the genomic regions responsible for the functionality of the genotype. Identification of SSRs from ESTs (EST–SSR), referred to as genetic microsatellites, can be directly linked to the genes having agronomic significance (Varshney et al. 2002; Gupta et al. 2003). EST–SSRs have been successfully developed through data mining in various crop plants, including sorghum (Ramu et al. 2009; Srinivas et al. 2008, 2009). This marker system has been employed effectively in studying diversity in rice (Chao et al. 2000), wheat (Leigh et al. 2003; Gupta et al. 2003), and barley (Varshney et al. 2008). About 55 SSRs were developed and mapped through data mining of sorghum EST sequence by Ramu et al. (2009). Recently, Ramu et al. (2013) effectively used 40 reported EST–SSRs to study the functional diversity across the reference set developed by Billot et al. (2013). They identified 360 polymorphic alleles, out of which 88.9 % were represented by cultivated sorghum accessions, while wild genotypes contributed 71.4 % of alleles. The number of alleles ranged from 3 to 39 with an average of 9 per locus. Observed average alleles per locus were much lower than what was reported using neutral genomic SSR markers (19.2) on the same set of material by Billot et al. (2013). This suggested that EST–SSRs are under selective pressure. Cultivated and wild sorghum genotypes accounted for 28.6 % and 11.1 % unique alleles, respectively. Maximum numbers of alleles (58.9 %) were detected in the *bicolor* race, followed by *caudatum* (57.8 %), *durra* (49.2 %), and *guinea* (50 %), excluding *guinea margaritiferum* (*Gma*). *Kafir* race detected least number of alleles (29.7 %). Among *guinea* race, *Gma* contributed 25.3 % of detected alleles. African genotypes contributed 88.6 % of alleles, whereas those from the remaining parts of the world contributed 85.0 % of alleles. Central Africa contributed 50.8 % of alleles, followed by the East Africa (69.7 %), West Africa (62.5 %), and Southern Africa (58.9 %). Accessions from Asia were grouped into three, viz., East Asia, the Middle East, and India. Indian accessions contributed 48.3 % of alleles, whereas those from the Middle East contributed 38.6 % of alleles and East Asia contributed 33.1 % of alleles. They found that grouping of accessions was identical in both distance-based and model-based clustering methods. Genotypes were clustered within 8 clusters primarily based on race within geographic origins (Fig. 2.1). Earlier studies failed to put accessions belonging to *bicolor* race distinctly in one group. However, EST–SSRs successfully grouped *bicolor* accessions into two major groups, suggesting the utility of EST–SSRs in defining functional grouping among accessions. Sub-race *Gma* from East Africa formed as a separate cluster in close proximity to wild accessions. This suggested an independent domestication event for *Gma*. It was also found that *guinea*s from India and West Africa formed two distinct groups, while *kafir* accessions formed the most homogeneous group, which support the observations using anonymous SSR markers (Billot et al. 2013). Accessions originating from India were mainly grouped into two clusters. Out of the 40 EST–SSR markers used in the study, 33 followed stepwise mutation model, while seven fitted to the infinite allele model. This study clearly demonstrated the utility of EST–SSRs in defining functional diversity among accessions as compared to other genomic SSRs.

Functional diversity analysis in sorghum is not extensive. In recent past, Zheng et al. (2011) have re-sequenced two sweet sorghum and one grain sorghum genotypes using NGS techniques to identify a large number of SNPs, indels (insertions and deletions), PAVs (presence/absence variations), and CNVs (copy number variations). Similar SNP variations across 8 accessions have been reported through short-read genome sequencing of 8 diverse accessions of sorghum (Nelson et al. 2011). Recently, Jiang et al. (2013) adapted an integrative approach by using computational and experimental analyses to study the expression diversity between grain and sweet sorghum lines. It was observed that the genome sequences of grain and sweet sorghum exhibited considerable differences, but only limited divergence was observed in their functional genes, though more than 3,000 differentially expressed genes were detected between the grain and sweet sorghum varieties. It was concluded that such expression divergence resulted from mutations in expression regulatory sequences and DNA methylation, which was genetically determined by functionally divergent genes between the two genomes. Tandemly and segmentally duplicated as well as expanded sorghum genes by mobile elements contributing expression diversity were identified. A higher expression divergence in segmentally duplicated genes than tandemly duplicated genes was also observed. These duplicated genes in the grain sorghum experienced higher ratio of expression divergence when compared to those in the sweet sorghum.

2.6 Establishment of Core and Mini-Core Collections Using Molecular Markers

Most of the diversity studies have indicated that geographic origin and racial classification are associated with the organization of genetic diversity. With the realization of the utility of core collection, Prasada Rao and Ramanatha Rao (1995) first discussed the utility of characterization of data in establishing a core collection in sorghum. Grenier et al. (2000) studied the genetic diversity in the ICRISAT sorghum collection using morpho-agronomic and passport data and constituted three subsets of it. An initial study with limited microsatellite markers suggested that the sorghum collection at ICRISAT was highly structured genetically (Dje et al. 2000). Grenier et al. (2001a) first made an effort to assemble a core collection from the ICRISAT sorghum germplasm collection, using ecogeographical data, and stratified the col-

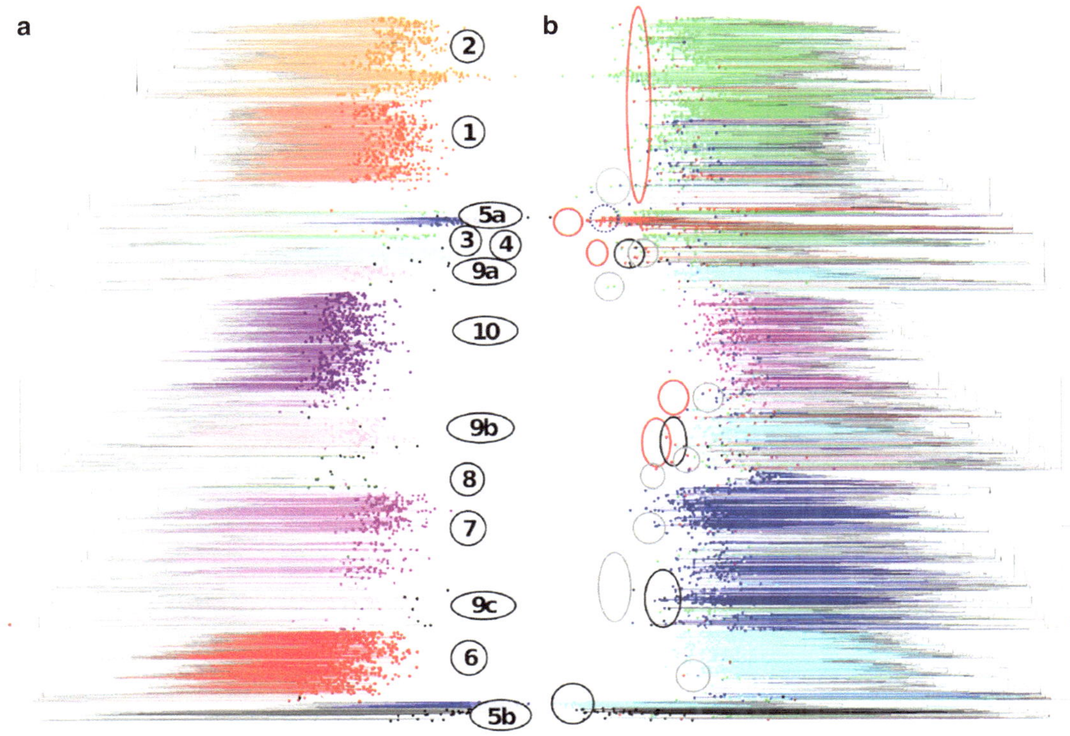

Fig. 2.1 Hierarchical NJ cluster analysis of 3,367 sorghum accessions of a global composite germplasm collection based on allelic data from 41 SSR markers (simple matching distance). (**a**) Accessions grouped by Bayesian analysis (Fig. 2.3, $K = 10$) are represented in color, corresponding to Group 1 in *orange*, Group 2 in *light orange*, Group 3 in *light green*, Group 4 in *light blue*, Group 5 in *dark blue*, Group 6 in *red*, Group 7 in *light purple*, Group 8 in *dark green*, Group 9 in *pink*, and Group 10 in *purple*. NJ clustering enabled finer resolution of these groups, leading to sub-divisions into Group 5a and Group 5b in *dark blue* and Group 9a, Group 9b, and Group 9c in *pink*. Unassigned accessions are presented in *gray*. Wild accessions are presented in *black*. (**b**) Accessions colored according to their classification in various taxonomic components: bicolor in *red*, caudatum in *dark blue*, durra in *green*, guinea in *light blue*, kafir in *purple*, unclassified in *gray*, and wild in *black*. The dendrogram sectors including dispersed components accessions (wild/weedy, bicolor, and unclassified) are highlighted by circles of the corresponding colors (*black*, *red*, *gray*) doi:10.1371/journal.pone.0059714.g004 (Adapted from Billot et al. 2013)

lection into four clusters, which act as the basis for a random sampling to establish a core collection of 225 accessions by using logarithmic sampling strategy (Grenier et al. 2001b). Subsequently, Deu et al. (2006) developed a refined core collection of 210 accessions by retaining 128 accessions of the core set of 225 and removing the *kafir* types. Further, 49 landrace accessions from a collection representing racial and geographic diversity (Deu et al. 1994) comprising of 16 accessions from the ICRISAT world sorghum collection and 17 from the CIRAD collection were added, and the structure of genetic diversity of this refined core collection was analyzed using morphological traits and RFLP markers. Subsequently, Upadhyaya et al. (2007) developed a mini-core collection comprising of 242 accessions at ICRISAT based on 20 morpho-agronomic characters of 2246 ICRISAT core collection, which explained over 90 % genetic diversity in the core collection. This mini-core collection is now being used in different studies. Recently, at the Directorate of Sorghum Research, Hyderabad, India, under the National Agriculture Innovation Project (NAIP), this mini-core collection along with some Indian *rabi* landraces were evaluated for their post-flowering moisture stress response and genotyped with 39 SSR markers from sorghum SSR diversity kit reported by Billot et al.

(2012). Based on their post-flowering moisture stress response and molecular diversity, a reference set of 96 genotypes have been developed for further use in the mining of alleles for moisture stress response in sorghum (Rakshit et al. unpublished data). Recently, Billot et al. (2013) have developed a reference set of 384 genotypes based solely on molecular diversity among 3,365 global composite germplasm collection (GCGC) of sorghum. This reference set represents 78.3 % alleles present in the GCGC collection, and this was further evaluated by Ramu et al. (2013) using EST–SSRs.

Besides the collection at ICRISAT, similar efforts have also been made elsewhere. Dahlberg et al. (2004) developed a refined subset from Sudan using over 40,000 sorghum accessions of the US National Plant Germplasm System. Using this set, unique information about the relative heat tolerance of the Sudanese sorghum accessions was obtained. Casa et al. (2008) reported a community resource of 377 sorghum accessions, which was subjected to population structure and kinship study using 47 SSR markers and phenotyped for eight morphological traits. This resource was reported to be a starting point for sorghum researchers to begin association studies of traits and markers or candidate genes of interest.

2.7 Applications of Diversity Analysis for Genetic Improvement in Sorghum

It is evident that diversity analysis in sorghum using different marker systems is quite extensive. Most of the studies have identified diverse genotypes and grouping of the genotypes mostly congruent to their racial and geographical lineages. However, rarely the generated information has been used in the practical sorghum improvement program. Mostly breeders in their regular breeding program prefer elite × elite crosses, where pedigree-based diversity is considered more, than any other information. In India, two distinct adaptive types of sorghum are under cultivation, viz., *kharif* (rainy season) and *rabi* (post-rainy season). The *kharif* season cultivars are predominantly *caudatum*, *kafir*, and *bicolor* races, while *rabi* cultivars are mainly *durra* types (Ganapathy et al. 2012). One of the most popular *rabi* cultivars in India is Maldandi, which was released in 1930. Subsequently, a selection from Maldandi, M35-1, was released for this region in 1969. Over the last 80 years, many selections have been made from Maldandi and are being cultivated as landraces across sorghum-growing regions. A diversity study across 82 Maldandi landraces was conducted by Rakshit et al. (2012) using both morphological and SSR markers (Fig. 2.2). Considerable diversity was recorded among these accessions, and 13 promising accessions were selected based on diversity data. Few landraces with better *roti* making quality than M35-1 were also identified. These selections are now being used in the *rabi* sorghum improvement program. Ganapathy et al. (2012) succeeded in distinguishing parental lines of hybrids based on their fertility groups, especially lines from rainy season sorghum using SSR markers (Fig. 2.3). They also demonstrated the presence of low genetic diversity among *rabi* genotypes, as compared to *kharif* accessions. This indicated the need to increase the diversity within *rabi* genotypes through introgression program. Genetic diversity among 31 sorghum parents was studied by Madhusudhana et al. (2012) using 413 SSR markers resulting in the identification of diverse parents for developing mapping population for important traits like shoot fly (296B × IS18551, 27B × IS2122), grain mold (296B × B58586), drought and charcoal rot (M35-1 × B35; CSV216R × B35), ergot (2219B × IS8525), and foliar diseases (IS10284 × IS26866). Using such diverse parents, important QTLs for shoot fly resistance were successfully identified (Satish et al. 2009) and validated (Aruna et al. 2011) at DSR.

Though not many studies are available in the use of molecular diversity information directly in the breeding program, it is being extensively used in association mapping of different traits in the recent past. A genome-wide association study (GWAS) of a diverse panel of 300 accessions with 1,290 SNPs was conducted by Sukumaran et al. (2012), which resulted in the identification of eight significant marker-trait associations after the association analysis between 333 SNPs in candidate genes and/or loci and grain quality traits. An SNP in starch synthase *IIa* (*SSIIa*)

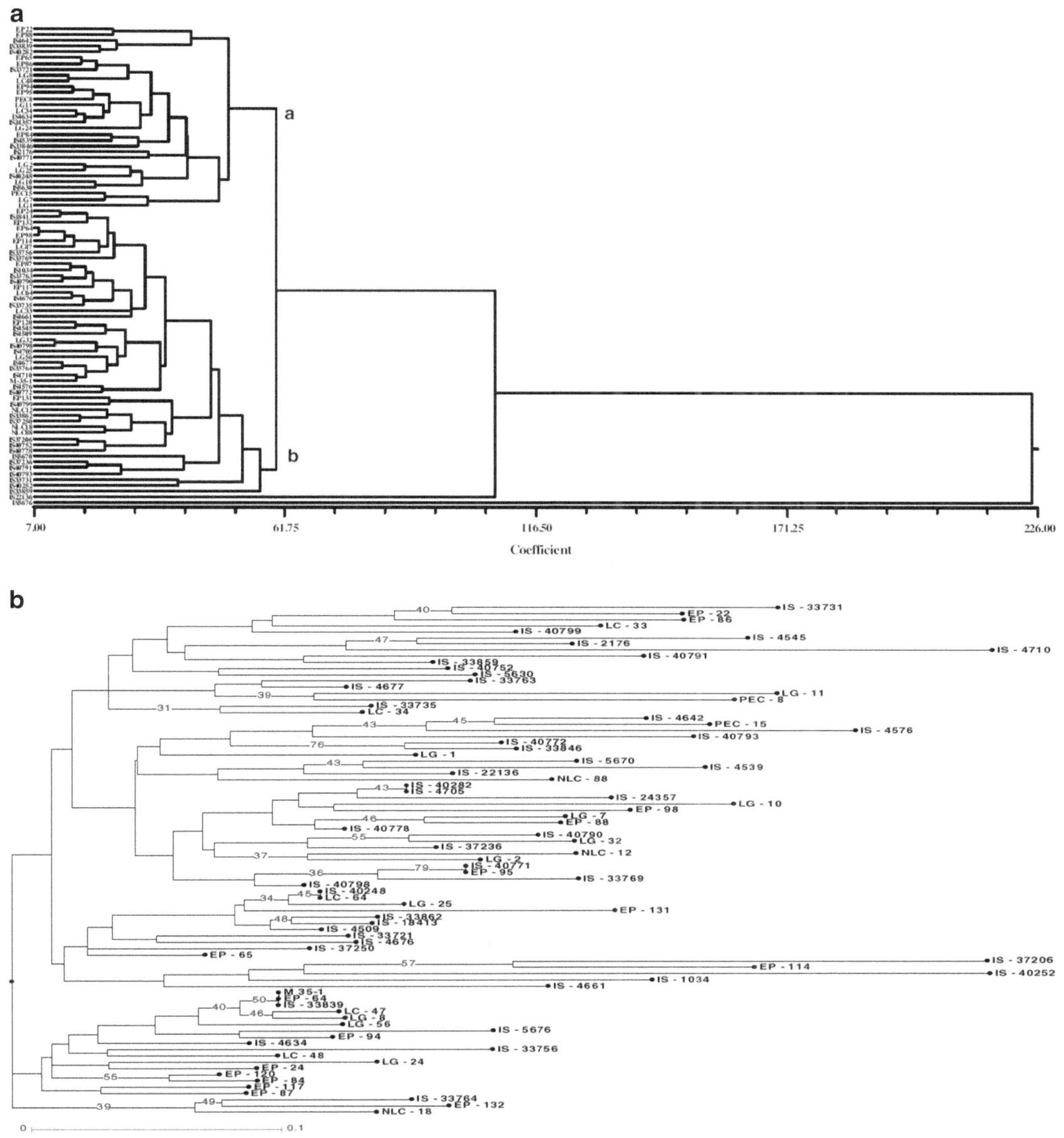

Fig. 2.2 (**a**) Sequential agglomerative hierarchical nested clustering of the 82 Maldandi landraces based on Euclidian distances using 17 quantitative traits. (**b**) Unweighted neighbor joining clustering of 82 Maldandi accessions using genotyping data of 16 SSR markers (bootstrap values above 30 are indicated) (Adapted from Rakshit et al. 2012)

gene was associated with kernel hardness (KH), while an SNP in starch synthase (*SSIIb*) gene and *pSB1120* locus was associated with starch content. Similarly, Morris et al. (2013) conducted GWAS for plant height and candidate genes for inflorescence architecture using ~265,000 single nucleotide polymorphisms (SNPs) in 971 worldwide accessions resulting in mapping of several classical loci for the two traits. In addition, the independent spread of multiple haplotypes carrying alleles for short-stature or long-inflorescence branches was also traced.

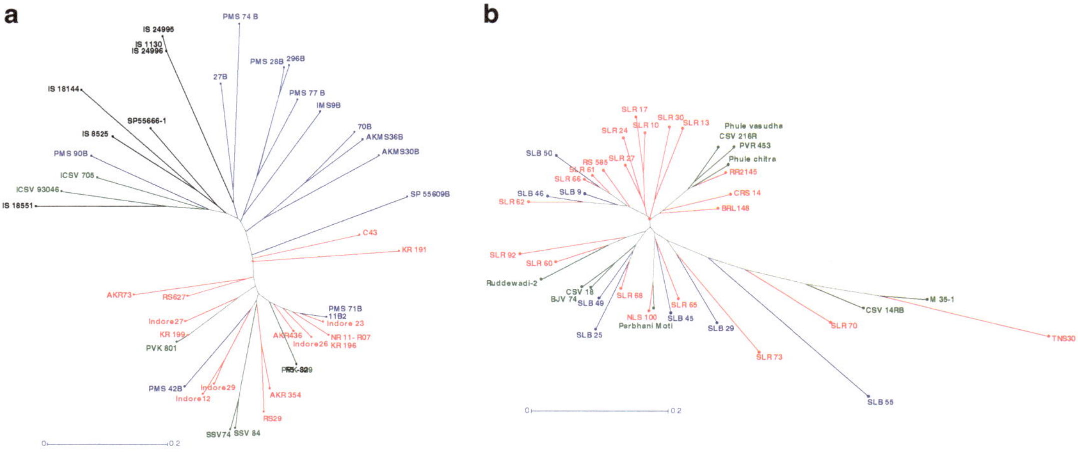

Fig. 2.3 Unrooted neighbor joining tree showing genetic relationship among 43 *kharif* sorghum genotypes (**a**) and 39 rabi sorghum genotypes (**b**) using 35 SSR markers The different working groups are identified by specific colors (*blue* for maintainers, *red* for restorers, *green* for varieties, and *black* for germplasm lines) (Adapted from Ganapathy et al. 2012)

2.8 Conclusions

The strength of any breeding program is on the availability of a well-characterized germplasm collection with highest trait diversity. Germplasm size indeed plays a very important role, but the mere size of the collection has no meaning unless it is characterized and the diversity well documented. In this regard, the establishment of the reference collection and core and mini-core collections retaining much of the variability in the original collection plays a very important role. Excellent bioresources in the form of reference collection and mini-core collection are already available with sorghum. However, in many cases, though they represent much of the diversity across the world collection, constituent lines of these collections may not be adapted to a specific climatic condition. Under those situations it is of paramount importance to replace the unadapted types with some diverse local collections. Indian sorghum workers have made an incredible contribution in collecting sorghum germplasm across the country. Currently, the size of the germplasm collection of the sorghum gene bank at the Directorate of Sorghum Research is 31,742. Sporadically selected entries of this collection have been used in the breeding program, and some of them have been characterized morphologically. However, till now no systematic approach has been made to characterize the whole collection. Recently, the DSR has taken an initiative toward this direction. However, morphological diversity analysis of this collection needs to be complemented with molecular diversity analysis in light of the recent work of Billot et al. (2013). Information already generated in this study needs to be supplemented with the characterization data of Indian germplasm collection, and then India-specific reference set may be created. The country-specific reference set having additional entries from world composite collections will prove to be an important resource to initiate association mapping for important traits. In this regard, economic botanists, plant breeders, plant protection scientists, physiologists, and biotechnologists should join hands to bring the progress in genomics research toward practical improvement of sorghum crop.

References

Abdel-Mawgood AL (2012) DNA-based techniques for studying genetic diversity. In: Genetic diversity in microorganisms, Prof. Mahmut Caliskan (ed), InTech, ISBN: 978-953-51-0064-5, pp 95–122

Agrama HA, Tuinstra MR (2003) Phylogenetic diversity and relationships among sorghum accessions using SSRs and RAPDs. Afr J Biotechnol 10:334–340

Aldrich PR, Doebley J (1992) Restriction fragment variation in the nucleus and chloroplast genomes of cultivated and wild *Sorghum bicolor*. Theor Appl Genet 85:293–302

Anderson EC (1957) A semi-graphical method for the analysis of complex problem. Proc Natl Acad Sci U S A 43:923–927

Aremu CO (2011) Genetic diversity: A review for need and measurements for intra-species crop improvement. J Microbiol Biotechnol Res 1:80–85

Arif IA, Khan HA, Bahkali AH, Al Homaidan AA, Al Farhan AH, Al Sadoon M, Shobrak M (2011) DNA marker technology for wild life conservation. Saudi J Biol Sci 18:219–225

Aruna C, Bhagwat VR, Madhusudhana R, Sharma V, Hussain T, Ghorade RB, Khandalkar HG, Audilakshmi S, Seetharama N (2011) Identification and validation of genomic regions that affect shoot fly resistance in sorghum [*Sorghum bicolor* (L.) Moench]. Theor Appl Genet 122:1617–1630

Arya L, Verma M, Sandhia GS, Singh SK, Lakhanpaul S (2008) Pattern of genetic relationship as revealed by AFLP markers in Indian sorghum [*Sorghum bicolor* (L.) Moench]. Indian J Genet 68:139–144

Ayana A, Bekele E, Bryngelsson T (2000a) Genetic variation in wild sorghum (*Sorghum bicolor* ssp. *Verticilliflorum* L. Moench) germplasm from Ethiopia assessed by random amplified polymorphic DNA (RAPD). Hereditas 132:249–254

Ayana A, Bryngelsson T, Bekele E (2000b) Genetic variation of Ethiopian and Eritrean sorghum (*Sorghum bicolor* L. Moench) germplasm assessed by random amplified polymorphic DNA (RAPD). Genetic Res Crop Evol 47:471–482

Beaumont MA, Ibrahim KM, Boursot P, Bruford MW (1998) Measuring genetic distance. In: Karp AP, Isaac I, Ingram DS (eds) A molecular tools for screening biodiversity. Chapman and Hall, London, pp 315–325

Bernardo R (2008) Molecular markers and selection for complex traits in plants: learning from the last 20 years. Crop Sci 48:1649–1664

Billot C, Rivallan R, Sall MN, Fonceka D, Deu M, Glaszmann JC, Noyer JL, Rami JK, Risterucci AM, Wincker P, Ramu P, Hash CT (2012) A reference microsatellite kit to assess for genetic diversity of *Sorghum bicolor* (Poaceae). Am J Bot 99:e245–260

Billot C, Ramu P, Bouchet S et al (2013) Massive sorghum collection genotyped with SSR markers to enhance use of global genetic resources. PLoS One 8:e59714

Borevitz JO, Liang D, Plouffe D, Chang HS, Zhu T, Weigel D, Berry CC, Winzeler E, Chory J (2003) Large-scale identification of single-feature polymorphisms in complex genomes. Genome Res 13:513–523

Botstein D, White RL, Skolnick M, Davis RW (1980) Construction of a genetic linkage map in man using restriction fragment length polymorphisms. Am J Hum Genet 32:314–331

Bouchet S, Pot D, Deu M et al (2012) Genetic structure, linkage disequilibrium and signature of selection in sorghum: lessons from physically anchored DArT markers. PLoS One 7:e33470

Brown SM, Hopkins MS, Mitchell SE, Senior ML, Wang TY, Duncan RR, Gonzalez-Candelas F, Kresovich S (1996) Multiple methods for the identification of polymorphic simple sequence repeats in sorghum [*Sorghum bicolor* (L.) Moench]. Theor Appl Genet 93:190–198

Casa AM, Mitchell SE, Hamblin MT, Sun H, Bowers JE, Paterson AH, Aquadro CF, Kresovich S (2005) Diversity and selection in sorghum: simultaneous analyses using simple sequence repeats. Theor Appl Genet 111:23–30

Casa AM, Pressoir G, Brown PJ, Mitchell SE, Rooney WL, Tuinstra MR, Franks CD, Kresovich S (2008) Community resources and strategies for association mapping in sorghum. Crop Sci 48:30–40

Chao YG, Ishii T, Temnykh S, Chen X, Lipovich L, Mc Couch SR, Park WD, Ayres N, Cartinhour S (2000) Diversity of microsatellites derived from genomic libraries and GenBank sequences in rice (*Oryza sativa* L.). Theor Appl Genet 100:713–722

Chen X, Sullivan PF (2003) Single nucleotide polymorphism genotyping: biochemistry, protocol, cost and throughput. Pharmacogenomics J 3:77–96

Cui YX, Xu GW, Magill CW, Schertz KF, Hart GE (1995) RFLP-based assay of *Sorghum bicolor* (L.) Moench genetic diversity. Theor Appl Genet 90:787–796

Dahlberg JA, Burke JJ, Rosenow DT (2004) Development of sorghum core collection: refinement and evaluation of a subset from Sudan. Econ Bot 58:556–557

de Oliveira AC, Richter T, Bennetzen JL (1996) Regional and racial specifities in sorghum germplasm assessed with DNA markers. Genome 39:579–587

Deu M, Gonzalez-de-Leon D, Glazsmann JC, Degremont I, Chantereau J, Lanaud C, Hamon P (1994) RFLP diversity in cultivated sorghum in relation to racial differentiation. Theor Appl Genet 88:838–844

Deu M, Rattunde F, Chantereau J (2006) A global view of genetic diversity in cultivated sorghums using core collection. Genome 49:168–180

Deu M, Sagnard F, Chantereau J et al (2008) Niger- wide assessment of *in situ* sorghum genetic diversity with microsatellite markers. Theor Appl Genet 116:903–913

Dhaliwal MS, Singh S, Singh K (2009) Detection of genotype specific fingerprints and assessment of genetic diversity in elite genetic stocks of tomato (*Solanum lycopersicum* L.) using RAPD primers and agronomic traits. Indian J Genet 69:44–49

Dje Y, Heuertz M, Lefebre C, Vekemans X (2000) Assessment of genetic diversity within and among germplasm accessions in cultivated sorghum using microsatellite markers. Theor Appl Genet 100:918–925

Efron B (1979) Bootstrap methods: another look at the jackknife. Ann Stat 7:1–26

Elshire RJ, Glaubitz JC, Sun Q, Poland JA, Kawamoto K (2011) A robust, simple genotyping-by-sequencing

(GbS) approach for high diversity species. PLoS One 6:e19379
Everitt B (1980) Cluster analysis, 2nd edn. Halstead Press, New York
Felsenstein J (1985) Confidence limits on phylogenies: an approach using the bootstrap. Evolution 39:783–791
Franco J, Crossa J, Villasenor J, Taba S, Eberhart SA (1997) Classifying Mexican maize accessions using hierarchical and density search methods. Crop Sci 37:972–980
Ganapathy KN, Gomashe SS, Rakshit S, Prabhakar B, Ambekar SS, Ghorade RB, Biradar BD, Saxena U, Patil JV (2012) Genetic diversity revealed utility of SSR markers in classifying parental lines and elite genotypes of sorghum (*Sorghum bicolor* L. Moench). Aust J Crop Sci 6:1486–1493
Geleta N, Labuschangne MT, Chris D, Viljoen CD (2006) Genetic diversity analysis in sorghum germplasm as estimated by AFLP, SSR and morpho-agronomical markers. Biodivers Conserv 15:3251–3265
Ghebru B, Schmidt RJ, Bennetzen JL (2002) Genetic diversity of Eritrean sorghum landraces assessed with simple sequence repeat (SSR) markers. Theor Appl Genet 105:229–236
Gower JC (1971) A general coefficient of similarity and some of its properties. Biometrics 27:857–874
Grenier C, Bramel-Cox PJ, Noirot M, Prasada Rao KE, Hamon P (2000) Assessment of genetic diversity in three subsets constituted from the ICRISAT sorghum collection using random vs. non-random sampling procedures. A. Using morpho-agronomical and passport data. Theor Appl Genet 101:190–196
Grenier C, Hamon P, Bramel-Cox PJ (2001a) Core collection of sorghum: I. Stratification based on eco-geographical data. Crop Sci 41:234–240
Grenier C, Hamon P, Bramel-Cox PJ (2001b) Core collection of sorghum: II. Comparison of three random sampling strategies. Crop Sci 41:241–246
Gupta PK, Rusotgi S, Sharma S, Singh R, Kumar N, Balyan HS (2003) Transferable EST-SSR markers for the study of polymorphism and genetic diversity in bread wheat. Mol Genet Genomics 270:315–323
Hair JR, Anderson RE, Tatham RL, Black WC (1995) Multi-variate data analysis with readings, 4th edn. Prentice-Hall, Englewood Cliffs
Harlan JR, de Wet JML (1972) A simplified classification of cultivated sorghum. Crop Sci 12:172–176
Hillis DM, Bull JJ (1993) An empirical test of bootstrapping as a method for assessing confidence in phylogenetic analysis. Syst Biol 42:182–192
Hoshino AA, Bravo JP, Nobile PM, Morelli KA (2012) Microsatellites as tools for genetic diversity analysis. In: Caliskan M (ed) Genetic diversity in microorganisms. INTECH, Croatia, pp 149–170. ISBN 978-953-51-0064-5
Jaccard P (1908) Nouvelles recherches sur la distribution florale. Bull Soc Vaud Sci Nat 44:223–270
Jeya Prakash SP, Biji KR, Michael Gomez S, Ganesa Murthy K, Chandra Babu R (2006) Genetic diversity analysis of sorghum (*Sorghum bicolor* L. Moench) accessions using RAPD markers. Indian J Crop Sci 1:109–112
Jiang SY, Ma Z, Vanitha J, Ramachandran S (2013) Genetic variation and expression diversity between grain and sweet sorghum lines. BMC Genomics 14:18
Johnson AR, Wichern DW (1992) Applied multivariate statistical analysis, 3rd edn. Prentice-Hall, Englewood Cliffs
Jolliffe IT (1986) Principal component analysis. Springer, Berlin
Kantety RV, Zeng X, Jeffrey LB, Zehr BE (1995) Assessment of genetic diversity in dent and popcorn (*Zea mays* L.) inbred lines using inter-simple sequence repeat (ISSR) amplification. Mol Breed 1:365–373
Kilian A, Huttner E, Wenzl P, Jaccoud D (2005) The fast and the cheap: SNP and DArT-based whole genome profiling for crop improvement. In: Tuberosa R, Phillips RL, Gale M (ed) Proceedings of the international congress in the wake of the double helix: from the green revolution to the gene revolution May 27–31, Bologna, pp 443–461
Kumar B, Rakshit S, Singh RD, Gadag RN, Nath R, Paul AK, Wasialam (2008) Diversity analysis of early maturing elite Indian maize (*Zea mays* L.) inbred lines using simple sequence repeats. J Plant Biochem Biotechnol 17:133–140
Leigh F, Lea V, Law J, Wolters P, Powell W, Donini P (2003) Assessment of EST- and genomic microsatellite markers for variety discrimination and genetic diversity studies in wheat. Euphytica 133:359–366
Lessa EP (1990) Multidimensional analysis of geographic genetic structure. Syst Zool 39:242–252
Link W, Dickens C, Singh M, Schwall M, Melchinger AE (1995) Genetic diversity in European and Mediterranean faba bean germplasm revealed by RAPD markers. Theor Appl Genet 90:27–32
Lombard V, Baril CP, Dubreuil P, Blouet F, Zhang D (2000) Genetic relationships and fingerprinting of rapeseed cultivars by AFLP: consequences for varietal registration. Crop Sci 40:1417–1425
Lynch M, Milligan BG (1994) Analysis of population genetic structure with RAPD markers. Mol Ecol 3:91–99
Mace ES, Xia L, Jordan DR, Halloran K, Parh DK, Huttner E, Wenzl P, Kilian A (2008) DArT markers: diversity analyses and mapping in *Sorghum bicolor*. BMC Genomics 9:26
Madhusudhana R, Balakrishna D, Rajendrakumar P, Seetharama N, Patil JV (2012) Molecular characterization and assessment of genetic diversity of sorghum inbred lines. Afr J Biotechnol 11:15626–15635
Medraoui L, Ater M, Benlhabib O, Msikine D, Filali-Maltouf A (2007) Evaluation of genetic variability of sorghum (*Sorghum bicolor* L. Moench) in northwestern Morocco by ISSR and RAPD markers. CR Biol 330:789–797
Melchinger AE (1993) Use of RFLP markers for analyses of genetic relationships among breeding materials and prediction of hybrid performance. In: Buxton DR (ed)

Proceedings of the international crop science congress, 1st, Ames, IA. July 1992, CSSA, Madison, pp 621–628

Menkir A, Golsbrough P, Ejeta G (1997) RAPD-based assessment of genetic diversity in cultivated races of sorghum. Crop Sci 37:564–569

Menz MA, Klein RR, Unruh NC, Rooney WL, Klein PE, Mullet JE (2004) Genetic diversity of public inbreds of sorghum determined by mapped AFLP and SSR markers. Crop Sci 44:1236–1244

Miller FR (1968) Some diversity in the world sorghum collection. In: Sutherland JI, Falasca RJ (ed) Proceedings of the 23rd annual Corn and Sorghum research conference, Chicago, IL, 10–12 December 1968, American Seed Trade Association, Washington, DC, pp 120–128

Miller MR, Atwood TS, Eames BF, Eberhart JK, Yan YL, Postlethwait JH, Johnson EA (2007a) RAD marker microarrays enable rapid mapping of zebrafish mutations. Genome Biol 8:R105

Miller MR, Dunham JP, Amores A, Cresko WA, Johnson EA (2007b) Rapid and cost-effective polymorphism identification and genotyping using restriction site associated DNA (RAD) markers. Genome Res 17:240–248

Milligan GW, Cooper M (1985) An examination of procedures for determining the number of clusters in a data set. Psychometrika 50:159–179

Mohammadi SA, Prasanna BM (2003) Analysis of genetic diversity in crop plants – salient statistical tools and considerations. Crop Sci 43:1235–1248

Morris GP, Ramu P, Deshpande SP et al (2013) Population genomic and genome-wide association studies of agroclimatic traits in sorghum. Proc Natl Acad Sci U S A 110:453–458

Mullis KB, Faloona FA (1987) Specific synthesis of DNA *in vitro* via a polymerase-catalyzed chain reaction. Methods Enzymol 155:335–350

Murty BR, Govil JN (1967) Description of 70 groups in genus sorghum based on a modified Snowden's classification. Indian J Genet 27:75–91

Namkoong G (1988) Sampling for germplasm collections. Hortic Sci 23:79–81

Nei M, Chesser RK (1983) Estimation of fixation indices and gene diversities. Ann Hum Genet 47:253–259

Nei M, Li WH (1979) Mathematical model for studying genetic variation in terms of restriction endonucleases. Proc Natl Acad Sci U S A 76:5269–5273

Nelson JC, Wang S, Wu Y, Li X, Antony G, White FF, Yu J (2011) Single nucleotide polymorphism discovery by high-throughput sequencing in sorghum. BMC Genomics 12:352

Ollitrault P (1987) Evaluation of genetique des sorghos cultives (*Sorghum bicolor* Moench) par l'analyse conjointe des diversites enzymatiques et morphophysiologique. Relation avec les sorghos sauvages: Universite Paris XI, France

Panchen AL (1992) Classification, evolution and the nature of biology. Cambridge University Press, Cambridge

Peeters JP, Martinelli JA (1989) Hierarchical cluster analysis as a tool to manage variation in germplasm collections. Theor Appl Genet 78:42–48

Prasada Rao KE, Ramanatha Rao V (1995) The use of characterization data in developing a core collection of sorghum. In: Hodgkin T, Brown AHD, van Hintum TJL, Morales EAV (eds) Core collections of plant genetic resources. IPGRI/Wiley, New York, pp 109–115

Rakshit S, Zaidi PH, Mishra SK (2002) Molecular markers and tagging of genes in crop plants. In: Hemantaranjan A (ed) Advances in plant physiology, vol 4. Scientific Publisher (India), Jodhpur, pp 205–223

Rakshit S, Gomashe SS, Ganapathy KN, Elangovan M, Ratnavathi CV, Seetharama N, Patil JV (2012) Morphological and molecular diversity reveal wide variability among sorghum Maldandi landraces from south India. J Plant Biochem Biotechnol 21:145–156

Ramu P, Kassahun B, Senthilvel S, Kumar CA, Jayashree B, Folkertsma RT, Reddy LA, Kuruvinashetti MS, Haussmann BI, Hash CT (2009) Exploiting rice-sorghum synteny for targeted development of EST-SSRs to enrich the sorghum genetic linkage map. Theor Appl Genet 119:1193–1204

Ramu P, Billot C, Rami JF, Senthilvel S, Upadhyaya HD, Ananda Reddy L, Hash CT (2013) Assessment of genetic diversity in the sorghum reference set using EST-SSR markers. Theor Appl Genet 126:2051–2064

Rincon F, Johnson B, Crossa J, Taba S (1996) Cluster analysis, an approach to sampling variability in maize accessions. Maydica 41:307–316

Ritter KB, Lynne McIntyre C, Godwin ID, Jordan DR, Chapman SC (2007) An assessment of the genetic relationship between sweet and grain sorghums, within *Sorghum bicolor* ssp. *bicolor* (L.) Moench, using AFLP markers. Euphytica 157:161–176

Rogers JS (1972) Measures of genetic similarity and genetic distance. In: Studies in Genetics VII, University of Texas Publication 7213. University of Texas, Austin, pp 145–153

Rohlf FJ (1972) An empirical comparison of three ordination techniques in numerical taxonomy. Syst Zool 21:271–280

Rohlf FJ (1992) NTSYS-pc numerical taxonomy and multivariate analysis system. Version 1.70. Exeter, Setauket, NY

Sagnard F, Deu M, Dembélé D et al (2011) Genetic diversity, structure, gene flow and evolutionary relationships within the *Sorghum bicolor* wild-weedy-crop complex in a western African region. Theor Appl Genet 123:1231–1246

Saiki RK, Scharf S, Faloona F, Mullis KB, Horn GT, Erlich HA, Arnheim N (1985) Enzymatic amplification of beta-globin genomic sequences and restriction site analysis for diagnosis of sickle cell anemia. Science 230:1350–1354

Satish K, Srinivas G, Madhusudhana R, Padmaja PG, Nagaraja Reddy R, Murali Mohan S, Seetharama N (2009) Identification of quantitative trait loci for resistance to shoot fly in sorghum [*Sorghum bicolor* (L.) Moench]. Theor Appl Genet 119:1425–1439

Schlotterer C (2004) The evolution of molecular markers – just a matter of fashion? Nat Rev Genet 5:63–69

Schulman AH (2007) Molecular markers to assess genetic diversity. Euphytica 158:313–321

Smith JSC, Smith OS, Boven SL, Tenborg RA, Wall SJ (1991) The description and assessment of distances between inbred lines of maize. III: a revised scheme for the testing of distinctiveness between inbred lines utilizing DNA RFLPs. Maydica 36:213–226

Smith JSC, Kresovich S, Hopkins MS, Mitchell SE, Dean RE, Woodman WL, Lee M, Porter K (2000) Genetic diversity among elite sorghum inbred lines assessed with simple sequence repeats. Crop Sci 40:226–232

Sneath PHA, Sokal RR (1973) Numerical taxonomy. The principles and practice of numerical classification, XV. Freeman, San Francisco, p 573

Snowden JD (1936) The cultivated races of sorghum. Adlard and Son, Ltd., London

Sokal RR, Michener CD (1958) A statistical method for evaluating systematic relationships. Univ Kansas Sci Bull 38:1409–1438

Sokal R, Rohlf J (1994) Biometry: the principles and practice of statistics in biological research, 3rd edn. Freeman & Co, New York

Souza E, Sorrells ME (1991) Relationships among 70 North American oat germplasms: I. Cluster analysis using quantitative characters. Crop Sci 31:599–605

Srinivas G, Satish K, Murali Mohan S, Nagaraja Reddy R, Madhusudhana R, Balakrishna D, Venkatesh Bhat B, Howarth CJ, Seetharama N (2008) Development of genic-microsatellite markers for sorghum staygreen QTL using a comparative genomic approach with rice. Theor Appl Genet 117:283–296

Srinivas G, Satish K, Madhusudhana R, Reddy RN, Murali Mohan S, Seetharama N (2009) Identification of quantitative trait loci for agronomically important traits and their association with genic-microsatellite markers in sorghum. Theor Appl Genet 118:1439–1454

Sukumaran S, Xiang W, Bean SR, Pedersen JF, Kresovich S, Tuinstra MR, Tesso TT, Hamblin MT, Yu J (2012) Association mapping for grain quality in a diverse sorghum collection. Plant Genome 5:126–135

Swofford DL, Olsen GJ, Waddell PJ, Hillis DM (1996) Phylogenetic inference. In: Hillis DM, Moritz C, Mable BK (eds) Molecular systematics, 2nd edn. Sinauer Associates, Sunderland, pp 407–514

Tao Y, Manners JM, Mm L, Henzell RG (1993) DNA polymorphism in grain sorghum [*Sorghum bicolor* (L.) Moench]. Theor Appl Genet 86:679–688

Taramino G, Tarchini R, Ferrario S, Lee M, Pe' ME (1997) Characterization and mapping of simple sequence repeats (SSRs) in *Sorghum bicolor*. Theor Appl Genet 95:66–72

Upadhyaya HD, Gopal Reddy V, Gowda CLL, Singh S (2007) A minicore collection of sorghum [*Sorghum bicolor* (L.) Moench] for enhancing utilization of germplasm in crop improvement. In: The ASA-CSSA-SSSA international annual meetings, 4–8 November 2007, New Orleans, Louisiana

Uptmoor R, Wenzel W, Friedt W, Donaldson G, Ayisi K, Ordon F (2003) Comparative analysis on the genetic relatedness of *Sorghum bicolor* accessions from Southern Africa by RAPDs, AFLPs and SSRs. Theor Appl Genet 106:1316–1325

Varshney RK, Thiel T, Stein N, Langridge P, Graner A (2002) *In silico* analysis on frequency and distribution of microsatellites in ESTs of some cereal species. Cell Mol Biol Lett 7:537–546

Varshney RK, Thiel T, Sretenovic-Rajicic T, Baum M, Valkoun J, Guo OP, Grando S, Ceccarelli S, Graner A (2008) Identification and validation of a core set of informative genic SSR and SNP markers for assaying functional diversity in barley. Mol Breed 22:1–13

Vos P, Hogers R, Bleeker M et al (1995) AFLP: a new technique for DNA fingerprinting. Nucleic Acids Res 23:4407–4414

Wang ML, Zhu C, Barkley NA, Chen Z, Erpelding JE, Murray SC, Tuinstra MR, Tesso T, Pederson GA, Yu J (2009) Genetic diversity and population structure analysis of accessions in the US historic sweet sorghum collection. Theor Appl Genet 120:13–23

Ward JH Jr (1963) Hierarchical grouping to optimize an objective function. J Am Stat Assoc 58:236–244

Weir BS (1990) Genetic-data analysis: method for discrete population genetic data. Sinauer Associates, Sunderland

Wiley EO (1981) Phylogenetics: the theory and practice of phylogenetics and systematics. Wiley, New York

Williams JGK, Kubelik AR, Livak KJ, Rafalski JA, Tingey SV (1990) DNA polymorphisms amplified by arbitrary primers are useful as genetic markers. Nucleic Acids Res 18:6531–6535

Wu YQ, Huang Y, Tauer CG, Porter DR (2006) Genetic diversity of sorghum accessions resistant to greenbugs as assessed with AFLP markers. Genome 49:143–149

Zheng LY, Guo XS, He B et al (2011) Genome-wide patterns of genetic variation in sweet and grain sorghum (*Sorghum bicolor*). Genome Biol 12:R114

Zietkiewicz E, Rafalski A, Labuda D (1994) Genome fingerprinting by simple sequence repeat (SSR)-anchored polymerase chain reaction amplification. Genomics 20:176–183

3 Linkage Mapping

R. Madhusudhana

Contents

3.1	**Introduction**	48
3.2	**Linkage Map**	48
3.3	**Mapping Populations**	48
3.3.1	Criteria for Developing Mapping Population in Linkage-Based Analysis	49
3.3.2	Populations for Linkage-Based QTL Mapping	49
3.3.3	Populations to Exploit Linkage Disequilibrium	51
3.3.4	Population to Exploit Both Linkage and Linkage Disequilibrium	51
3.3.5	Mapping Populations for Cross-Pollinating Species	52
3.4	**Marker Systems**	52
3.5	**Genetic Linkage Maps of Sorghum**	54
3.6	**Trait Mapping**	60
3.6.1	Mapping Major Genes	60
3.6.2	Mapping Quantitative Trait Loci (QTL)	61
3.7	**Software Resources**	64
3.7.1	Resources for Linkage-Based Mapping	64
3.7.2	Resources for Association Mapping	65
3.8	**Conclusions**	65
References		65

R. Madhusudhana (✉)
Plant Breeding, ICAR-Indian Institute of Millets Research, Rajendranagar, Hyderabad, Telangana 500 030, India
e-mail: madhu@millets.res.in

Abstract

The past three decades have seen the development and use of DNA marker systems in linkage mapping of genes/QTLs for economic traits in almost all crop plants. Mapping populations of different types and sizes, DNA marker systems, and genotyping technologies were developed along with advances in statistical analyses for linkage mapping. The construction of genetic maps in sorghum started during 1990s using RFLP markers. Early genetic maps were based on F_2 populations and lacked resolution due to less number of markers and smaller population size. Later maps were developed with recombinant inbred line populations with more number of markers and population size. New classes of PCR-based markers like SSRs, AFLPs were used in the construction of maps. The availability of whole sorghum genome sequence resulted in the development of thousands of SSRs and identification of millions of SNPs leading to the construction of high-density linkage maps. Saturated genetic maps contribute substantially to the fine mapping and positional cloning of important genes and offer a tool for gene discovery, allele mining, etc. Linkage maps covering whole of sorghum genome were developed, and comprehensive maps with molecular, cytological, and physical elements established integrating inputs from several mapping efforts. Linkage maps are useful for elucidation of complex biological processes

R. Madhusudhana et al. (eds.), *Sorghum Molecular Breeding*, DOI 10.1007/978-81-322-2422-8_3

directly related to superior agronomic performance and in identifying gene/QTL marker associations for gene pyramiding, marker-assisted breeding of crop plants including sorghum.

Keywords
Sorghum • Mapping populations • Markers • Linkage mapping

3.1 Introduction

Ever since Gregor Mendel formulated the laws of inheritance (Mendel 1865), associating genetic factors with trait phenotype variations became an important activity in plant breeding. Initially, this was restricted to morphological traits, which could be visually seen and measured at plant morphology. The traits of pea plant, which Gregor Mendel studied to form the principles of Genetics, were in fact the genetic markers that formed the basis for the concept of genetic linkage put forth by de Vilmorin and Bateson during 1911. However, the concept of linkage groups representing the chromosomes was not clear until the establishment of the first genetic map with 37 markers distributed over seven linkage groups in *Pisum* (Lamprecht 1948). Later, with the advancement made in molecular biology and other allied sciences, more and more genetic markers (protein markers, DNA markers) were found for their application in construction of linkage maps and in establishing marker-trait associations.

3.2 Linkage Map

The central idea behind the construction of a linkage map is that the frequency of recombination between two markers can be used as the measure of distance between them on a chromosome. The nearer the two markers on a chromosome, less is the frequency of recombination between them. Therefore, the markers in close proximity on a chromosome are said to be linked. Each DNA marker occupies a specific position called marker locus on a chromosome. A linkage map is the linear order of DNA markers indicating the positions and relative genetic distances between them on a chromosome. The most important use of linkage maps is to identify genomic locations associated with the expression of genes or QTLs. In gene/QTL mapping studies, construction of a linkage map for a species is the first step for which development of a mapping population is essential. A suitable mapping population, appropriate marker system, and software for data analyses are key to the construct of a genetic map. Map construction requires (1) selecting the most appropriate mapping population(s); (2) genotyping with polymorphic markers and calculating pair-wise recombination frequencies; (3) establishing linkage groups and estimating map distances; and (4) determining loci order. Since large mapping populations are often characterized by hundreds of markers, maps are constructed using various computer software, viz., Mapmaker (Lander et al. 1987), GMendel (Echt et al. 1992), JoinMap (Stam 1993), and others. They utilize genotypic data of a mapping population to estimate the recombination frequency to determine the linear arrangement of genetic markers on a chromosome. The linkage map and marker data have applications for finding out marker-trait associations involving trait phenotypic data.

3.3 Mapping Populations

Mapping population is a group of individuals used for gene mapping. Selection of appropriate mapping population is fundamental to the success of a gene mapping project. Broadly, mapping populations can be of two types, one that involves the development of new experimental population with related individuals (linkage based) while the other is based on the exploitation of the existing natural or breeding populations through linkage disequilibrium analysis. In genetics and breeding, mapping populations are the genetic tools used to identify the genetic loci controlling measurable phenotypic traits. They consist of individuals of one species, or in some cases, derived from crosses among related species where the parents differ in the traits of interest. The trait to be studied/mapped needs to be polymorphic between the parents in linkage-based analysis or should have diversity in the natural or breeding population. Additionally, a significant trait heritability is essential.

3.3.1 Criteria for Developing Mapping Population in Linkage-Based Analysis

Selection of parents for developing a mapping population is critical to successful map construction and mapping genes/QTLs. It is always advisable to screen a panel of parents for their trait phenotype to identify the extremes of the phenotypic distribution before choosing the parents of a mapping population. It is expected that the more the parental lines differ at phenotypic traits, there will be more genetic factors controlling the trait in the segregating generations and hence easier for their genetic mapping. However, the parents should not be genetically too distant to cause sterility in their progeny and very high levels of segregation distortion. It is also possible that if the parents are too diverse, the chances of getting transgressive segregants become less.

Other important factors that need to be considered while developing a mapping population are the reproductive mode of the crop species, time available for the development of population, and the map resolution needed. Predominantly, crop species can either self naturally (rice, tomato, pea, etc.) or can be manually selfed (maize, sorghum, etc.) or self-incompatible, inbreeding sensitive (potato etc.). Thus, the type of mapping population that can be developed depends on their reproductive mode. In crops where selfing is natural or done manually, it is possible to develop progenies with maximum homozygosity, while it is not possible in crops, which are self-incompatible and show maximum inbreeding depression. Thus, we may develop mapping populations suitable for self-fertilizing plants and mapping populations for cross-pollinating species.

The precision with which genetic distances are measured in a genetic map is directly related to the size of the mapping population. If only 30 individuals are studied and no recombination found between two markers, then one would find the genetic distance between the two markers as 0 cM. On the other hand, 100 individuals might reveal recombination between the same markers to indicate some distance in cM. Thus, the higher the number involved, the better is the resolution and more precise will be the order and linkage distance between markers in a linkage map. Thus, one needs to consider the level of precision, cost involved, etc., in the development of a linkage map. It is essential to develop a mapping population with at least 250 individuals to identify QTLs with major and minor effects.

3.3.2 Populations for Linkage-Based QTL Mapping

If pure lines are possible without any loss in vigor, different types of mapping populations can be developed for linkage map-based QTL studies. They consist of F_2 population, backcrosses (BC), doubled haploids (DHs), recombinant inbred lines (RILs), near-isogenic lines (NILs), etc.

F_2 Population

This is the simplest type of mapping population developed from selfing or inter-mating F_{1s} of a cross involving two homozygous and genetically diverse parents differing for the trait of interest. The advantage of F_2 population is that the development of population is easy, as it requires less time and efforts. It is good for estimating both additive and dominance effects and well suited for preliminary mapping and also for fine mapping a QTL/gene as thousands of F_2 plants can be grown easily. The expected segregation ratio in a F_2 population differs for dominant (3:1) and for a codominant (1:2:1) marker. However, there are some limitations on the use of F_2 mapping population. Since it is the result of one cycle of meiosis, the map resolution is low. In QTL mapping, even a marker that is far from the QTL remains linked. Secondly, the G×E effects of a trait expression cannot be estimated since the F_2 population is a temporary/mortal population and cannot be replicated over time and space.

Backcross Population

Backcross (BC) population can also be easily developed by backcrossing the F_1 to one of its parent. The expected genotypic ratio in a BC depends on the parent type and marker used. With dominant parent (BC_1), the ratio segregates in 1:0 for a dominant marker and 1:1 with a codominant marker. However, the segregation remains same (1:1) irrespective of marker type if backcrossing is made with recessive parent (BC_2). Similar to F_2 population, BC population also requires less time for its development. It is also a mortal population and hence cannot be replicated over time and space.

Doubled Haploids

Haploids contain a single complete set of chromosomes and can be derived from anther culture from F_1 plant. Doubled haploid (DH) lines are produced by chromosome doubling of haploid plants and hence they are complete homozygotes and contain two identical sets of chromosomes/genes. The genotypic ratio expected in a DH population is 1:1 irrespective of whether a marker is dominant or codominant. Generally, five methods are used for the production of haploids in plants for genetics and breeding programs (Palmer and Keller 2005). They are chromosome or genome elimination, ovary culture or gynogenesis, anther culture or androgenesis, semigamy, and haploid inducer lines. Since DH lines are complete homozygotes, they are permanent mapping populations and are ideal for estimating QTL × E interactions as the population can be replicated over locations and years. The production of DH is instantaneous and saves a lot of time. The disadvantages of DH lines are that the dominance effects cannot be estimated and recombination from male side alone is accounted. Due to the involvement of in vitro techniques, DH production requires more technical skills. Suitable culturing methods/haploid production methods are usually not available for a number of crops, and many crops are recalcitrant in their tissue culture response.

Recombinant Inbred Lines

Recombinant inbred lines (RILs) are the products of successive inbreeding. They are developed by continuous selfing or sib mating the progeny of individual members of an F_2 population until complete homozygosity is achieved. Single-seed descent (SSD) method (Goulden 1939) is followed for the development of RILs. Starting from F_2, one seed from each F_2 plant is advanced to next generation. Again, in F_3 one seed from each family is taken to advance to F_4. This process is continued until all the lines become homozygous and show no further segregation of traits. RILs developed though selfing in a self-pollinated crop requires less time than those developed through sibmating as done in cross-pollinated crops. Selfing is an intensified form of inbreeding than sibmating and thus results in more rapid achievement of homozygosity. In RILs, the genetic segregation ratio for both dominant and codominant marker would be 1:1. As the genetic constitution of each RIL is fixed and does not change upon inbreeding, RILs are permanent/immortal mapping population, which can be easily multiplied, shared with other groups, and replicated over locations and years. RILs show a higher degree of recombination than F_2 as they have undergone several cycles of meiosis and therefore are very useful in identifying tightly linked markers. Since RILs are homozygous, only additive effects of trait expression can be estimated. Development of RILs takes considerable time and their development is also difficult in crops with a high degree of inbreeding depression.

Near-Isogenic Lines

Near-isogenic lines (NILs) derived from inbreeding are usually the products of successive backcrossing. They are obtained by repeated backcrossing of recurrent parent simultaneously following selection for the trait of interest that is introgressed from the donor parent. The donor genome is progressively diluted in the introgression lines until only a short segment of chromosome is retained containing the gene from donor.

When the NILs are compared with molecular markers, only those that are linked to the gene of interest are expected to be polymorphic. Irrespective of the nature of marker, the expected segregation ratio is 1:1. NILs are also permanent mapping population, suitable for tagging gene for a trait, and are very useful in functional genomics. However, like RILs, NILs also require several generations to develop. Linkage drag is a potential problem in the development of NILs.

3.3.3 Populations to Exploit Linkage Disequilibrium

Association Mapping Population

Association mapping (AM) is a promising approach for detecting marker-trait association using existing populations. The approach exploits the linkage disequilibrium (LD) that exists between trait and markers. In this, no segregating population is developed afresh, but the existing diverse germplasm lines, genetic stocks, landraces, etc., which have undergone several generations of recombination, will constitute the AM population. The genotypes selected for constituting the AM population should be genetically diverse to avoid population structure. A non-structured population is preferred for trait-marker association which otherwise may lead to spurious marker-trait associations. It differs from classical linkage map-based QTL mapping in that no segregating progeny have to be developed. Phenotyping efforts may be reduced as existing phenotypic data from national trials, adaptive trials, etc., can be used in addition to or as a replacement of new trials. Association mapping can greatly accelerate positional cloning of QTL approaches since the trait-loci association can be mapped with more precision and with high resolution.

Multi-parent Advanced Generation Inter-cross

Multi-parent advanced generation inter-cross (MAGIC) is a method to increase the precision with which genetic markers are linked to QTL. It is an extension of the advanced inter-cross method (Cavanagh et al. 2008). MAGIC involves two extensions to traditional methods of searching for marker-trait correlations among segregating progeny of a cross between two parents. First, the mapping population is established by inter-crossing multiple founder lines. A MAGIC population is therefore more genetically diverse than a conventional biparental mapping population and more QTLs can be detected. Second, the MAGIC population can be cycled through several extra generations of forced inter-mating. Each extra generation mills the genetic contribution from the founder lines finer. If "n" founder lines are available, they are inter-crossed for "n/2" generations until all the founders are combined with equal genetic proportions. Once the inter-crossing is over, RILs may be derived from them upon selfing. In comparison to QTL mapping in biparental crosses, the increased recombination and diversity of MAGIC gives greater precision in QTL location and greater opportunity to detect more QTLs for multiple traits. The larger number of parental accessions increases the allelic and phenotypic diversity over traditional RILs, potentially increasing the number of QTLs that segregate in the population. The successive rounds of recombination cause LD to decay, thereby increasing the precision of QTL location (Mackay and Powell 2007). The major limitation of this MAGIC population is that the inter-crossing cycles also increase proportionately with increase in founder size. Another limitation of MAGIC populations is that they are likely to show extensive segregation for developmental traits, like maturity and plant height. Segregation for such traits may influence the overall performance for complex traits like yield or drought tolerance thereby limiting their utility (Gupta et al. 2010; Varshney et al. 2009).

3.3.4 Population to Exploit Both Linkage and Linkage Disequilibrium

The Nested Association Mapping (NAM) strategy was proposed by Yu et al. (2008), which combines the advantages of both linkage and association mapping strategies. The procedure for the development of NAM population involves selecting diverse founders and crossing them with a common parent for developing a large set of related mapping progenies, preferably RILs. Thus, with 25 diverse founders crossed with a common parent, there will be 25 RIL populations developed, each with at least 200 RILs, totaling 5,000 nested RILs. The RILs are "nested" in the sense that they all share a common parent, but each RIL population has a different alternate parent.

3.3.5 Mapping Populations for Cross-Pollinating Species

Development of mapping populations like RILs, NILs, etc., through selfing is difficult in several crop species due to high-inbreeding depression (carrot, potato, alfalfa, etc.) or self-incompatibility (*Brassica*, radish, tobacco, etc.). In such cases, heterozygous parental plants are used to derive the mapping populations such as F_1 or backcross lines (BC). In the foundation cross population, different alleles are contributed from either parent to the individual F_1 plants. The linkage between markers is assessed by the production of a genetic map for either parent.

3.4 Marker Systems

Genetic markers, also called as DNA markers, are the biological features that are determined by allelic forms of genes or genetic loci, which can be transmitted from one generation to another, and thus can be used as experimental probes or tags to keep track of an individual, a tissue, a cell, a nucleus, a chromosome, or a gene (Xu 2010). Broadly, markers can be classified into two categories, viz., classical markers and DNA markers. Classical markers are early generation markers and represent markers at morphological, cytological, or biochemical levels. However, the DNA markers represent the sequence differences that can be visualized at the DNA level. The differences can be visualized using either autoradiography or fluorescence or chemical staining of the DNA. DNA markers have several advantages over the classical markers as they are numerous, highly polymorphic, represent greater genome coverage, no epistasis, neutral to stage and environment, amenable for fast and easy assay, easy access, transferability across species and labs, etc. Several marker systems have been developed, but only few of them are extensively used in crop systems. A brief review of these markers is given below.

RFLPs

Restriction fragment length polymorphism (RFLP) markers are the first-generation markers, which were extensively used earlier in several crop species for linkage mapping. Genome mapping was revolutionized by the development of RFLPs (Beckmann and Soller 1983), which are codominant and locus-specific markers with wide genome coverage. They are highly reproducible and are powerful tools for comparative and syntenic mapping across species due to their easy transferability across species. Alteration or elimination of restriction enzyme recognition sites results in varied restriction products, which can be visualized using Southern blotting technique. Although RFLPs have played an important role in developing early generation linkage maps in several crops, they are not ideally suited to large-scale MAS applications (Rafalski and Tingey 1993). This is primarily because the RFLP system is labor-intensive, time-consuming, not amenable for high-throughput genotyping, requires suitable probes and relatively a large amount of quality DNA for the analysis.

RAPDs

Randomly amplified polymorphic DNA (RAPD) is a PCR-based technique that uses single, short random primers of arbitrary sequence to produce random amplification of DNA fragments from the whole genomes (Williams et al. 1990). Here, the primer anneals to the genomic DNA at two different sites on complementary strands of DNA template. If the two primer binding sites are within an amplifiable range, a discrete DNA product is formed, which can be visualized as presence or absence of band in gel electrophoresis. Enormous attraction for RAPDs was due to the fact that there is no requirement for cloning DNA for probe synthesis or any sequence information for the design of specific primers. Moreover, the procedure does not involve blotting or hybridizing steps for their detection, requires small amounts of DNA (10 ng per reaction), and can achieve quite high-sample throughput for their analysis. Therefore, this technique is quick, simple, and efficient in principle. The major limitations of RAPDs are that they are dominant markers and the technique is highly sensitive to changes in external PCR conditions resulting in low reproducibility (Nagaoka and Ogihara 1997). RAPDs have been used for a variety of purposes including the construction of genetic linkage maps, gene tagging, identification of cultivars, and genetic diversity studies.

AFLPs

The development of amplified fragment length polymorphisms (AFLPs) created another source of markers with the potential to rapidly saturate genetic maps (Vos et al. 1995). It can be applied to any organism and does not require prior sequence information. AFLP is a multi-locus marker system, which combines the power of RFLP and the flexibility of PCR-based markers. The technique is based on selective PCR amplification of restriction fragments generated by specific restriction enzymes and oligonucleotide adapters of few nucleotide bases. A typical AFLP fingerprint contains 50–100 amplified fragments, of which up to 80 % may serve as genetic markers. Although AFLP is a highly powerful and reliable marker technique, its usage in marker-assisted selection is limited as it is a dominant marker and requires high-quality DNA, higher cost for genotyping and complexities in scoring, and interpretation of amplicons.

SSRs

Simple sequence repeats (SSRs), also called as microsatellites or short tandem repeats (STRs) or sequence tagged microsatellites (STMS), are one of the most important categories of DNA markers. SSRs are tandem repetitions of short nucleotide stretches of 2–6 base pairs in length, usually repeated about 15–30 times. SSRs are widely used for DNA fingerprinting, genetic mapping, MAS, and studies of genetic diversity and population genetics (Hearne et al. 1992; Zietkiewicz et al. 1994). SSR markers are abundant, codominant and show high level of polymorphism, which makes them markers of choice for detailed mapping of genomes. The number and composition of microsatellite repeats differ in plants and animals. The flanking sequences of SSRs are often unique, allowing primers to be designed that result in tagged SSR markers representing a single locus. The majority of the allelic variation of SSRs is thought to arise as a result of slipped strand mispairing. Unlike RAPDs, SSRs as a marker system have been shown to be highly reproducible (Jones et al. 1997). In addition, they are easily transferable between laboratories as the sequence information can be distributed. In the recent times, with the availability of genome sequence for several crops, the identification of SSR motif and design of primer pairs has become very simple. Thousands of SSR markers have been developed in several plant species using genome sequence.

SSRs derived from Expressed Sequence Tags (ESTs) or genes are called genic SSRs and are found to be less polymorphic compared to SSRs derived from non-coding regions due to the sequence conservation in transcribed regions of DNA (Eujayl et al. 2004). EST-SSR markers are an important class of marker since they can contribute to "direct allele selection," if they are shown to be completely associated or even responsible for a targeted trait (Sorrells and Wilson 1997). These markers can be used for accurate assaying functional diversity in the natural populations, germplasm collections as well as used in comparative mapping, and evolutionary studies as anchor markers. Another most important feature of the genic SSR markers over genomic SSRs is their transferability among distantly related species. Recently, the potential use of EST-SSRs developed for barley and wheat has been demonstrated for comparative mapping in wheat, rye, and rice (Yu et al. 2004).

A major disadvantage of the EST-derived microsatellites is the sequence redundancy that yields multiple sets of markers at the same locus. But nowadays, the random EST sequences are assembled into unique gene sequences called unigenes that circumvents the problem of redundancy in EST databases (Parida et al. 2006). The unigene-based SSRs show unique identity and positions in the transcribed regions of the genome. The availability of large unigene databases make it possible to systematically search for microsatellites in the unigenes.

ISSRs
Inter-Simple Sequence Repeats (ISSR) is a RAPD-like marker system that does not require any prior knowledge of genome sequence (Godwin et al. 1997). The technique combines most of the benefits of AFLP and microsatellite analysis with the universality of RAPD. ISSR can access variation in the numerous microsatellite regions by using primers that are anchored at the 5′ or 3′ end of a repeat region and extend into the flanking region. This technique allows amplification of the genomic segments between inversely oriented repeats (ISSRs). Generally, a series of single primers are used to generate a series of fragments that are size separated on either an agarose gel or a polyacrylamide gel (Nagaraju et al. 2002).

DArT
Diversity Arrays Technology (DArT) markers are a class of DNA markers with a potential of high-throughput whole genome profiling based on hybridization-based technology (Jaccoud et al. 2001). It is a sequence-independent marker system and generates genome-wide marker fingerprints. DArT markers are biallelic and inherit in dominant manner. DArT can overcome the limitations of genome coverage, reproducibility, time, etc., of RFLP, AFLP, RAPD, and SSR marker systems. It is a non-gel-based technology amenable for high-throughput automation, providing high-quality datasets of hundreds of markers in a single assay. Although DArT is a powerful genotyping tool, the technology is not affordable by less resourced laboratories as it requires high initial investment. DArT is a dominant marker and requires higher technical skills.

SNPs
Currently, Single Nucleotide Polymorphisms (SNPs) are the marker of choice due to their availability in large numbers in virtually all crop species. SNP is the ultimate marker representing a single nucleotide difference between two individuals. SNPs may be present within coding or non-coding sequences of genes or in the intergenic regions between genes at different frequencies in different chromosome regions (Jiang 2013). Once developed, SNP is the most prolific, highly efficient, and inexpensive marker technology for use in plant breeding. The technology holds enormous potential for multiplexing and high throughput resulting in less operational cost. SNPs are increasingly used for germplasm characterization, linkage mapping and QTL studies, population structure and genome-wide association analysis, map-based cloning, and functional genomics.

3.5 Genetic Linkage Maps of Sorghum

Initial Maps
Development of linkage map is a prerequisite for the detailed genetic analysis of a trait and its manipulation through MAS (Tanksley et al. 1989). DNA markers are preferred for genetic mapping of important traits (agronomic, pest-resistance, stress-tolerance, and quality traits) besides addressing several issues in genetic diversity, phylogeny, etc. Dense and saturated linkage maps substantially contribute to the positional cloning of important genes and form the base for gene discovery and isolation, germplasm characterization, allele mining, etc. Several molecular marker systems have been used for the development of genetic maps in various crop species (rice, wheat, maize, tomato, soybean, etc.) including sorghum.

Mapping efforts in sorghum involving DNA markers began in the early 1990s and then on several linkage maps have been reported using different DNA markers (Table 3.1). The first linkage map in sorghum was constructed with 36 RFLPs with a length of 283 cM wherein the map represented 8 of the 10 linkage groups (Hulbert et al. 1990). Further efforts involving more RFLPs resulted in several linkage maps with increased map length, marker density, and resolution (Berhan et al. 1993; Binelli et al. 1992; Chittenden et al. 1994; Dufour et al. 1997; Lin et al. 1995; Pereira et al. 1994; Whitkus et al. 1992; Xu et al. 1994), but were far from representing the complete genome of sorghum. These maps in fact were linkage group segments representing a set of linked markers. For the first time, Pereira et al. (1994) and Chittenden et al. (1994) developed complete sorghum linkage maps using maize and sorghum RFLP probes with increased map density and map length. These earlier maps with endogenous and exogenous RFLP probes were useful in comparative genomic studies in establishing the homology and synteny between the members of *Poaceae*. Probes derived from genomic DNA (gDNA) and cDNA libraries specific to sorghum were added to linkage maps along with exogenous probes developed from other related genomes. A map constructed using 38 sorghum and 33 maize gDNA probes contained 15 linkage groups and spanned a map length of 633 cM with an average marker distance of 8.9 cM (Ragab et al. 1994). Since alignment and integration of early maps was necessary, Subudhi and Nguyen (2000) aligned five major RFLP maps (Boivin et al. 1999; Chittenden et al. 1994; Pereira et al. 1994; Ragab et al. 1994; Xu et al. 1994) with ten linkage groups by integrating with the map of a RIL population. This was helpful in assessing the accuracy of available maps and the assessment of linkage of QTL markers in a particular genomic region.

High-Density Maps

With the advent of new marker systems, later sorghum maps were generated with PCR-based marker systems like RAPDs, AFLPs, and SSRs. Though RAPDs were used in the initial mapping efforts, they have not been used extensively in sorghum due to the problems of their reproducibility. The map of Tuinstra et al. (1996) had the highest number of RAPDs (150), followed by Agrama et al. (2002) and Knoll et al. (2008) with 75 and 67 markers, respectively. Further efforts were made in saturating the available maps with SSRs and AFLPs as these markers show high level of polymorphism, marker stability, and repeatability across labs. Use of more common SSR markers in different maps allowed comprehensive comparison and their integration. In many of these maps, RFLPs were used as anchor probes for linkage group identification, alignment, and for comparative genetic studies. AFLPs being powerful, consistent, and efficient for genetic mapping were extensively utilized for saturation of linkage maps. The RFLP map of Dufour et al. (1997) was further saturated with AFLPs by Boivin et al. (1999) which resulted in increased map length (from 977 to 1,899 cM) and map density. The combined map of two RIL populations (Haussmann et al. 2002) consisted of 249 AFLPs, and the distribution of AFLPs on the genome was found to be clustered similar to the one observed earlier (Boivin et al. 1999). Several of the maps also included AFLP markers for map saturation and QTL identification in sorghum (Mace et al. 2008; McIntyre et al. 2005; Murray et al. 2008; Ramu et al. 2009; Ritter et al. 2008; Shiringani and Friedt 2011).

SSR marker technology has proven to be a dependable, rapid, and inexpensive tool for plant genotyping (Yang et al. 1996) and is preferred over RFLP and AFLP markers in terms of their technical simplicity, throughput level, and automation (Varshney et al. 2005). Taramino et al. (1997) developed 13 sorghum SSRs and mapped 7 SSRs on to the RFLP map of Pereira et al. (1994). Similarly, eight new SSR markers were assigned to an RFLP linkage map of cross QL39 × QL41 (Tao et al. 2000; Tao et al. 1998). Kong et al. (2000) and Bhattramakki et al. (2000) added 31 and 113 new SSR markers on a previously developed RFLP linkage map (Peng et al. 1999). With the availability of new marker systems and rapid genotyping facilities, existing linkage maps

Table 3.1 Genetic linkage maps of sorghum

Reference	Parents	Population	Marker type	Map length (cM)	Linkage group
Hulbert et al. (1990)	Shanqui Red × M 91051	55 F_2	37 RFLPs	283	8
Binelli et al. (1992)	IS 18729 × IS 24756	149 F_2	35 RFLPs	440	5
Whitkus et al. (1992)	IS 2482C × IS 18809	81 F_2	91 RFLPs, 7 isozymes	949	13
Berhan et al. (1993)	IS 18729 × IS 24756	55 F_2	96 RFLPs	709	15
Ragab et al. (1994)	BSC 35 × BTx 623	93 $F_{2:3}$	71 RFLPs	633	15
Chittenden et al. (1994)	BTx 623 × *S. propinquum*	56 F_2	276 RFLPs	1,445	10
Pereira et al. (1994)	CK 60 × PI 229828	78 F_2	201 RFLPs	1,530	10
Xu et al. (1994)	IS 3620C × BTx 623	50 F_2	190 RFLPs	1,789	14
Lin et al. (1995)	BTx 623 × *S. propinquum*	370 F_2	202 RFLPs	935	11
Pereira and Lee (1995)	CK 60 × PI 229828	152 F_2	111 RFLPs	1,299	10
Tuinstra et al. (1996)	TX 7078 × B 35	98 $F_{5:7–8}$ RIL	20 RFLPs,150 RAPD	1,580	17
Dufour et al. (1997)	IS 2807 × 379	110 F_5 RIL	145 RFLPs, 4 cloned genes, 2 morphological markers	977	
	IS 2807 × 249	91 F_5 RIL	133 RFLPs, 4 cloned genes, 1 morphological marker	878	12
	Composite map of above two populations		183 RFLPs, 3 cloned genes, 2 morphological markers	1,095	13
Taramino et al. (1997)	CK 60 × PI 229828	68 F_2	7 SSRs	1,575	
Tao et al. (1998)	QL 39 × QL 41	128 F_5 RIL	155 RFLPs, 8 SSRs	1,400	21
Boivin et al. (1999)	IS 2807 × 379	110 F_5 RIL	298 RFLPs, 137 AFLPs	1,899	11
Crasta et al. (1999)	B 35 × Tx 430	96 F_5 RIL	142 RFLPs	1,602	14
Peng et al. (1999)	BTx 623 × IS 3620C	137 $F_{6–8}$ RIL	321 RFLPs	1,364	10
Subudhi and Nguyen (2000)	B 35 × Tx 7000	98 F_7 RIL	214 RFLPs, 3 SSRs, 7 RAPD	1,200	10
Kong et al. (2000)	BTx 623 × IS 3620C	138 $F_{6–8}$ RIL	114 RFLPs, 31 SSRs	1,287	10
Bhattramakki et al. (2000)	BTx 623 × IS 3620C	139 $F_{6–8}$ RIL	323 RFLPs, 143 SSRs	1,406	10
Tao et al. (2000)	QL 39 × QL 41	152 F_5 RIL	281 RFLPs, 25 SSRs, 5 morphological markers	>2,750	10
Xu et al. (2000)	B 35 × Tx 7000	98 F_7 RIL	162 RFLPs	837	10
Bennetzen et al. (2001)	Framework map derived from comparison of maps from Kong et al. (2000), Peng et al. (1999), Berhan et al. (1993) and Pereira et al. (1994)		154 RFLPs, 34 SSRs, 10 morphological markers	1,450	10
Kebede et al. (2001)	SC 56 × Tx 7000	125 F_7 RIL	144 RFLPs	1,355	10

(continued)

Table 3.1 (continued)

Reference	Parents	Population	Marker type	Map length (cM)	Linkage group
Klein et al. (2001)	RTx 430 × Sureno	125 F_5 RIL	85 RFLPs, 44 SSRs, 1 morphological marker	970	10
Hart et al. (2001)	BTx 623 × IS 3620C	139 F_{6-8} RIL	145 RFLPs and SSRs	1,278	10
Menz et al. (2002)	BTx 623 × IS 3620C	137 F_{6-8} RIL	336 RFLPs, 136 SSRs, 2454 AFLPs	1,713	10
Huassmann et al. (2002)	IS 9830 × E36-1	225 $F_{3:5}$ RIL	14 RFLPs, 125 AFLPs, 45 SSRs, 3 RAPD	1,265	10
	N 13 × E 36-1	226 $F_{3:5}$ RIL	16 RFLPs, 158 AFLPs, 55 SSRs, 3 RAPD	1,410	12
	Composite map of above two populations		339 RFLP, AFLP, 55 SSR, RAPD markers	1,424	11
Agrama et al. (2002)	GB 1 K × Redlan	93 RIL	38 SSR, 75 RAPD	1,530	12
Tao et al. (2003)	ICSV 745 × 90562	120 RIL	269 RFLP, 8 SSRs	1,472	12
Bowers et al. (2003)	BTx 623 × *S. propinquum*	65 F_2	2512 RFLPs	1,059.2	10
McIntyre et al. (2005)	31945-2-2 × *S. arundinaceus*	120 F_2	169 RFLPs, 30 RGAs, 28 SSRs 86 AFLPs	–	14
Deu et al. (2005)	Malisor 84–7 × S34	218 F_2	92 RFLP, SSR markers	1,160	13
Brown et al. (2006)	BTx 623 × IS 3620C	119 F_{9-10} RIL	336 RFLPs, 136 SSRs, 2454 AFLPs	1,713	10
Bian et al. (2006)	Early Folger × N 32B	207 F_{2-3}	31 RFLPs, 254 AFLPs, 42 SSRs	983.5	20
Feltus et al. (2006)	BTx 623 × IS 3620C	137 F_{6-8}	177 RFLPs and SSRs	1,438.8	10
	BTx 623 × *S. propinquum*	370 F_2	96 RFLPs, 36 SSRs	1,409.8	10
Wu et al. (2007)	Westland A × PI 550610	277 F_2	188 SSRs	997.5	16
Knoll et al. (2008)	Shan Qui Red × SRN 39	153 RIL	17 RFLPs, 56 SSRs, 67 RAPD, 3 morphological markers	2,128	14
Ritter et al. (2008)	R 9188 × R 9403463-2-1	184 F_6 RIL	42 sorghum SSRs,10 sugarcane SSRs, 247 AFLPs	2,012.9	16
Mace et al. (2008)	R 931945-2-2 × IS 8525	146 RIL	358 DArTs, 47 SSRs, 188 AFLPs, 2 morphological markers	1,413.6	10
Murray et al. (2008)	Rio × BTx 623	176 F_{4-5}	191 AFLP, 68 SSRs	1,836	10
Parh et al. (2008)	R 931945-2-2 × IS 8525	146 RIL	148 DArTs, 36 SSRs, 117 AFLPs, 2 morphological markers	1,625.2	10
Fernandez et al. (2008)	KS 115 × Macia	312 RIL	85 SSRs, 27 INDELs	1,364.6	11

(continued)

Table 3.1 (continued)

Reference	Parents	Population	Marker type	Map length (cM)	Linkage group
Duan et al. (2009)	B2 V4 × 1383-2	150 F_2	122 MSAP, 22 SSRs	483.6	11
Srinivas et al. (2009a)	296B × IS 18551	168 RIL	100 SSRs, 28 EST-SSRs	1,074.5	15
Srinivas et al. (2009b)	296B × IS 18551	168 RIL	100 SSRs, 38 EST-SSRs, 10 Unigene SSRs, 2 morphological markers	1,098.5	15
Satish et al. (2009)	296B × IS 18551	168 RIL	107 SSRs, 38 EST-SSRs, 10 Unigene SSRs, 3 morphological markers	1,143	16
Ramu et al. (2009)	N 13 × E 36-1	94 RIL	14 RFLPs, 125 AFLPs, 45 genomic SSRs, 55 EST-SSRs, 3 RAPD markers	2,838	10
Mace et al. (2009)	BTx 623 × IS 3620C	137 F_{6-8} RIL	303 DArTs, 259 RFLPs, 226 SSRs, 2 morphological markers	1,528	10
	R 890592 × ICSV 745	119 RIL	234 DArTs, 10 SSRs, 244 RFLPs	1,433	10
	R 931945-2-2 × IS 8525	146 RIL	357 DArTs, 47 SSRs, 188 AFLPs, 2 morphological markers	1,453	10
	B 923296 × SC 170-6-8	88 RIL	170 DArTs, 13 SSRs	1,138	10
	BTx 642 × QL 12	94 RIL	117 DArTs	910	10
	SAR 10 × SSM 249	183 RIL	627 DArTs, 131 SSRs, 47 RFLPs, 2 morphological markers	1,227	10
	Consensus map of above six populations		1190 DArTs, 839 Non-DArT markers	1,603.5	10
Shiringani et al. (2010)	M 71 × SS 79	188 RIL	102 AFLP, 49 SSR, and 6 EST-SSR	1,029	11
Mace and Jordan (2011)	B 923296 × SC 170-6-8	141 RIL	377 DArTs	2,259	10
Sabadin et al. (2012)	BR 007 × SC 283	90 RIL	255 DArTs, 83 SSRs, 5 sequence-tagged site (STS), and one RFLP marker	2,034.9	10
Shiringani and Friedt (2011)	SS 79 × M7 1	188 RIL	102 AFLP, 49 SSR, and 6 EST-SSR	1,029	11
Guan et al. (2011)	Shihong 137 × L-Tian	186 F_2	118 SSRs	1,884.6	15
Aruna et al. (2011)	27B × IS 2122	210 RIL	127 SSRs, 21 genic-SSRs, and one morphological marker	700	10

(continued)

Table 3.1 (continued)

Reference	Parents	Population	Marker type	Map length (cM)	Linkage group
Nagaraja Reddy et al. (2012, 2014)	M35-1 × B 35	245 RIL	228 SSRs and 3 morphological markers	1,235.5	10
Zou et al. (2012)	654 × LTR 108	244 RIL	3418 bin markers, SNPs	1,591.4	10
Takai et al. (2012)	MS 138B × 74LH 3213	136 RIL	247 SSRs	1,697.2	10
Kong et al. (2013)	BTx 623 × *S. propinquum*	161 RIL	141 SSRs	773.1	10
Alam et al. (2014)	ICSV 745 × R 890562	119 RIL	234 DArTs, 10 SSRs, and 244 RFLPs	1,487.96	12
	IS 8525/R 931945-2-2	146 RIL	357 DArTs, 51 SSRs, and two morphological	1,390.20	13
	B 923296/SC 170-6-8	141 RIL	337, major DArT	2,264.45	17
	R 931945-2-2*2 × *S. bicolor subsp. verticilliflorum*	214 BC_1F_4	467, major DArT	–	–

were increasingly saturated to develop high-density maps with the addition of more markers. Bhattramakki et al. (2000) generated a high-density linkage map of sorghum involving 323 RFLPs and 143 SSRs with an average distance of 3.1 cM between the markers. Menz et al. (2002) saturated this map further with AFLPs markers to construct extra high-density sorghum linkage map with 2,926 markers (2,454 AFLPs, 136 SSRs, and 336 RFLPs from rice, barley, oat, and maize cDNA and genomic clones) with a map length of 1,713 cM, and an average marker distance of 0.5 cM. Another high-density map with 2,512 RFLP loci from 2,050 endogenous and exogenous probes developed (Bowers et al. 2003) had the highest marker density with a least average distance of 0.4 cM between markers. The probes included 1,189 from sorghum cDNA and gDNA clones, others from maize, sugarcane, wheat, barley, rice, millet, oat, rye, and *Arabidopsis* genomes. Thus, the two high-density maps (Bowers et al. 2003; Menz et al. 2002) were highly useful for undertaking comparative genomic investigations in various members of *Poaceae*.

Several linkage maps exclusively with microsatellite markers have been developed (Moens et al. 2006; Nagaraja Reddy et al. 2012; Srinivas et al. 2009a). Although the utility of SSR markers has been well established, their de novo development can be costly, complex, and time-consuming on a locus-by-locus basis (Bhattramakki et al. 2000; Brown et al. 1996). Consequently, their development has been made much easier by the generation of large numbers of EST libraries (Pratt et al. 2005) and by the availability of complete sorghum genome sequence (Paterson et al. 2009). Several maps (Ramu et al. 2009; Srinivas et al. 2009a) therefore involved EST-SSRs, which were very useful in comparative genomic analysis with other cereals like rice, maize, etc. Looking into the problem of sequence redundancy, unique gene sequences called unigenes (Parida et al. 2006) were used for deriving SSRs. The unigene-based SSRs have the advantages of unique identity and positions in the transcribed regions of the genome and can be used for accurately assaying the functional diversity in natural populations and germplasm collections as well as for comparative mapping and evolutionary studies as anchor markers. Several unigene-based SSR markers were developed and their map positions were determined in a RIL population along with the genomic SSRs (Srinivas et al. 2009b), which was further strengthened with the mining of 1,519 unigene SSRs and construction of linkage map with 228 SSRs in sorghum (Nagaraja Reddy et al. 2012). An important milestone was the availability of the complete DNA sequence of sorghum (Paterson et al. 2009). Li et al. (2009) developed 1,758 new genomic SSR primers from the sorghum sequence, and 1,692 of the SSRs were in silico mapped on to the 10 sorghum chromosomes. Similarly, Yonemaru et al. (2009) designed 5599 non-redundant SSR markers, including regions flanking the SSRs, in whole-genome shotgun sequences of sorghum line ATx623. Of them, chromosomal locations of 5012 SSR markers were determined in silico. This strategy is convenient, saves labor, and involves low cost compared to conventional genetic maps and will serve as reference map for quick saturation of genetic map and comparative mapping.

DArT markers show the potential of high-throughput whole genome profiling (Jaccoud et al. 2001). DArT is developed as an alternative to marker types reported earlier to overcome several of the limitations (lower genome coverage and discrimination power, reproducibility, etc.). Sorghum linkage map developed by Mace et al. (2008) consisted of 358 DArTs, 47 genomic-SSR, and 188 AFLP markers distributed over 10 chromosomes, which spanned a genetic distance of 1431.6 cM. The average distance between adjacent markers was 2.39 cM. Later, Mace et al. (2009) developed a consensus map based on six populations comprising of 1,190 DArT markers and 839 non-DArT markers distributed on 10 chromosomes. This map is currently used as a reference map resource for various genetic studies, besides providing a framework for transferring genetic information between different marker technologies and for integrating DArT markers with other genomic resources.

SNPs are the ultimate markers to represent polymorphism. In future, they are the markers of choice for several genetic analyses in trait dissection and marker-assisted selection. With the advent of newer, efficient, and cheaper genome-sequencing technologies, the detection of SNPs in several crops is progressing at a great speed. Genotyping-by-sequencing (GBS) is one such technology (Elshire et al. 2011) where sequencing and genotyping of the individual are done simultaneously. The utility of re-sequencing in sorghum was established with the development of an ultra-high-density linkage map based on high-quality SNPs generated from low-coverage sequences (~0.07 genome sequence) involving 244 RILs of a sorghum cross (Zou et al. 2012). This map consisted of 3,418 bin markers and spanned 1,591.4 cM of genome size with an average distance of 0.5 cM between adjacent bins. The relationship between the genetic bin map and the physical position of SNP was consistent, and therefore it was easy to anchor the physical interval and find the putative genes for several agronomic traits in the target regions.

3.6 Trait Mapping

3.6.1 Mapping Major Genes

Traits with qualitative inheritance are usually controlled by a single major gene. Such traits are scored on a binary scale as "1" or "0." This data is treated as a maker and inserted as a locus in linkage map construction to identify the map position of the single major gene controlling a gene expression. Besides this approach, the following two approaches are also effectively used for mapping major genes.

Bulked Segregant Analysis (BSA)
BSA has been successfully used in mapping single major gene (Barua et al. 1993). In situations where development of a molecular map is not the objective or possible due to high cost and time factors, BSA can be effectively used. In this procedure, DNA pools from both the extremes of phenotype from segregating population can be effectively used for mapping the major gene involved in contrast trait expression. Each of these bulk DNA samples will contain a random sample of all the loci, except for the loci that are in the region of the gene. Therefore, any differences in marker pattern between these two bulks should be linked to the locus upon which the bulk was developed. This powerful technique has gained wide acceptance for mapping major gene traits.

Selective Genotyping
Genotyping only the part of the population that represents extreme phenotypes of the target traits is called as selective genotyping (Xu et al. 2008). It involves genotyping of only the selected individuals that are selected based on the individual phenotype (with low and high trait value). Since only the selected individuals are genotyped and not the entire population, genotyping cost is greatly reduced. The method is highly useful for mapping genes/QTLs with major effects. Selective genotyping can be bidirectional if the two tails of the distribution are considered or unidirectional if only one tail is considered. The latter is more suitable for traits that have been subjected to strong negative or lethal selection in unfavorable environments (Xu et al. 2008). Selective genotyping analysis has been widely used in genetic mapping in plants with numerous reports for single major gene validation.

3.6.2 Mapping Quantitative Trait Loci (QTL)

Any trait, which shows continuous distribution of trait values in a segregation population, is called as a quantitative trait. Since many genes are involved in its expression, it is also called as a polygenic trait. Unlike the qualitative traits where measurement is made on "counts," these quantitative traits are measured in statistical measurements like mean, variance, and range. Since there are many genes involved in the quantitative trait expression, it is likely that the genes are present on different chromosomes. The trait may be controlled by many genes with small effects or with few genes with large effects. These loci controlling quantitative trait expression are referred to as QTL (quantitative trait loci), and the procedure to detect and locate them on a chromosome is called as QTL mapping. QTL mapping aims to discover and locate probable QTLs (or genes) controlling a quantitative trait. Mapping of QTLs is based on a systematic search for linkage disequilibrium between marker loci and QTLs. In other words, QTL analysis is detecting an association between phenotype and the genotype of markers.

3.6.2.1 Linkage Analysis-Based QTL Mapping

The statistical analyses of associations between phenotype and genotype in a linkage-based population to detect QTLs include single-marker analysis (Edwards et al. 1987; Luo and Kearsey 1989), interval mapping (Lander and Botstein 1989), and composite interval mapping (Zeng 1993, 1994) and multiple trait mapping (Jiang and Zeng 1995; Ronin et al. 1995).

Single-Marker Analysis (SMA)
The simplest method for QTL mapping is single-marker analysis, implemented as a simple *t*-test, or analysis of variance (ANOVA) or simple linear regression using any statistical software, which assess the segregation of a phenotype with respect to a marker genotype (Soller et al. 1976). Progeny are classified based on marker genotype, and then phenotypic mean between the two classes is compared for statistical significance using *t*-test or ANOVA. A significant trait mean difference between the two marker classes indicates that the marker is linked to a QTL. The estimate of QTL effect on the trait is measured as the difference between the phenotypic class means. Linear regression is most commonly used because the coefficient of determination (R^2) from the marker explains the phenotypic variation arising from the QTL linked to the marker. The merit of this procedure is that the method does not require linkage map, is simple, and can be analyzed by any statistical software. However, this method cannot estimate the QTL position precisely, cannot distinguish a small effect QTL close to the marker and large effect QTL far from the marker (both give same probability), and underestimates the effect of a linked QTL due to recombination between marker and QTL and confounding of the effect of one QTL by many others that influence the trait.

Interval Mapping (IM)
A more powerful QTL mapping method, known as interval mapping, also called as simple interval mapping (SIM) was developed by Lander and Botstein (1989). This requires a linkage map for the estimation of position and effect of a

QTL. The basic principle is to test a QTL model for the presence of a QTL at many positions in a marker interval between two mapped marker loci and searches for QTL parameters that give the best approximation for quantitative trait distributions that are observed for each marker class in the marker interval. The procedure is based on maximum likelihood or regression that maximizes the likelihood of a single-gene model by averaging over the possible states of the unknown genotype at each possible QTL location. The method has its own drawbacks. The identification and estimation of QTL effect and position through SIM can be biased as the method considers only one QTL at a time in the model even though there could be other QTLs present (Zeng 1994). Second, QTLs outside the interval under consideration can affect the ability to find a QTL within it (Zeng 1993). Third, a false identification of QTL (false positive or "ghost peak") can arise if two QTLs are located in small intervals and act in the same direction (coupling).

Composite Interval Mapping (CIM)

CIM was developed by Jansen and Stam (1994) and Zeng (1994) which is an extension of IM that places certain markers into the model as cofactors. The approach of CIM assesses the probability that an interval between two markers is associated with a QTL affecting the trait of interest and is as well controlling for the effects of other background markers on the trait (Li and Börner 2004). In theory, CIM gives more power and accuracy than SIM because the effects of other QTLs are not present as residual variance. Four major limitations of CIM are (1) effect of uneven distribution of markers in the genome (i.e., the test statistics in a marker-rich region may not be comparable to that in a marker-poor region), (2) difficulty in estimating the joint contribution to the genetic variance of multiple-linked QTLs, (3) not directly extendable for analyzing epistasis, and (4) reduction in the statistical power to detect QTLs due to the use of tightly linked markers as cofactors (Zeng et al. 1999).

Multiple Interval Mapping (MIM)

MIM was proposed and implemented by Kao et al. (1999) to overcome the limitations of CIM and to map multiple QTLs simultaneously. It uses multiple marker intervals simultaneously to fit multiple putative QTLs directly in the model. The MIM model is based on Cockerham's model for interpreting genetic parameters and the method of maximum likelihood for estimating genetic parameters. The precision and power of QTL mapping could be improved by the MIM approach. Also, epistasis between QTLs, genotypic values of individuals, and heritabilities of quantitative traits can be readily estimated and analyzed.

MIM consists of four components: (1) an evaluation procedure designed to analyze the likelihood of the data given a genetic model (number, positions, and epistatic terms of QTL), (2) a search strategy optimized to select the best genetic model (among those sampled) in the parameter space, (3) an estimation procedure for all parameters of the genetic architecture of the quantitative traits (number, positions, effects, and epistasis of QTL; genetic variances and covariances explained by QTL effects), and (4) a prediction procedure to estimate or predict the genotypic values of individuals and their offspring based on the selected genetic model and estimated genetic parameter values (Xu 2010).

Limitations of linkage-based QTL mapping methods: Linkage-based QTL mapping approaches are associated with several limitations. First is the time involved in the development of a mapping population. If the targeted mapping population is a RIL, then one has to wait for 6–7 filial generations, which also involves lot of labor and other resources. Secondly, the precision of QTL mapping largely depends on the genetic variation (or genetic background) covered by a mapping population, its size, and the number of marker loci used to represent the whole genome. The linkage-based QTL mapping strategy has low resolution with simultaneous evaluation of only a few alleles (Abdurakhmonov and Abdukarimov 2008). Third, the resolution with which a QTL is identified usually spans 5–10 cM that is due to limited recombination events that occurred during the development of a mapping population. Besides these, an identified

QTL needs to be validated in different genetic backgrounds before taking a MAS program. It should also be noted that an increased number of markers may pose problems for linkage analysis in providing correct marker order and can lead to erroneous QTL mapping (Collard et al. 2009).

3.6.2.2 Association Mapping (AM)

AM is a population-based approach to exploit natural variation. It relies on the historical, unrecorded sources of disequilibrium to create population-wide marker-phenotype associations (Jannink et al. 2001). This approach overcomes some of the demerits of linkage-based QTL mapping. The terms AM and linkage disequilibrium (LD) have been used interchangeably, but AM is an application of LD. AM refers to a significant association of a molecular marker with a phenotypic trait, while LD is a non-random association of two markers (alleles at different loci), between two genes or QTLs, and between gene/QTL and a marker (Gupta et al. 2005). LD refers to the correlation between alleles in a population (Flint-Garcia et al. 2003) but not necessarily on the same chromosome. There are two strategies through which AM is applied, viz., candidate-gene association mapping and the genome-wide association mapping. For candidate gene-based strategy, the information on the biochemical, genetic, and physiological pathways in the development of a trait along with the location and functions of genes involved is necessary (Mackay 2001). This also requires the identification of SNPs between lines in specific gene/genes since SNPs offer ultimate QTL map resolution as they are potentially in LD with the causative polymorphism (Rafalski 2002). On the other hand, genome-wide association mapping scans for the LD between marker and trait value all along the genome length. Therefore, this approach requires facilities for genotyping SNPs in high density. AM can greatly accelerate positional cloning of QTL approaches, since the trait-associated loci can be mapped with more precision and with high resolution. Some of the earlier AM studies were done in crops like maize (Bar-Hen et al. 1995), rice (Virk et al. 1996), and oat (Beer et al. 1997).

AM also suffers with some limitations. Firstly, it usually requires a large number of markers for genotyping, since the number of markers depends on the LD decay and the genome size. More numbers of markers are required for a small LD like in out-crossing species compared to self-fertilizing species where the LD is larger. Secondly, AM has low power to detect rare alleles in populations as the power to detect the marker-trait association depends on the allele frequency in the population. Thirdly, the confounding effects of population structure (Yu and Buckler 2006; Zhao et al. 2007) arises when phenotypic traits are correlated with the underlying population structure at non-causal loci. In such cases, even loci that are unrelated to the trait will show varying degrees of association because of the confounding effects of population structure. Association of trait and loci is considered realistic only if the population structure is homogeneous. Choice of appropriate sample, control of population structure, and correct interpretation of results are essential to avoid misleading inference from AM. Finally, it requires thorough statistical assessment to investigate the relatedness of the lines and the overall population structure.

3.6.2.3 Nested Association Mapping (NAM)

NAM strategy addresses the complex trait dissection at a fundamental level through the generation of a common mapping resource that enables researchers to efficiently exploit genetic, genomic, and systems biology tools (Yu et al. 2008). It combines the advantages of both linkage and association mapping. NAM has the advantages of lower sensitivity to genetic heterogeneity and higher power as well as higher efficiency in using the genome sequence or dense markers while still maintaining high allele richness due to diverse founders. While previous joint linkage and linkage disequilibrium studies focused on mining existing mapping population in pedigrees or heterogeneous stocks, NAM is an integrated mapping population specifically designed for a full genome scan with high power for QTL with effects of different sizes.

NAM population uses RIL populations derived from several crosses of parental inbreds.

Due to diminishing chances of recombination over short genetic distance and a given number of generations, the genomes of these RILs are mosaics of chromosomal segments of their parental genomes. Consequently, within the chromosomal segments, the linkage LD information across the parental inbreds is maintained. Thus, if diverse parental inbreds are used, LD decays within the chromosomal segments of the RILs over a short physical distance (Wilson et al. 2004). Therefore, the NAM strategy allows to exploit both recent and ancient recombination and, thus, will show a high mapping resolution (Yu et al. 2008). Furthermore, due to the balanced design underlying the proposed mapping strategy as well as the systematic reshuffling of the genomes of the parental inbreds during RIL development, NAM populations are expected to show a high power to detect QTL in genome-wide approaches (Buckler et al. 2009).

3.7 Software Resources

3.7.1 Resources for Linkage-Based Mapping

MapMaker/QTL

Several software are available for establishing the marker-trait associations in linkage-based mapping populations. Of them, MapMaker/QTL was first to be developed (Lincoln et al. 1993b) and widely used. This works in DOS, UNIX, or Mac operating systems. Estimation of linkage between a marker and phenotype is carried out based on maximum likelihood using interval mapping method. This software uses the input files from its sister software, MapMaker/EXP (Lincoln et al. 1993a). The software is available freely at http://www.broadinstitute.org/ftp/distribution/software/mapmaker3/ and ftp://ftp-genome.wi.mit.edu/distribution/software/newqtl/.

PLABQTL

PLABQTL (Utz and Melchinger 2007) runs in DOS or Windows computing environments. It works on multiple regressions and handles populations from F_2 until RIL, or test cross generations. It estimates QTL using SIM or CIM methods. It can also estimate QTL × environment interactions. Similar to MapMaker/QTL, PLABQTL also requires the marker data, linkage map, and trait values for accomplishing QTL analysis. The software is freely available at https://plant-breeding.uni-hohenheim.de/software.html#jfmulticontent_c110647-2.

QGene

QGene 4.0 (Joehanes and Nelson 2008; Nelson 1997) is a free, open-source Java program that runs on any operating system. It requires a data file containing marker data, trait data, and a map file with all the markers. It is a user-friendly program for QTL analysis, NIL selection, and many other functions. It is especially good for advanced backcross populations. The software is available to download at http://www.qgene.org/qgene/index.php.

Map Manager QTX

Map Manager QTX (Meer et al. 2004) analyzes the results of experimental genetic crosses to map QTL. QTX is available for both Windows and Mac operating systems at http://www.mapmanager.org/. QTX functions for mapping mendelian markers and uses regression to detect and map QTLs by four methods: association with single-marker loci, simple interval mapping, composite interval mapping, and a search for pairs of interacting QTLs. QTX supports advanced backcross, advanced inter-cross, and recombinant inbred inter-cross designs and offers weighted regression for designs in which an estimate of the trait variance is available for each line. It is a very user-friendly program for QTL analysis, NIL selection, and many other functions.

MapQTL

MapQTL (van Ooijen 2011) is a commercial software for detecting QTL in diploid populations. It can analyze data from various experimental populations like F_2, backcross, RILs, DH, and out-breeder full-sib family. Two algorithms, maximum likelihood and regression, are used in interval mapping and composite interval mapping for the mapping of QTL. It also performs single-marker analysis using non-parametric statistics of Kruskal-Wallis rank sum test. The software is easy to use, very fast, and presents the results in tables and charts. The software requires locus files with segregation data for all markers, map order for each linkage groups, trait values, and trait names. The commercial software is available at http://www.kyazma.nl/index.php/mc.MapQTL.

WinQTL Cartographer
WinQTL Cartographer (Basten et al. 2003) maps QTL in cross populations from inbred lines. This includes a powerful graphic tool for presenting and summarizing mapping results and can import and export data in a variety of formats. It analyzes the data with the following statistical methods: IM, SIM, CIM, Bayesian interval mapping, MIM, multiple trait analysis, and categorical trait analysis. It uses a graphical interface to depict the mapping results. The software can import and export data in a variety of formats and is available at http://statgen.ncsu.edu/qtlcart/WQTLCart.htm.

3.7.2 Resources for Association Mapping

TASSEL
Trait Analysis by aSSociation, Evolution and Linkage (TASSEL) is a software to evaluate traits associations, evolutionary patterns, and linkage disequilibrium (Bradbury et al. 2007). It performs a variety of genetic analyses including LD mapping, diversity estimation, and calculations of LD (http://sourceforge.net/projects/tassel/; Zhang, et al. 2006). It provides a number of powerful statistical approaches to association mapping such as a general linear model (GLM) and mixed linear model (MLM). It can handle a wide range of indels (insertion and deletions). The software is available at http://www.maizegenetics.net/index.php?option=com_content&task=view&id=89&Itemid=119.

GAPIT
Genome Association and Prediction Integrated Tool (Lipka et al. 2012) is an R package that performs genome-wide association study (GWAS) and genome prediction (or selection). This program uses state-of-the-art methods developed for statistical genetics, such as the unified mixed model, EMMA, the compressed mixed linear model, and P3D/EMMAx. The software is available at http://www.maizegenetics.net/gapit.

3.8 Conclusions

Using the concept of linkage along with morphological and biochemical markers, earlier studies detected major genes involved in the trait expression. However, due to the limitations in the use of conventional markers, development and use of DNA markers in trait mapping is a significant step in unraveling the genetic architecture of traits. Many of the traits, which are economically important, are controlled by many genes. Genetic mapping is the most valuable approach to identify genetic factors that underlie quantitatively inherited traits. Linkage and linkage disequilibrium-based mapping approaches have made great strides in mapping genes/QTLs in several crops. High-throughput genotyping techniques based on SNPs allow the development of high-density linkage maps and form significant tools in trait mapping. Genetically linked markers are used in marker-assisted breeding to identify individuals with desirable traits.

References

Abdurakhmonov IY, Abdukarimov A (2008) Application of association mapping to understanding the genetic diversity of plant germplasm resources. Int J Plant Genom 2008:574927

Agrama H, Widle G, Reese J, Campbell L, Tuinstra M (2002) Genetic mapping of QTLs associated with greenbug resistance and tolerance in *Sorghum bicolor*. Theor Appl Genet 104:1373–1378

Alam MM, Mace ES, van Oosterom EJ, Cruickshank A, Hunt CH, Hammer GL, Jordan DR (2014) QTL analysis in multiple sorghum populations facilitates the dissection of the genetic and physiological control of tillering. Theor Appl Genet 127:2253–2266

Aruna C, Bhagwat VR, Madhusudhana R, Vittal S, Hussain T, Ghorade RB, Khandalkar HG, Audilakshmi S, Seetharama N (2011) Identification and validation of genomic regions that affect shoot fly resistance in sorghum [*Sorghum bicolor* (L.) Moench]. Theor Appl Genet 122:1617–1630

Bar-Hen A, Charcosset A, Bourgoin M, Guiard J (1995) Relationship between genetic markers and morphological traits in a maize inbred lines collection. Euphytica 84:145–154

Barua U, Chalmers K, Hackett C, Thomas W, Powell W, Waugh R (1993) Identification of RAPD markers linked to a *Rhynchosporium secalis* resistance locus in barley using near-isogenic lines and bulked segregant analysis. Heredity 71:177–184

Basten C, Weir B, Zeng Z (2003) WinQTL Cartographer. v. 2.0. Dept. of Statistics, North Carolina State University, Raleigh

Beckmann J, Soller M (1983) Restriction fragment length polymorphisms in genetic improvement: methodologies, mapping and costs. Theor Appl Genet 67:35–43

Beer SC, Siripoonwiwat W, O'donoughue LS, Souza E, Matthews D, Sorrells ME (1997) Associations between molecular markers and quantitative traits in an oat germplasm pool: can we infer linkages. J Agric Genom 3:1–16

Bennetzen JL, Subramanian V, Xu J, Salimath SS, Subramanian S, Bhattramakki D, Hart GE (2001) A framework genetic map of sorghum containing RFLP, SSR and morphological markers. In: Phillips RL, Vasil IK (eds) DNA-based markers in plants. Springer, Netherlands, pp 347–355

Berhan AM, Hulbert S, Butler L, Bennetzen J (1993) Structure and evolution of the genomes of *Sorghum bicolor* and *Zea mays*. Theor Appl Genet 86:598–604

Bhattramakki D, Dong J, Chhabra AK, Hart GE (2000) An integrated SSR and RFLP linkage map of *Sorghum bicolor* (L.) Moench. Genome 43:988–1002

Bian YL, Yazaki S, Inoue M, Cai HW (2006) QTLs for sugar content of stalk in sweet sorghum (*Sorghum bicolor* L. Moench). Agric Sci China 5:736–744

Binelli G, Gianfranceschi L, Pe M, Taramino G, Busso C, Stenhouse J, Ottaviano E (1992) Similarity of maize and sorghum genomes as revealed by maize RFLP probes. Theor Appl Genet 84:10–16

Boivin K, Deu M, Rami J-F, Trouche G, Hamon P (1999) Towards a saturated sorghum map using RFLP and AFLP markers. Theor Appl Genet 98:320–328

Bowers JE, Abbey C, Anderson S, Chang C, Draye X, Hoppe AH, Jessup R, Lemke C, Lennington J, Li Z (2003) A high-density genetic recombination map of sequence-tagged sites for sorghum, as a framework for comparative structural and evolutionary genomics of tropical grains and grasses. Genetics 165:367–386

Bradbury PJ, Zhang Z, Kroon DE, Casstevens TM, Ramdoss Y, Buckler ES (2007) TASSEL: software for association mapping of complex traits in diverse samples. Bioinformatics 23:2633–2635

Brown S, Hopkins M, Mitchell S, Senior M, Wang T, Duncan R, Gonzalez-Candelas F, Kresovich S (1996) Multiple methods for the identification of polymorphic simple sequence repeats (SSRs) in sorghum [*Sorghum bicolor* (L.) Moench]. Theor Appl Genet 93:190–198

Brown PJ, Klein PE, Bortiri E, Acharya CB, Rooney WL, Kresovich S (2006) Inheritance of inflorescence architecture in sorghum. Theor Appl Genet 113:931–942

Buckler ES, Holland JB, Bradbury PJ, Acharya CB, Brown PJ, Browne C, Ersoz E, Flint-Garcia S, Garcia A, Glaubitz JC (2009) The genetic architecture of maize flowering time. Science 325:714–718

Cavanagh C, Morell M, Mackay I, Powell W (2008) From mutations to MAGIC: resources for gene discovery, validation and delivery in crop plants. Curr Opin Plant Biol 11:215–221

Chittenden L, Schertz K, Lin Y, Wing R, Paterson A (1994) A detailed RFLP map of *Sorghum bicolor*×*S. propinquum*, suitable for high-density mapping, suggests ancestral duplication of sorghum chromosomes or chromosomal segments. Theor Appl Genet 87:925–933

Collard B, Mace E, McPHAIL M, Wenzl P, Cakir M, Fox G, Poulsen D, Jordan D (2009) How accurate are the marker orders in crop linkage maps generated from large marker datasets? Crop Pasture Sci 60:362–372

Crasta OR, Xu WW, Rosenow DT, Mullet J, Nguyen HT (1999) Mapping of post-flowering drought resistance traits in grain sorghum: association between QTLs influencing premature senescence and maturity. Mol Gen Genet 262:579–588

de Vilmorin P, Bateson W (1911) A case of gametic coupling in Pisum. Proc R Soc Lond B Biol Sci 84:9–11

Deu M, Ratnadass A, Hamada M, Noyer J, Diabate M, Chantereau J (2005) Quantitative trait loci for head-bug resistance in sorghum. Afr J Biotechnol 4:247–250

Duan Y, Qian J, Sun Y, Yi Z, Yan M (2009) Construction of methylation linkage map based on MSAP and SSR markers in *Sorghum bicolor* (L.). IUBMB Life 61:663–669

Dufour P, Deu M, Grivet L, D'hont A, Paulet F, Bouet A, Lanaud C, Glaszmann J, Hamon P (1997) Construction of a composite sorghum genome map and comparison with sugarcane, a related complex polyploid. Theor Appl Genet 94:409–418

Echt C, Knapp S, Liu B (1992) Genome mapping with non-inbred crosses using GMendel 2.0. Maize Genet Coop Newsl 66:27–29

Edwards M, Stuber C, Wendel J (1987) Molecular-marker-facilitated investigations of quantitative-trait loci in maize. I. Numbers, genomic distribution and types of gene action. Genetics 116:113–125

Elshire RJ, Glaubitz JC, Sun Q, Poland JA, Kawamoto K, Buckler ES, Mitchell SE (2011) A robust, simple genotyping-by-sequencing (GbS) approach for high diversity species. PLoS One 6:e19379

Eujayl I, Sledge M, Wang L, May G, Chekhovskiy K, Zwonitzer J, Mian M (2004) Medicago truncatula EST-SSRs reveal cross-species genetic markers for *Medicago* spp. Theor Appl Genet 108:414–422

Feltus FA, Hart GE, Schertz KF, Casa AM, Kresovich S, Abraham S, Klein PE, Brown PJ, Paterson AH (2006) Alignment of genetic maps and QTLs between inter- and intra-specific sorghum populations. Theor Appl Genet 112:1295–1305

Fernandez MGS, Hamblin MT, Li L, Rooney WL, Tuinstra MR, Kresovich S (2008) Quantitative trait loci analysis of endosperm color and carotenoid content in sorghum grain. Crop Sci 48:1732–1743

Flint-Garcia SA, Thornsberry JM IV, Iv B (2003) Structure of linkage disequilibrium in plants. Annu Rev Plant Biol 54:357–374

Godwin ID, Aitken EA, Smith LW (1997) Application of inter simple sequence repeat (ISSR) markers to plant genetics. Electrophoresis 18:1524–1528

Goulden C (1939) Problems in plant selection. In: Proceedings of 7th international genetics congress. Edinburgh Cambridge University Press, Cambridge, UK, pp 132–133

Guan YA, Wang HL, Qin L, Zhang HW, Yang YB, Gao FJ, Li RY, Wang HG (2011) QTL mapping of bio-energy related traits in sorghum. Euphytica 182:431–440

Gupta PK, Rustgi S, Kulwal PL (2005) Linkage disequilibrium and association studies in higher plants: present status and future prospects. Plant Mol Biol 57:461–485

Gupta P, Langridge P, Mir R (2010) Marker-assisted wheat breeding: present status and future possibilities. Mol Breed 26:145–161

Hart GE, Schertz KF, Peng Y, Syed NH (2001) Genetic mapping of *Sorghum bicolor* (L.) Moench QTLs that control variation in tillering and other morphological characters. Theor Appl Genet 103:1232–1242

Haussmann B, Mahalakshmi V, Reddy B, Seetharama N, Hash C, Geiger H (2002) QTL mapping of stay-green in two sorghum recombinant inbred populations. Theor Appl Genet 106:133–142

Hearne CM, Ghosh S, Todd JA (1992) Microsatellites for linkage analysis of genetic traits. Trends Genet 8:288–294

Hulbert SH, Richter TE, Axtell JD, Bennetzen JL (1990) Genetic mapping and characterization of sorghum and related crops by means of maize DNA probes. Proc Natl Acad Sci U S A 87:4251–4255

Jaccoud D, Peng K, Feinstein D, Kilian A (2001) Diversity arrays: a solid state technology for sequence information independent genotyping. Nucleic Acids Res 29:e25

Jannink J-L, Bink MC, Jansen RC (2001) Using complex plant pedigrees to map valuable genes. Trends Plant Sci 6:337–342

Jansen RC, Stam P (1994) High resolution of quantitative traits into multiple loci via interval mapping. Genetics 136:1447–1455

Jiang GL (2013) Molecular markers and marker-assisted breeding in plants. In: Andersen SB (ed) Plant breeding from laboratories to fields. InTech. ISBN 978-953-51-1090-3

Jiang C, Zeng ZB (1995) Multiple trait analysis of genetic mapping for quantitative trait loci. Genetics 140:1111–1127

Joehanes R, Nelson JC (2008) QGene 4.0, an extensible Java QTL-analysis platform. Bioinformatics 24:2788–2789

Jones C, Edwards K, Castaglione S, Winfield M, Sala F, Van de Wiel C, Bredemeijer G, Vosman B, Matthes M, Daly A (1997) Reproducibility testing of RAPD, AFLP and SSR markers in plants by a network of European laboratories. Mol Breed 3:381–390

Kao CH, Zeng Z-B, Teasdale RD (1999) Multiple interval mapping for quantitative trait loci. Genetics 152:1203–1216

Kebede H, Subudhi PK, Rosenow DT, Nguyen HT (2001) Quantitative trait loci influencing drought tolerance in grain sorghum (*Sorghum bicolor* L. Moench). Theor Appl Genet 103:266–276

Klein R, Rodriguez-Herrera R, Schlueter J, Klein P, Yu Z, Rooney W (2001) Identification of genomic regions that affect grain-mould incidence and other traits of agronomic importance in sorghum. Theor Appl Genet 102:307–319

Knoll J, Gunaratna N, Ejeta G (2008) QTL analysis of early-season cold tolerance in sorghum. Theor Appl Genet 116:577–587

Kong L, Dong J, Hart G (2000) Characteristics, linkage-map positions, and allelic differentiation of *Sorghum bicolor* (L.) Moench DNA simple-sequence repeats (SSRs). Theor Appl Genet 101:438–448

Kong W, Jin H, Franks CD et al (2013) Genetic analysis of recombinant inbred lines for *Sorghum bicolor* × *Sorghum propinquum*. G3 (Bethesda) 3:101–108

Lamprecht H (1948) The variation of linkage and the course of crossing over. Agric Hort Genet 6:10–48

Lander ES, Botstein D (1989) Mapping mendelian factors underlying quantitative traits using RFLP linkage maps. Genetics 121:185–199

Lander ES, Green P, Abrahamson J, Barlow A, Daly MJ, Lincoln SE, Newburg L (1987) MAPMAKER: an interactive computer package for constructing primary genetic linkage maps of experimental and natural populations. Genomics 1:174–181

Li JZ, Börner PDA (2004) Mapping of new microsatellite markers and molecular identification of quantitative trait locus (QTL) for agronomically important traits in barley. Ph.D. dissertation, Der Landwirtschaftlichen Fakultat der Martin-Luther-Universitat Halle-Wittenberg. http://sundoc.bibliothek.uni-halle.de/diss-online/04/04H319/prom.pdf

Li M, Yuyama N, Luo L, Hirata M, Cai H (2009) *In silico* mapping of 1758 new SSR markers developed from public genomic sequences for sorghum. Mol Breed 24:41–47

Lin YR, Schertz KF, Paterson AH (1995) Comparative analysis of QTLs affecting plant height and maturity across the Poaceae, in reference to an interspecific sorghum population. Genetics 141:391–411

Lincoln S, Daly M, Lander E (1993a) Constructing genetic linkage maps with MAPMAKER/EXP. Version 3.0. Whitehead Institute for Biomedical Research Technical Report, 3rd edn

Lincoln S, Daly M, Lander E (1993b) Mapping genes controlling quantitative traits using MAPMAKER/QTL. Version 1.1. Whitehead Institute for Biomedical Research Technical Report, 2nd edn

Lipka AE, Tian F, Wang Q, Peiffer J, Li M, Bradbury PJ, Gore MA, Buckler ES, Zhang Z (2012) GAPIT: genome association and prediction integrated tool. Bioinformatics 28:2397–2399

Luo Z, Kearsey M (1989) Maximum likelihood estimation of linkage between a marker gene and a quantitative locus. Heredity 63:401–408

Mace ES, Jordan DR (2011) Integrating sorghum whole genome sequence information with a compendium of sorghum QTL studies reveals uneven distribution of QTL and of gene-rich regions with significant implications for crop improvement. Theor Appl Genet 123:169–191

Mace E, Xia L, Jordan D, Halloran K, Parh D, Huttner E, Wenzl P, Kilian A (2008) DArT markers: diversity analyses and mapping in *Sorghum bicolor*. BMC Genom 9:26

Mace ES, Rami JF, Bouchet S, Klein PE, Klein RR, Kilian A, Wenzl P, Xia L, Halloran K, Jordan DR (2009) A consensus genetic map of sorghum that integrates multiple component maps and high-throughput Diversity Array Technology (DArT) markers. BMC Plant Biol 9:13

Mackay TF (2001) The genetic architecture of quantitative traits. Annu Rev Genet 35:303–339

Mackay I, Powell W (2007) Methods for linkage disequilibrium mapping in crops. Trends Plant Sci 12:57–63

McIntyre CL, Casu R, Drenth J, Knight D, Whan V, Croft B, Jordan D, Manners J (2005) Resistance gene analogues in sugarcane and sorghum and their association with quantitative trait loci for rust resistance. Genome 48:391–400

Meer JM, Cudmore JRH, Manly K (2004) MapManager/QTX: software for complex trait analysis. http://www.mapmanager.org/mmQTX.html

Mendel G (1865) Experiments in plant hybridization (Verhandlungen des naturforschenden Vereins Brünn). Available online: www.mendelweb.org/Mendel.html. Accessed 16 May 2015

Menz M, Klein R, Mullet J, Obert J, Unruh N, Klein P (2002) A high-density genetic map of *Sorghum bicolor* (L.) Moench based on 2926 AFLP®, RFLP and SSR markers. Plant Mol Biol 48:483–499

Moens P, Wu Y, Huang Y (2006) An SSR genetic map of *Sorghum bicolor* (L.) Moench and its comparison to a published genetic map. Genome 50:84–89

Murray SC, Sharma A, Rooney WL, Klein PE, Mullet JE, Mitchell SE, Kresovich S (2008) Genetic improvement of sorghum as a biofuel feedstock: I. QTL for stem sugar and grain non-structural carbohydrates. Crop Sci 48:2165–2179

Nagaoka T, Ogihara Y (1997) Applicability of inter-simple sequence repeat polymorphisms in wheat for use as DNA markers in comparison to RFLP and RAPD markers. Theor Appl Genet 94:597–602

Nagaraja Reddy R, Madhusudhana R, Mohan SM, Chakravarthi D, Seetharama N (2012) Characterization, development and mapping of Unigene-derived microsatellite markers in sorghum [*Sorghum bicolor* (L.) Moench]. Mol Breed 29:543–564

Nagaraja Reddy R, Madhusudhana R, Murali Mohan S, Seetharama N, Patil JV (2014) Detection and validation of stay-green QTL in post-rainy sorghum involving widely adapted cultivar, M35-1 and a popular stay-green genotype B35. BMC Genomics 15:909

Nagaraju J, Kathirvel M, Kumar RR, Siddiq E, Hasnain SE (2002) Genetic analysis of traditional and evolved Basmati and non-Basmati rice varieties by using fluorescence-based ISSR-PCR and SSR markers. Proc Natl Acad Sci U S A 99:5836–5841

Nelson JC (1997) QGENE: software for marker-based genomic analysis and breeding. Mol Breed 3:239–245

Palmer CED, Keller WA (2005) Overview of haploidy. In: Palmer CE, Keller WA, Kasha KJ (eds) Haploids in crop improvement II. Springer, Berlin, pp 3–9

Parh DK, Jordan DR, Aitken EAB, Mace ES, Jun-Ai P, McIntyre CL, Godwin ID (2008) QTL analysis of ergot resistance in sorghum. Theor Appl Genet 117:369–382

Parida SK, Kumar KAR, Dalal V, Singh NK, Mohapatra T (2006) Unigene derived microsatellite markers for the cereal genomes. Theor Appl Genet 112:808–817

Paterson AH, Bowers JE, Bruggmann R et al (2009) The *Sorghum bicolor* genome and the diversification of grasses. Nature 457:551–556

Peng Y, Schertz K, Cartinhour S, Hart G (1999) Comparative genome mapping of *Sorghum bicolor* (L.) Moench using an RFLP map constructed in a population of recombinant inbred lines. Plant Breed 118:225–235

Pereira MG, Lee M (1995) Identification of genomic regions affecting plant height in sorghum and maize. Theor Appl Genet 90:380–388

Pereira M, Lee M, Bramel-Cox P, Woodman W, Doebley J, Whitkus R (1994) Construction of an RFLP map in sorghum and comparative mapping in maize. Genome 37:236–243

Pratt LH, Liang C, Shah M, Sun F, Wang H, Gingle AR, Paterson AH, Wing R, Dean R, Klein R (2005) Sorghum expressed sequence tags identify signature genes for drought, pathogenesis, and skotomorphogenesis from a milestone set of 16,801 unique transcripts. Plant Physiol 139:869–884

Rafalski JA (2002) Novel genetic mapping tools in plants: SNPs and LD-based approaches. Plant Sci 162:329–333

Rafalski J, Tingey S (1993) RFLP map of soybean (*Glycine max*) 2 N=40. Cold Spring Harbor Laboratory, Cold Spring Harbor, New York, pp 149–156

Ragab R, Dronavalli S, Maroof MS, Yu Y (1994) Construction of a sorghum RFLP linkage map using sorghum and maize DNA probes. Genome 37:590–594

Ramu P, Kassahun B, Senthilvel S, Kumar CA, Jayashree B, Folkertsma R, Reddy LA, Kuruvinashetti M, Haussmann B, Hash C (2009) Exploiting rice–sorghum synteny for targeted development of EST-SSRs to enrich the sorghum genetic linkage map. Theor Appl Genet 119:1193–1204

Ritter KB, Jordan DR, Chapman SC, Godwin ID, Mace ES, McIntyre CL (2008) Identification of QTL for sugar-related traits in a sweet × grain sorghum (*Sorghum bicolor* L. Moench) recombinant inbred population. Mol Breed 22:367–384

Ronin Y, Kirzhner V, Korol A (1995) Linkage between loci of quantitative traits and marker loci: multi-trait analysis with a single marker. Theor Appl Genet 90:776–786

Sabadin PK, Malosetti M, Boer MP et al (2012) Studying the genetic basis of drought tolerance in sorghum by managed stress trials and adjustments for phenological and plant height differences. Theor Appl Genet 124:1389–1402

Satish K, Srinivas G, Madhusudhana R, Padmaja PG, Reddy RN, Mohan SM, Seetharama N (2009) Identification of quantitative trait loci for resistance to shoot fly in sorghum [*Sorghum bicolor* (L.) Moench]. Theor Appl Genet 119:1425–1439

Shiringani AL, Friedt W (2011) QTL for fibre-related traits in grain × sweet sorghum as a tool for the enhancement of sorghum as a biomass crop. Theor Appl Genet 123:999–1011

Shiringani AL, Frisch M, Friedt W (2010) Genetic mapping of QTLs for sugar-related traits in a RIL population of *Sorghum bicolor* L. Moench. Theor Appl Genet 121:323–336

Soller M, Brody T, Genizi A (1976) On the power of experimental designs for the detection of linkage between marker loci and quantitative loci in crosses between inbred lines. Theor Appl Genet 47:35–39

Sorrells ME, Wilson WA (1997) Direct classification and selection of superior alleles for crop improvement. Crop Sci 37:691–697

Srinivas G, Satish K, Madhusudhana R, Reddy RN, Mohan SM, Seetharama N (2009a) Identification of quantitative trait loci for agronomically important traits and their association with genic-microsatellite markers in sorghum. Theor Appl Genet 118:1439–1454

Srinivas G, Satish K, Madhusudhana R, Seetharama N (2009b) Exploration and mapping of microsatellite markers from subtracted drought stress ESTs in *Sorghum bicolor* (L.) Moench. Theor Appl Genet 118:703–717

Stam P (1993) JoinMap: a computer package to construct integrated genetic linkage maps. Plant J 3:739–744

Subudhi PK, Nguyen HT (2000) Linkage group alignment of sorghum RFLP maps using a RIL mapping population. Genome 43:240–249

Takai T, Yonemaru JI, Kaidai H, Kasuga S (2012) Quantitative trait locus analysis for days-to-heading and morphological traits in an RIL population derived from an extremely late flowering F_1 hybrid of sorghum. Euphytica 187:411–420

Tanksley S, Young N, Paterson A, Bonierbale M (1989) RFLP mapping in plant breeding: new tools for an old science. Biotechnology 7:257–264

Tao Y, Jordan D, Henzell R, McIntyre C (1998) Identification of genomic regions for rust resistance in sorghum. Euphytica 103:287–292

Tao Y, Henzell R, Jordan D, Butler D, Kelly A, McIntyre C (2000) Identification of genomic regions associated with stay green in sorghum by testing RILs in multiple environments. Theor Appl Genet 100:1225–1232

Tao YZ, Hardy A, Drenth J, Henzell RG, Franzmann BA, Jordan DR, Butler DG, McIntyre CL (2003) Identifications of two different mechanisms for sorghum midge resistance through QTL mapping. Theor Appl Genet 107:116–122

Taramino G, Tarchini R, Ferrario S, Lee M (1997) Characterization and mapping of simple sequence repeats (SSRs) in *Sorghum bicolor*. Theor Appl Genet 95:66–72

Tuinstra MR, Grote EM, Goldsbroug PB, Ejeta G (1996) Identification of quantitative trait loci associated with pre-flowering drought tolerance in sorghum. Crop Sci 36:1337–1344

Utz F, Melchinger A (2007) PLABQTL: a computer program to map QTL. Institute of Plant Breeding, Seed Science, and Population Genetics, University of Hohenheim, Stuttgart

van Ooijen J (2011) MapQTL 6: software for the mapping of quantitative trait loci in experimental populations of diploid species. Kyazma B. V., Wageningen

Varshney RK, Graner A, Sorrells ME (2005) Genic microsatellite markers in plants: features and applications. Trends Biotech 23:48–55

Varshney RK, Nayak SN, May GD, Jackson SA (2009) Next-generation sequencing technologies and their implications for crop genetics and breeding. Trends Biotech 27:522–530

Virk PS, Ford-Lloyd BV, Jackson MT, Pooni HS, Clemeno TP, Newbury HJ (1996) Predicting quantitative variation within rice germplasm using molecular markers. Heredity 76:296–304

Vos P, Hogers R, Bleeker M, Reijans M, van De Lee T, Hornes M, Friters A, Pot J, Paleman J, Kuiper M (1995) AFLP: a new technique for DNA fingerprinting. Nucleic Acids Res 23:4407–4414

Whitkus R, Doebley J, Lee M (1992) Comparative genome mapping of sorghum and maize. Genetics 132:1119–1130

Williams JG, Kubelik AR, Livak KJ, Rafalski JA, Tingey SV (1990) DNA polymorphisms amplified by arbitrary primers are useful as genetic markers. Nucleic Acids Res 18:6531–6535

Wilson LM, Whitt SR, Ibáñez AM, Rocheford TR, Goodman MM, Buckler ES (2004) Dissection of maize kernel composition and starch production by candidate gene association. Plant Cell 16:2719–2733

Wu YQ, Huang Y, Porter DR, Tauer CG, Hollaway L (2007) Identification of a major quantitative trait locus conditioning resistance to greenbug biotype E in sorghum PI 550610 using simple sequence repeat markers. J Econ Entomol 100:1672–1678

Xu Y (2010) Molecular plant breeding. CABI, Wallingford/Cambridge, MA. ISBN 9781845933920

Xu GW, Magill C, Schertz K, Hart G (1994) A RFLP linkage map of *Sorghum bicolor* (L.) Moench. Theor Appl Genet 89:139–145

Xu W, Subudhi PK, Crasta OR, Rosenow DT, Mullet JE, Nguyen HT (2000) Molecular mapping of QTLs conferring stay-green in grain sorghum (*Sorghum bicolor* L. Moench). Genome 43:461–469

Xu Y, Wang J, Crouch J (2008) Selective genotyping and pooled DNA analysis: an innovative use of an old concept. In: Recognizing past achievement, meeting future needs. Proceedings of the 5th International Crop Science Congress, 14–18 Apr 2008, International Crop Science Society, Jeju, The Republic of Korea

Yang W, de Oliveira AC, Godwin I, Schertz K, Bennetzen JL (1996) Comparison of DNA marker technologies in characterizing plant genome diversity: variability in Chinese sorghums. Crop Sci 36:1669–1676

Yonemaru J, Ando T, Mizubayashi T, Kasuga S, Matsumoto T, Yano M (2009) Development of genome-wide simple sequence repeat markers using whole-genome shotgun sequences of sorghum (*Sorghum bicolor* (L.) Moench). DNA Res 16:187–193

Yu J, Buckler ES (2006) Genetic association mapping and genome organization of maize. Curr Opin Biotechnol 17:155–160

Yu JK, La Rota M, Kantety R, Sorrells M (2004) EST derived SSR markers for comparative mapping in wheat and rice. Mol Genet Genomics 271:742–751

Yu J, Holland JB, McMullen MD, Buckler ES (2008) Genetic design and statistical power of nested association mapping in maize. Genetics 178:539–551

Zeng ZB (1993) Theoretical basis for separation of multiple linked gene effects in mapping quantitative trait loci. Proc Natl Acad Sci U S A 90:10972–10976

Zeng ZB (1994) Precision mapping of quantitative trait loci. Genetics 136:1457–1468

Zeng ZB, Kao CH, Basten CJ (1999) Estimating the genetic architecture of quantitative traits. Genet Res Camb 74:279–289

Zhang Z, Btadbury PJ, Kroon DE, Casstevens TM, Buckler ES (2006) TASSEL 2.0: a software package for association and diversity analysis in plants and animals. Plant and Animal Genomes XIV conference, Poster P956/CP012, San Diego

Zhao H, Fernando R, Dekkers J (2007) Effects of population structure on power and precision of regression-based linkage disequilibrium mapping of QTL. J Anim Sci 85:63

Zietkiewicz E, Rafalski A, Labuda D (1994) Genome fingerprinting by simple sequence repeat (SSR)-anchored polymerase chain reaction amplification. Genomics 20:176–183

Zou G, Zhai G, Feng Q, Yan S, Wang A, Zhao Q, Shao J, Zhang Z, Zou J, Han B (2012) Identification of QTLs for eight agronomically important traits using an ultra-high-density map based on SNPs generated from high-throughput sequencing in sorghum under contrasting photoperiods. J Exp Bot 63:5451–5462

Application of DNA Markers for Genetic Improvement

4

R. Madhusudhana

Contents

4.1	**Introduction**	71
4.2	**Applications of Marker-Assisted Selection**	72
4.3	**Prerequisites for MAS**	73
4.4	**Selection Schemes for MAS**	73
4.4.1	Marker-Assisted Backcrossing	73
4.4.2	Gene Pyramiding	74
4.4.3	Marker-Assisted Recurrent Selection	75
4.4.4	Genomic Selection	75
4.4.5	Advanced Backcross QTL	76
4.4.6	Breeding by Design	76
4.5	**MAS in Sorghum**	76
4.5.1	Agronomic Traits	76
4.5.2	Biotic Stresses	79
4.5.3	Abiotic Stresses	87
4.6	**Major Effect Genes**	89
4.7	**DNA Markers for Seed Purity**	89
4.8	**MAS: Challenges**	93
References		93

Abstract

The advent of DNA markers has heralded a new era in genetics and plant breeding. DNA markers have provided valuable tools in various genetic analyses ranging from diversity, development of molecular maps, and gene/QTL mapping to the positional cloning of genes. Marker-assisted selection has the potential to pyramid favorable gene combinations for improved trait performance. Over the past two decades, considerable progress has been made in the development of genomic resources in sorghum. Development and use of new marker systems and dense maps has resulted in tagging and mapping of major genes and several quantitative traits of economic importance. QTL mapping of various agronomic, biotic, abiotic stress traits has resulted in identification of many QTL spread across the genome. Validation and fine mapping of QTL provide an opportunity to employ MAS for sorghum improvement.

Keywords

Sorghum • Marker-assisted selection • Gene pyramiding • Major effect genes

4.1 Introduction

DNA markers are nothing but DNA sequence variations between individuals, which can be detected using tools like Southern blotting, PCR-

R. Madhusudhana (✉)
Plant Breeding, ICAR-Indian Institute of Millets Research, Rajendranagar, Hyderabad, Telangana 500 030, India
e-mail: madhu@millets.res.in

R. Madhusudhana et al. (eds.), *Sorghum Molecular Breeding*, DOI 10.1007/978-81-322-2422-8_4

based techniques, microarray, and sequencing. Markers assist the plant breeders in selecting a plant with desirable trait(s) directly or indirectly, and different marker systems have been used for the selection of desired plants. Morphological markers are the first genetic markers used for this purpose. These are the visible plant traits like color of flower, anther, stigma, pericarp, seed, leaf shape, awn, plant height, seed size, pubescence, etc. The first association of a simply inherited major effect gene with a quantitative trait in plants was reported several decades ago (Sax 1923), and the phenomenon has since been observed for a range of traits in many crops. Selection of semi-dwarf plants in rice and wheat is the most successful example of use of major effect genes in modern plant breeding. However, the availability of such morphological markers is limited; many are not linked with economic traits, interact with environmental conditions, are less polymorphic, etc. The next classes of markers used with limited application in plant breeding are the biochemical or protein markers. Isozyme/proteins with their alternate forms and mobility have been utilized as molecular markers in seed purity testing. These markers also suffer from the limitations similar to those of morphological markers.

Molecular breeding is the application of molecular biology tools for the genetic improvement of traits in crop or animal species. DNA markers have become important tools for genetic analysis and crop improvement since 1980s. Unlike morphological and biochemical markers, DNA markers are abundant, available across the length of the genome, phenotypically neutral, stage and time independent, and therefore are considered as ideal marker systems for application in molecular breeding of crop plants. During the past two to three decades, several marker systems such as RFLP (Restriction Fragment Length Polymorphism), RAPD (Random Amplified Polymorphic DNA), AFLP (Amplified Fragment Length Polymorphism), SSR (Simple Sequence Repeats), DArTs (Diversity Array Technique), and SNPs (Single Nucleotide Polymorphisms) have been developed. However, not all these marker systems are equally preferred for molecular breeding. SSR markers characterized by their hyper-variability, reproducibility, codominant nature, locus specificity, and genome-wide distribution make them the markers of choice for high-throughput genotyping, high-density linkage map construction, and useful for gene mapping and marker-assisted selection. A number of linkage maps based on SSRs have been developed in many cereal species including sorghum. Thousands of SSRs have been developed and used for linkage map construction and QTL analysis in sorghum. However, with the developments in economical genome sequencing technologies characterized by speed and efficiency, millions of SNPs are identified in several crop plants. Typically, SNP frequencies are in a range of one SNP every 100–300 bp in plants (Edwards et al. 2007). SNPs are codominant, often linked to genes, and have become very attractive and potential genetic markers in genetic analysis and molecular breeding. Since they are available in millions per genome, easily detectable, amenable for automation, and cover whole genome, it is expected that the SNPs will be increasingly used in most of the crop species for several genetic analyses including marker-assisted selection.

4.2 Applications of Marker-Assisted Selection

Molecular marker-assisted selection (MAS) is a plant breeding approach that has been developed to circumvent the problems associated with conventional plant breeding. In this approach, the selection is not based on trait phenotype but on the DNA markers linked to trait gene/QTL. The use of markers for trait improvement becomes advantageous over conventional phenotypic selection for higher genetic gains in several situations like when the trait of interest is:

- Expressed in the later stage of plant development (e.g., male sterility, seed traits)
- Recessive (e.g., brown midrib)
- Depending on its expression on specific environment (e.g., cold, disease, etc.)
- Difficult to measure (e.g., moisture stress)

- Controlled by two or more unlinked genes (e.g., multiple genes)
- Quantitative trait with low heritability (e.g., grain yield)
- With linkage drag
- Measured involving difficult assays (e.g., biochemical traits)
- Also in gene pyramiding (e.g., genes for resistance)
- Locating a major effect gene

4.3 Prerequisites for MAS

The success of MAS depends on the strength of marker-trait associations established for a given trait. Therefore, before the start of a MAS program, there is a need to identify, validate, and establish a stable marker-trait association. This can either be done by using conventional QTL detection methods in a segregating biparental population or through an association-mapping approach involving a diverse genotype panel. For successful MAS, critical information on component traits, accurate phenotyping, identification of candidate genes and quantitative trait loci, the relationship between QTL and genes, the contribution of individual QTL to the phenotype, and their variability across different locations and different crop seasons are essentially required. Besides this, there are several essential requirements for MAS (Jiang 2013). Some of the important ones are:

- *Suitable marker system and its reliability*: Marker should be simple, high-throughput, low-cost, codominant, highly reproducible, show high levels of polymorphism, and reliable.
- *DNA extraction method*: Needs to be very quick and on high-throughput basis.
- *Saturated genetic maps*: Linkage maps provide the basis for the detection of marker-trait associations. Therefore, the maps should be highly saturated. A marker at every 3–5 cM is highly desirable in QTL analysis and for MAS.
- *Knowledge of marker-trait associations*: Tightly linked markers should be available for successful MAS.
- *Quick and efficient data processing and management*: In MAS, since the decision on the selection of plants is time bound, there should be a system for quick and efficient data analysis and its management.

4.4 Selection Schemes for MAS

4.4.1 Marker-Assisted Backcrossing

Backcross method is one of the important breeding procedures used in plant breeding for the transfer of a target gene from a donor (non-recurrent parent) line into a recipient (recurrent) line. The method was proposed to transfer simply inherited traits from a donor to a recipient and is one of the most widely used methods for line improvement in plant breeding across crops. The goal of backcrossing is to obtain a line as identical as possible to the recurrent parent with the addition of the traits/gene of interest from donor. Therefore, the elite variety used as a recurrent parent usually has most desirable agronomic traits except for one or two undesirable traits. However, this method has some limitations as it involves 4–5 years of time, scheme differs for dominant and recessive traits, concurrent transfer of unwanted genes/traits due to linkage drag from donor parent especially when the donor is an unadapted genotype. Under these circumstances, marker-assisted backcrossing (MABC) has great advantage in plant breeding activities.

MABC is the simplest form of MAS and is one of the most successfully used methods in molecular plant breeding. Instead of phenotypic selection in each round of backcrossing, in MABC, selection is made using DNA markers linked to the desired trait/QTL. With the availability of thousands of DNA markers and genetic and QTL maps, the MABC has become a potential method for the transfer of traits from a donor parent into the recurrent parent. In this method, the problems encountered in the traditional backcross breeding are largely overcome. The introgression of the target gene or trait is precisely and efficiently done in MABC using molecular markers. Three levels of MABC are described: foreground selection, recombinant selection, and background selection (Holland 2004).

- *Foreground selection*: Selection is made only for the marker alleles of donor parent at the target locus to maintain the target locus in heterozygous state until the final backcrossing is completed (Jiang 2013). This is also called as positive selection. Since the success of MABC depends on the strength of marker-trait association, it is always advised to use markers which are tightly linked to the trait/QTL. The most favorable situation is to have foreground marker within the gene of interest. Such a marker is considered as perfect marker/direct marker, since the selection of foreground marker guarantees the presence of the gene/QTL in the progeny. In situations where perfect markers are not available, use of tightly linked markers within 1 cM is always advised to minimize the recombination between QTL and marker. In the absence of such tightly linked markers, it is better to use foreground markers flanking the QTL so that the target control rate is improved (Hospital 2003).
- *Recombinant selection*: Selection of backcross progeny with the target gene and the recombination that happened between target locus and the linked flanking marker. The main purpose is to reduce the linkage drag around the target gene. It is often observed in conventional backcrossing that the length of linkage drag along with target gene is quite long in spite of several rounds of backcrossing. If any undesirable trait/genes are carried along with the donor segment, it may negatively affect the crop performance. Therefore, by using flanking markers on either side of the target gene (less than 5 cM), the linkage drag can be drastically reduced. Since the simultaneous occurrence of double recombination on either side of the target gene is extremely rare, recombination selection is usually performed for at least two backcross generations (Frisch et al. 1999).
- *Background selection*: Backcross progenies are selected with maximum representation of marker alleles from the recurrent parent in all genomic regions except the target locus of donor parent. This level of selection is advantageous to hasten the restoration of the recurrent genome at the quickest possible time. In conventional backcrossing, a minimum of six backcrosses is required to recover the recurrent parent genome, and there is still the possibility of some donor genomic segments unlinked to the target gene present in the recurrent parent. Using markers, the recovery of the recurrent genome can be achieved even at BC_3 or BC_4, thus saving two to three backcross generations.

4.4.2 Gene Pyramiding

Pyramiding of genes using DNA markers is one of the most successful applications of DNA markers in plant breeding. In this process, multiple genes/QTLs from different donor sources can be simultaneously combined in a single genotype using linked markers. The strategy of gene pyramiding has been proposed and effectively used to enhance resistance to diseases and insects in various crops. Genes conferring resistance to different races of pathogen or biotypes of an insect pest can be effectively pyramided in a single genotype using linked markers to provide horizontal resistance. The combined expression of pyramided genes/QTLs was found highly advantageous as it provided broader spectrum resistance through gene interactions and quantitative complementation (Yoshimura et al. 1995). Three main breeding approaches have been proposed for pyramiding of major genes/QTLs through MABC from different donors to an elite recurrent parent: stepwise transfer, simultaneous transfer, and simultaneous but stepwise transfer.

- *Stepwise transfer*: The recurrent parent is first introgressed with a specific gene/QTL using first donor parent in the first backcross program. Then the improved recurrent parent forms the base for second backcrossing with second donor parent for incorporating the second gene/QTL. This process continues until all the major genes from different donors are pyramided in the recurrent parent. Naturally, this approach takes maximum time since new backcrossing starts only after the completion of the earlier one with the incorporation of major gene into the recurrent parent. More precise transfer of genes, easier implementation, and requirement of less population and genotyping efforts are the advantages of this approach.

- *Simultaneous transfer*: The recurrent parent is crossed simultaneously but independently with different donors to develop different F_1s. The F_1s so produced are inter-crossed to produce double cross hybrids, which will be again inter-crossed to develop four-way cross hybrid. This is backcrossed with the recurrent parent simultaneously tracking the different genes from various donors. This method takes shorter time to develop recurrent parent with multiple genes. However, the main disadvantage is that it requires a large population and more genotyping since all the target genes/QTL are handled simultaneously.
- *Simultaneous but stepwise transfer*: This combines the advantages of both stepwise and simultaneous backcrossing strategies. In this approach, transfer of genes from different donor parents to a common recurrent parent is undertaken independently, but simultaneous backcrossing is followed to combine them in a single recurrent parent. This strategy is more often followed, as it is easy and assures gene fixation and pyramiding into the common recurrent parent.

4.4.3 Marker-Assisted Recurrent Selection

Several economically important agronomic traits are quantitatively inherited and influenced by the environment. Improvement of complex traits via phenotypic selection is generally possible through recurrent selection strategy involving cycles of selection and inter-mating, but this imposes restrictions on the practicability of this breeding method. With the use of DNA markers, recurrent selection can be accelerated considerably with precision and efficiency. Marker-assisted recurrent selection (MARS) includes identification and selection of several QTLs (up to 20 or even more) for complex traits within a single population. As defined by Ribaut et al. (2010), MARS is a recurrent selection scheme using molecular markers for the identification and selection of multiple genomic regions involved in the expression of complex traits to assemble the best-performing genotype within a single or across related populations. The scheme involves improvement of a F_2 segregating population by one cycle of marker-assisted selection (with both phenotypic data and marker scores) followed commonly by two or three cycles of marker-based selection (based on marker scores only). In continuous nursery programs, pre-flowering genotypic information is used for marker-assisted selection and controlled pollination. Thus, several selection-cycles are possible within 1 year, accumulating favorable QTL alleles in the breeding population (Eathington 2005).

4.4.4 Genomic Selection

Grain yield, quality, and resistance to biotic and abiotic stresses show complex inheritance involving several genes with small effect controlling their expression. Traditional phenotypic and the general marker-assisted selection have not been effective for genetic manipulation of such traits. Unlike identifying individual loci significantly associated with a trait as done in general QTL mapping approach, genomic selection (GS) uses whole genome marker data as predictors of performance and consequently delivers predictions of trait values that are more accurate for use in selection. GS is therefore a form of marker-assisted selection in which tens or hundreds of thousands of DNA markers covering the whole genome are selected so that all QTLs/genes are in linkage disequilibrium with at least one marker. This approach has become feasible due to the availability of a large number of single nucleotide polymorphisms (SNP) and new methods to efficiently genotype large number of SNP. GS uses a "training population" of individuals that have been both genotyped and phenotyped to develop a model that takes genotypic data from a "candidate population" of untested individuals and produces genomic estimated breeding values (GEBVs). Genomic prediction combines marker data with phenotypic and pedigree data (when available) in an attempt to increase the accuracy of the prediction of breeding and genotypic values. Selection of plants can be based on prediction values, potentially leading to more rapid and lower-cost gains from breeding.

4.4.5 Advanced Backcross QTL

Validation of marker-trait association is an important and essential step in any marker-assisted breeding scheme, which involves additional time and resources. However, validation can be avoided if genetic mapping and marker-assisted breeding scheme are combined. A molecular breeding method, referred to as advanced backcross QTL analysis (AB-QTL), was proposed by Tanksley and Nelson (1996) with the purpose of combining the process of QTL analysis with variety development, by simultaneously identifying and transferring favorable QTL alleles from unadapted (e.g., landraces, wild species) to cultivated germplasm. In this method, QTL detection is delayed until BC_2 or BC_3 generations, which offers more statistical power for detecting additive QTL and recovery of a QTL-line (QTL-NIL) with greater similarity to the recurrent parent. The QTL-NIL thus developed can be directly used as an improved cultivar or as a parent of an F_1 hybrid.

4.4.6 Breeding by Design

The advances in genomics, marker detection, and high-throughput genotyping for the development of extra dense linkage maps make it possible to identify the genetic basis of all agronomic traits. By understanding the genetic basis of all agronomically important characters and the allelic variation at those loci, the breeder would be able to design superior genotypes in silico and implement in plant breeding programs. This approach is called as Breeding by Design (Peleman and van der Voort 2003), which aims to control allelic variation for all genes of agronomic traits. Through the combination of precise genetic mapping, high-resolution chromosome haplotyping, and extensive phenotyping, allelic variations of all genes can be identified for precise breeding of an intended plant type. This approach, therefore, requires mapping loci involved in the expression of all agronomic traits, assessment of allelic variations at those loci, and then designing of superior plant type involving combination of favorable alleles. Since the outcome of a set of crosses based on molecular markers information is predicted in silico for precision breeding, this approach is also called as predictive breeding.

4.5 MAS in Sorghum

Sorghum genetic improvement through classical breeding approaches has been slow in addressing the major loss-causing and yield-destabilizing traits like susceptibility to insects and diseases, poor grain quality, drought, and striga menace, as these traits are greatly influenced by environment and no reliable genetic clues have been available for recombination breeding (Bhat et al. 2004). Therefore, the MAS approach is gaining importance in the improvement of sorghum crop. Several marker systems have been developed and used for tagging and mapping of major effect genes and quantitative traits of economic importance like grain yield and its component traits, resistance to insect pests, diseases, striga, drought, salinity, cold and nutritional quality traits, etc.

4.5.1 Agronomic Traits

Grain Yield

In sorghum, grain yield is a complex quantitative trait governed by several genes (Beil and Atkins

1967). It is dependent on the contributions from several reproductive, morphological, and phenological traits such as plant height, panicle length, panicle weight, number of primary branches/panicle, test weight, etc. Genetic improvement of grain yield is a challenging task as it involves accumulation of positive alleles involved in the expression of component traits. Over the last decade, few studies in sorghum have identified QTL for grain yield and its component traits (Brown et al. 2006; Hart et al. 2001; Nagaraja Reddy et al. 2013; Ritter et al. 2008; Srinivas et al. 2009). Eight QTLs involving different genetic backgrounds were identified on LG 2, 3, 6, 9, and 10. Three QTLs identified on LG 10 are meta-QTL indicating their consistent expression in different genetic backgrounds. Six of the eight QTLs are major effect QTL controlling >10 % of phenotypic expression for grain yield. A major QTL on LG 6 accounted highest trait variation (14.6 %) and is colocated with QTL for seven important grain yield traits (Srinivas et al. 2009). This QTL region harbored the major maturity gene, Ma_1, and major dwarfing gene, Dw_2. Clustering of grain yield and its component traits was reported at several genomic locations in sorghum (Nagaraja Reddy et al. 2013). Such consistent and major QTL can be considered for marker-assisted selection for grain improvement in sorghum.

Grain Traits

Grain weight is a trait of economic significance in sorghum as it contributes directly to grain yield and influences market price. Size, shape, luster, and color are some of the important grain quality traits that contribute to consumer preferences. Grain weight is a reliable index of grain size due to its high correlation with grain size. Grain size is positively correlated with grain yield, crude protein content, and staygreen trait (Borrell et al. 2014a). Since improvement in grain size has direct bearing on the grain yield, it is one of the most important traits for selection in increasing grain yield in sorghum.

Across genetic backgrounds, QTL analyses for grain weight have detected 28 QTL, of which nine are meta-QTLs accounting trait variance ranging from 4.8 to 35 %. Two QTLs on LG7 (35 %) and LG1 (20 %) contributed maximum to seed weight (Rami et al. 1998). A major QTL explaining 14.8 % of phenotypic variation for seed weight was detected near the genomic-microsatellite marker Xcup24 on SBI-01 (Srinivas et al. 2009; Tuinstra et al. 1998). Significantly, this QTL also corresponds to a seed weight QTL mapped in homologous regions on rice (chr. 3) and on maize (chr. 1) (Li et al. 2004). Sorghum gene Sb01g032830 (similar to grain length and weigh protein), an orthologous of rice (*GS3*) and maize gene (*ZmGS3*) for grain size, was present at this QTL region and was suggested to be associated with the domestication in sorghum similar to rice and maize. These important QTL could be the ideal candidates that may be utilized through marker-assisted breeding for improving seed size and grain yield in sorghum.

Plant Height

Sorghum produces excellent green and dry fodder (stover), demand for which is extremely high during the summer seasons in the arid and semi-arid regions. Sorghum fodder constitutes up to 45 % of the total dry weight of dairy animal feed during normal seasons and up to 60 % during dry seasons. Plant height is an important trait in sorghum and it mainly defines the total biomass of the plant. Tallness offer several advantages with thicker stem, higher juice content, ratoon ability, and grain yield regardless of its loose linkage with late maturity. Plant height is independent of stem structural composition, i.e., cellulose, hemicellulose, and lignin content (Murray et al. 2008), and therefore a variety with tall stem can be bred to contain more cellulose, stalk sugars, and less lignin which is ideal for bioethanol production. Since sorghum is gaining importance as a bioenergy crop, there is a lot of interest in the genetic dissection of plant height, which has direct relevance for improving total biomass at molecular level. Genetic control of plant height in sorghum has been characterized in terms of factors with qualitative effects (Quinby and Karper 1954). Four major loci affecting sorghum plant height (Dw_1, Dw_2, Dw_3, and Dw_4) have been reported. QTL studies have also identified several genomic regions associated with plant height in sorghum

(Feltus et al. 2006; Hart et al. 2001; Klein et al. 2001; Lin et al. 1995; Pereira and Lee 1995; Rami et al. 1998; Srinivas et al. 2009). QTL with major effects for plant height have been consistently identified in different genetic backgrounds and were related to two qualitative loci, *Dw_2* on SBI-06 and *Dw_3* on SBI-07 (Brown et al. 2006; Feltus et al. 2006; Klein et al. 2008; Mace and Jordan 2011). The map position of *Dw_2* on SBI-06 on the consensus map closely linked with DArT markers, sPb-7169 and sPB-1395, and *Dw_3* between a SSR marker, mcbCIR300 and a DArT marker, SSCIR57 on SBI-07 was reported by Mace and Jordan (2010). The map position for *Dw_1* was proposed on SBI-09, where a major height gene Sb.Ht9.1 has been reported earlier (Brown et al. 2008; Pereira and Lee 1995). So far, the map position of *Dw_4* is not reported. Of these major loci, *Dw_3* has been fine mapped, and the gene was identified (Sb07g023730), which codes for P-glycoprotein that regulates polar auxin transport and is orthologous to *br2* in maize (Multani et al. 2003). Recently, Morris et al. (2013) proposed that *Dw_2* phenotype is a result of loss of function in a sorghum histone deacetylase gene (Sb06g015420) analogous to its function in controlling plant height in other crops like maize, rice, and *Arabidopsis*. Similarly, plant height QTL SbHt9.1/*Dw_1* was proposed to be under the control of *GA2-oxidase* (Sb09g028360), a catabolic enzyme in the gibberellin pathway. Few other putative candidate genes such as *FtsZ*, *Ugt*, and GA 2-oxidase involved in the regulation of plant height have been reported in sorghum (Wang et al. 2012).

Panicle Traits

Like in rice, panicle is the most important determinant of grain yield in sorghum. Panicle length, width, weight, number of primary branches per panicle, number of seeds per panicle, and panicle harvest index are considered as most important components of panicle. Sorghum has been less studied for inflorescence architecture than other members of the *Poaceae* family. Five major QTL relating to primary and secondary seed branches have been identified in different genetic backgrounds (Brown et al. 2006; Pereira and Lee 1995; Srinivas et al. 2009) with 14–20 % of phenotypic variation. Panicle width is another trait related with grain yield for which seven major QTL have been detected (Hart et al. 2001). Four meta-QTL for panicle seed branch length, one each on LG 1, 3, 7, and 10, were consistent across genetic backgrounds (Mace and Jordan 2011). Similarly, 13 major QTL (>10 %) have been reported for panicle length, wherein 12 are meta-QTL with phenotypic variation ranging from 5.91 to 50.4 %. A major QTL on LG6 contributing more than 50 % of the variation was reported (Srinivas et al. 2009). Interestingly, this QTL is also colocated with the major QTL for plant height, harboring *Dw_2* gene. However, this QTL is negatively linked with grain yield implying selection for increased panicle length at this QTL region may lead to reduction in grain yield.

Maturity

Maturity is a key trait for adaptation of the plant to its environmental conditions. Six major effect genes (*Ma_1*, *Ma_2*, *Ma_3*, *Ma_4*, *Ma_5*, and *Ma_6*) influencing flowering time/maturity in sorghum have been reported (Quinby 1967; Rooney and Aydin 1999). Of these, only four genes (*Ma_1*, *Ma_3*, *Ma_4*, and *Ma_5*) were mapped on the linkage groups in sorghum. *Ma_1* has the largest impact on flowering time of all the six maturity genes. Mutations in *Ma_1* were critical for the early domestication and dispersal of sorghum from its center of origin across Africa and Asia (Quinby 1967). *Ma_1* is mapped on LG 6 flanked by the two SSR markers gap7 and gap72 (Mace and Jordan 2010). The other maturity genes are *Ma_3* mapped on LG1, *Ma_4* on LG10, and *Ma_5* on LG2. Both *Ma_1* and *Ma_3* have been cloned. *Ma_3* encodes a phytochrome B (Childs et al. 1997). The gene encoding pseudo-response regulator protein 37 (PRR37) was identified as a likely gene candidate for *Ma_1* based on the known roles of PRR genes in flowering of *Arabidopsis* (Murphy et al. 2011). *Ma_1* suppresses flowering by activating the floral inhibitor CONSTANS and repressing the floral activators Early Heading Date 1 and FLOWERING LOCUS T. Mutations in *Ma_1* produced early-maturing grain sorghum plants

(Murphy et al. 2011). Sorghum Ma_6, a strong repressor of flowering in long days, was identified as the CONSTANS, CO-like, and TOC1 (CCT)-domain protein encoded by SbGhd7 (Murphy et al. 2014). Sorghum *Ghd7* increases photoperiod sensitivity and delays flowering by inhibiting expression of the floral activator *SbEhd1* and genes encoding flowering time.

4.5.2 Biotic Stresses

Disease and insect management through host plant resistance has been an effective means of reducing losses in sorghum. Availability of markers for biotic stress resistance would do away with the need for phenotypic screening, and undesirable plants can be removed before flowering by marker analysis even at the seedling stage. In order to reduce the risk due to breakdown of resistance and to increase the levels of host resistance, new sources of resistance need to be explored, and alternative mechanisms of resistance should be incorporated, paving way for pyramiding different resistance genes into commercial cultivars. This goal cannot be achieved through conventional breeding technology alone and is best done by deploying MAS.

4.5.2.1 Insect Resistance

The development of sorghum cultivars resistant to insect pests is very important as insect pests cause maximum damage to sorghum from germination to grain maturity. One or the other insect pests attack every stage of the crop in sorghum. More than 150 insect pests have been reported to feed on sorghum. Controlling insect pests through insecticides is not economical and not a wise strategy in a long run. Plant resistance to insects is most often a quantitatively inherited trait. Strong effects of both the environment and the genetic variability within insect pest populations on the assessment of bioassays have resulted in a high degree of genotype-by-environment error (Smith et al. 1994). Therefore, marker-assisted host plant resistance breeding assumes greater significance for effective selection of resistant lines.

Shoot Fly

The sorghum shoot fly, *Atherigona soccata* (Rondani), is an economically important pest of grain, forage, and sweet sorghums in Asia, Mediterranean Europe, and Africa. It infests sorghum seedlings between the first and fourth week after emergence by laying eggs on the abaxial surface of the third to sixth basal leaves (Padmaja et al. 2010b). On hatching, the larva moves down between the leaf sheath and cuts through the central meristematic tissue of the developing leaf, resulting in withering of the central shoot known as deadheart (Deeming 1972). Shoot fly incidence is higher in sorghum crop sown late during the rainy season and in the early-sown crop during the post-rainy season (Jotwani et al. 1970). Sorghum is the main host plant of shoot fly, but populations may survive on wild graminaceous plants when sorghum is not available (Padmaja et al. 2010a). In India, the losses due to *A. soccata* damage have been estimated to reach as high as 90 % of grain and 45 % of fodder yield (Sukhani and Jotwani 1980). The annual economic losses in sorghum due to this pest have been estimated at US$337 million (ICRISAT 1992).

The development and release of new sorghum hybrids and varieties marked a genetic breakthrough in the otherwise stagnant grain yield levels in India during the late 1980s. Within a short period, it was realized that one of the important constraints in popularizing these high-yielding cultivars was their higher susceptibility to shoot fly. In view of the seriousness of the shoot fly problem in sorghum and the limitations (like high costs and toxicity hazards) of chemical control, it is necessary to develop new varieties, hybrid parents, and hybrids resistant to shoot fly. Extensive studies on various aspects of sorghum host plant resistance to shoot fly and the progress made in various areas, namely, screening techniques (natural and artificial methods), mechanisms and stability of resistance, morphological and biochemical factors of resistance, larval establishment in the plant whorl, and factors associated with resistance, are well documented (Padmaja et al. 2010a). Among the mechanisms of resistance, early vigor, glossiness, and trichomes are the most important. Screening technique was standardized and several landraces were identified as sources of resistance (Nwanze 1997). Mechanisms such as glossiness and trichomes were known to contribute to resistance in many of the source lines including IS18551.

Studies on the genetics of shoot fly resistance suggested that resistance to component traits is complex and polygenically inherited (Halalli et al. 1983) with predominantly additive gene effects. Shoot fly resistance in sorghum was classified into three components, viz., non-preference for oviposition, antibiosis, and tolerance (Soto 1974). Under field conditions, resistance to shoot fly is primarily due to non-preference for oviposition (also called antixenosis, observed as reduction in the number of eggs laid on the seedling). Many other important component traits like glossy leaves, leaf trichomes, seedling vigor, epicuticular wax, and biochemical factors are also associated with shoot fly resistance in sorghum. It was reported that trichome development is season dependent, indicating that hybrid parents need to be developed for rainy and post-rainy seasons separately and that both seed parents and restorers should have resistance to produce shoot fly-resistant hybrids (Dhillon et al. 2006). A smooth amorphous wax layer and sparse wax crystals characterized moderately resistant genotypes, while susceptible genotypes possessed a dense meshwork of crystalline epicuticular wax. Non-glossy plants showed a high density of star-shaped epicuticular waxes, whereas the glossy plants were characterized by a reduction in the number and type of waxes on leaves (Tarumoto 2005). It is generally assumed that sorghum seedlings emit volatiles that are specific to both adult fly oviposition attraction and larval orientation/migration (Padmaja et al. 2010b). Shoot fly females are attracted both to the volatiles emitted by the susceptible seedlings and phototactic (optical) stimuli that may facilitate orientation to its host for oviposition.

In the recent past, Satish et al. (2009) identified QTLs for shoot fly resistance in a recombinant inbred line (RIL) population of the cross, 296B × IS18551, wherein IS18551 originated from Ethiopia was the resistant parent. Twenty-nine QTLs were detected, viz., four each for leaf glossiness and seedling vigor, seven for oviposition, six for deadhearts, two for adaxial trichome density, and six for abaxial trichome density. LG SBI-10 hosts two QTL regions between SSR markers, Xgap1-Xnhsbm1011 and Xnhsbm1044-Xnhsbm1013. Similarly, SBI-05 carries a major gene for glossiness between SSR markers Xtxp65-Xtxp30. Major QTL regions identified correspond to QTL/genes for insect resistance in maize. Leaf glossiness QTL on SBI-05 and SBI-03 is syntenic to maize LG4 and LG3, respectively, carry genes *glossy3* and *glossy9* for leaf glossiness, and harbor long-chain Acyl-CoA synthetase and wax synthase genes involved in wax biosynthesis. Seedling vigor QTL on SBI-03 hosts a gene for indole-3 acetic acid-amino synthase GH3.5 that promotes plant growth, light, and stress adaptation. Similarly on SBI-10 where QTL for oviposition, deadhearts, and trichome density are colocated, genes, viz., cysteine protease *Mir1*, homogentisate phytyl transferase vte2, hydroxyproline-rich glycoprotein, NAC1, glossy15 and mh11 responsible for biotic and abiotic stress resistance and trichome density, have been identified. Some of these QTLs were validated in a second recombinant inbred population based on a different resistant (IS2122) and susceptible (27B) parents (Aruna et al. 2011). Efforts at ICRISAT in different genetic backgrounds also confirmed these QTL (Folkertsma et al. 2003). The key QTL on LG5 (for leaf glossiness) and LG10 (trichome density, oviposition, deadhearts) have been further saturated with new SSR markers (Satish et al. 2012b), and several putative candidate gene-linked markers have been identified. An orthologous insect resistance gene cysteine protease-*Mir1*(XnhsbmSFC34/SBI-10) involved in stalk borer resistance in maize was significantly associated with major QTL for all traits (except seedling vigor) explaining 22.1 % of the phenotypic variation for deadhearts%, a direct measure of shoot fly resistance. Similarly, a NBS-LRR gene (XnhsbmSFCILP2/SBI-10), involved in rice brown plant hopper resistance, was associated with deadhearts% and number of eggs per plant. Beta-1,3-glucanase (XnhsbmSFC4/SBI-10), involved in aphid and brown plant hopper resistance, was associated with deadhearts% and leaf glossiness. Comparative QTL analysis revealed the existence of common QTL for shoot fly and other important sorghum insect pests such as green bug, head bug, and midge. The associated candidate genes should aid in elucidating the molecular basis of resistance, high-resolution mapping, and map-based cloning of major QTL, besides providing powerful gene tags for marker-assisted selection of shoot fly resistance in sorghum.

Midge

Sorghum midge *Stenodiplosis sorghicola* (Coquillett) is one of the most damaging pests of grain sorghum worldwide (Harris 1976). It is widely distributed in Asia, Australia, Africa, the Americas, and Mediterranean Europe. The female midges lay their eggs into spikelets at anthesis, and the hatched larvae feed on the developing seed. Of the three insect resistance mechanisms (Painter 1951), ovipositional antixenosis is the most common (Henzell et al. 1994), while antibiosis to feeding larvae is also reported (Sharma 1985). One of the important morphological traits associated with midge resistance is glume size. Short and tight glumes make oviposition difficult for the midge and are therefore preferred trait for breeding resistance against midge. Faster rate of grain development and high tannin content in grain are also associated with midge resistance (Sharma et al. 1994). However, the genetic control of these mechanisms is poorly understood. Recessive to partial dominance were found to control resistance to sorghum midge. Both general and specific combining effects are significant, but resistance is controlled largely by additive gene action (Sharma et al. 1996).

QTL associated with two of the mechanisms of midge resistance, viz., antixenosis and antibiosis, were identified in a RIL population from the cross ICSV745 × 90562 (Tao et al. 2003). Two QTL on different linkage groups (SBI-03 and SBI-09) were found to be associated with antixenosis, explaining 12 and 15 % of the total variation in egg numbers/spikelet. One region on SBI-07 was significantly associated with antibiosis and explained 34.5 % of the variation of the difference of egg and pupal counts. The identification of DNA makers for both antixenosis and antibiosis mechanisms of midge resistance will be particularly useful for exploring new sources of midge resistance and for gene pyramiding of these mechanisms for achieving durable resistance through MAS.

Green Bug

Green bug, *Schizaphis graminum* (Rondani), is one of the major insect pests of sorghum causing significant economic damage. It is a sap-sucking insect, which removes the phloem sap and injects phytotoxins into the plants. The damage is characterized by a dark red spot at the feeding site, surrounded by an area of pale yellow discoloration. Apart from direct feeding damage, green bugs are also key vectors of viral pathogens (Harvey et al. 1996). The loss due to green bug damage was estimated to be $274 million annually (Eddleman et al. 1999). Several biotypes (C, E, I, and K) have been identified inflicting serious damage to sorghum grain.

Several studies revealing multiple QTL for green bug resistance in different genetic resistance sources have been conducted against green bug biotypes C, E, I, and K. Three loci present on SBI-05, SBI-06, and SBI-07 conferring resistance to green bug biotype I were identified by Katsar et al. (2002). Nine QTL affecting both resistance and tolerance to biotypes I and K of green bug have been identified (Agrama et al. 2002) with individual QTL accounting for 5.6–38.4 % of phenotypic variance. Four SSR and one RAPD marker were associated with the expression of all resistance and tolerance traits, and these markers were apparently linked to biotype non-specific resistance and tolerance. Four additional markers associated with biotype-specific resistance or tolerance traits were also identified. Nagaraj et al. (2005) detected three QTL on SBI-01 and SBI-04 for biotype I resistance and tolerance using chlorophyll loss as an indicator to green bug damage. Wu and Huang (2008) have shown a major QTL on SBI-09 for resistance to biotype I. It is evident from these studies that resistance to green bug damage is contributed by multiple genomic regions depending on the resistance source. Some of the alleles are biotype-specific, while others are biotype non-specific. Transcriptomic studies have shown the involvement of signaling compounds and defense-activated R-genes in defense response to green bug attack. Downregulation of cysteine proteinase inhibitors and the upregulation of genes such as *Xa1*, antimicrobial proteins (Park et al. 2006), and several other signaling compounds including an LRR-containing glycoprotein in response to green bug damage have been identified (Zhu-Salzman et al. 2004).

4.5.2.2 Disease Resistance

Sorghum is host to many diseases that are caused by fungi, bacteria, viruses, nematodes, and parasitic plants (Thakur et al. 2007). More than 50 diseases have been documented on sorghum occurring on different plant parts and at different crop stages. Only some of them are globally important, causing significant economic damage. Grain mold, root and stalk rots, ergot, smut, anthracnose, leaf blight, downy mildew, rust, leaf spots, and virus diseases (maize stripe and maize mosaic) cause substantial economic loss every year. The use of fungicides to control the disease decreases the profit margins of low cash-input sorghum production systems and increases the risk of residual effects from fungicide on forage. Therefore, host plant resistance is considered to be more practical and reliable for managing sorghum diseases. To develop disease-resistant cultivars, breeders require a detailed knowledge on the genetics and inheritance pattern of disease resistance. Resistance can be controlled by a single gene or complex with the involvement of several genes depending on the source of resistance, plant development stage, and the pathotype (Afanasenko et al. 1999). Molecular marker technology greatly facilitates the study of complex traits and has made it possible to dissect the polygenes controlling such traits into individual Mendelian factors (Paterson et al. 1988). DNA-based molecular markers delimiting disease resistance loci in sorghum have been reported for grain mold (Klein et al. 2001; Upadhyaya et al. 2013b), foliar diseases (Boora et al. 1999; Klein et al. 2001; McIntyre et al. 2005; Murali Mohan et al. 2009, 2010; Ramasamy et al. 2009; Singh et al. 2006; Tao et al. 1998; Upadhyaya et al. 2013b), charcoal rot (Reddy et al. 2008), head smut (Oh et al. 1994), and downy mildew (Agrama et al. 2002; Gowda et al. 1995).

Grain Mold

Grain mold is a major disease complex of sorghum that severely affects grain production and grain quality. The damage by grain mold is one of the main reasons for the drastic decline in the area of rainy sorghum in India. Due to the changes in the maturity patterns brought in through breeding efforts in improved rainy cultivars, susceptibility of sorghum to grain mold has greatly increased. The disease comes due to late rains and is favored by the conditions of high humidity and temperature prevalent during grain development. The fungi infect the developing seed from anthesis through grain maturity (grain mold) and also colonize the exposed sorghum grain mostly after physiological maturity (called weathering). Damage resulting from mold includes arrest of kernel development, discoloration, decrease in kernel mass and grain density, decrease in germination, decrease in seedling vigor, and contamination of grain with harmful mycotoxins. Such grains are not suitable either as food or for animal feed and thus fetch reduced market price to the growers (Somani and Indira 1999).

A complex of fungal pathogens, most of which are saprophytic, causes grain mold. However, *Fusarium* and *Curvularia* are parasitic fungi of the grain mold complex that can infect sorghum spikelet at anthesis itself. Of these, *Fusarium moniliforme* is the most predominant and most damaging fungal parasite of sorghum grain worldwide. Production loss ranges from 30 to 100 %, depending on cultivar, time of flowering, and prevailing weather conditions during flowering to harvesting (Singh and Bandyopadhyay 2000). It is difficult to estimate accurate losses due to grain mold disease as it involves losses from production to marketing and losses in utilization of the grain or seed. However, total annual loss due to grain mold in the semi-arid tropics has been estimated to be about US$ 130 million (ICRISAT 1992).

Genetic mechanisms governing grain mold resistance in sorghum are poorly understood. Resistance is complexly inherited with added complications from the involvement of several simply inherited, kernel-based traits influencing the level of grain mold resistance. Some of the kernel characteristics reported to enhance grain mold resistance are a pigmented testa, a red pericarp, a thin pericarp, corneous endosperm, increased flavan-4-ol content, reduced water uptake capacity in mature grain, open panicle structure, and taller plants (Bandyopadhyay and Mughogho 1988; Esele et al. 1993; Glueck and Rooney 1980; Harris and Burns 1973; Ibrahim et al. 1985; Menkir et al. 1996; Waniska et al. 1989). It was estimated that a minimum of four to ten genes control grain mold resistance (Rodriguez-Herrera et al. 2000) and polygenic nature of grain mold was also reported (Klein et al. 2001). Due to variation in the casual pathogen from location to location, resistance to grain mold should be evaluated in target environments (Audilakshmi et al. 2005). Because of the complex inheritance and the large environmental effects in disease development and expression, resistance to grain mold is influenced by large environmental and genotype-by-environment (G × E) interactions due to pathogen variability and variability in environmental conditions (Bhat et al. 2004). Identification of genetic markers linked to QTL for grain mold resistance would enable breeding efforts in developing sorghum cultivars with superior and stable host plant resistance.

Grain mold incidence was observed to be influenced by five QTL, each accounting for the phenotypic variance between 10 and 23 % (Klein et al. 2001). The effects and relative positions of QTL were in accordance with the QTL distribution of several agronomic traits correlated with grain mold incidence. Several genomic regions affected multiple traits including the one that affected grain mold incidence, plant height, panicle peduncle length, and grain-milling hardness, and the other that influenced grain mold and plant height. Collectively, QTL detected in the population explained between 10 and 55 % of the phenotypic variance. In a recent study, two SNP loci linked to grain mold resistance have been identified using an association-mapping panel of 242 mini-core sorghum genotypes (Upadhyaya et al. 2013b). Among these, one contained a NB-ARCLRR class of R-gene (Sb02g004900) that shares 37 % identity and 57 % similarity to the non-host resistance gene of maize, *Rxo1*. However, the map positions of the SNP markers did not overlap with the grain mold QTL from Klein et al. (2001). This could possibly be due to the differences in the pathogen among various environments causing differences in resistance expression (Audilakshmi et al. 2005; Little et al. 2012).

Foliar Diseases

Sorghum foliar diseases of fungal origin prevalent under warm humid conditions are highly destructive and drastically affect grain yield, fodder yield, and fodder quality by causing premature drying of leaves and defoliation in both grain and forage sorghums (Grewal 1988; Mathur and Bunker 2002). The estimated sorghum yield losses caused by foliar diseases in Asia, Africa, and the Americas range from 32 to 60 % (Frederiksen and Odvody 2000). Forage sorghum cultivars are quite susceptible to various foliar diseases [anthracnose (*Colletotrichum graminicola*), zonate leaf spot (*Gloeocercospora sorghi*), target leaf spot (*Bipolaris sorghicola*), *Drechslera* leaf blight (*Drechslera australiensis*), and rust (*Puccinia purpurea*)]. These diseases reduce the amount of green leaf area available for photosynthesis and affect the quality of fodder by reducing the protein, zinc, and in vitro dry matter digestibility (Rana et al. 1999). Little information is available about the genetic inheritance of resistance to zonate leaf spot wherein it is reported to be governed by duplicate epistasis (Grewal 1988) and its inheritance pattern is intermediate to recessive (Rosenow and Frederiksen 1982). A single dominant gene controlled anthracnose resistance, while the leaf midrib anthracnose was under the control of single recessive gene, both of which are unlinked (Erpelding and Wang 2007). The observed range in disease reaction from highly susceptible to highly resistant, in progeny derived from resistant × susceptible crosses, suggests that resistance to these diseases could be polygenic (Murali Mohan et al. 2009). Because quantitative resistance is difficult to assess phenotypically due to environmental interaction, markers linked to genes for resistance should be a very useful tool for foliar disease resistance breeding. Grain losses caused by foliar diseases in sorghum are usually low, but they are of more concern in the development of forage sorghums.

A major QTL on SBI-06 between SSR markers Xtxp95-Xtxp57 (Klein et al. 2001) influencing resistance against various unrelated pathogens causing foliar diseases was consistently detected with the phenotypic variation ranging from 32 % (bacterial leaf blight, zonate leaf spot) to 55 % (anthracnose) indicating involvement of a key gene for disease resistance. Disease-response QTL for other foliar disease like oval leaf spot was also found to colocate to this region on SBI-06. Consistent involvement of this QTL region in disease resistance against several foliar diseases was also reported (Murali Mohan et al. 2010). In a recent study, Upadhyaya et al. (2013a) identified eight SNP marker loci linked with anthracnose resistance across environments, of which, two SNPs were validated and were found to colocate with the two major QTL (*QAnt3* and *QAnt2*) reported by Murali Mohan et al. (2010). Genes known to be involved in plant defense mechanisms like NB-ARC class of R-genes, HR-related genes, a transcription factor that functions in the R-gene pathway, a gene that functions in the nonspecific host resistance, and a gene for antimicrobial compound production were identified as putative genes for anthracnose disease resistance in sorghum. These studies confirmed the strong correlation that was reported between plant color and foliar disease resistance. Tan plant color was associated with reduced foliar disease symptoms. Breeders have noted the relationship between tan plant color and apparent resistance to foliar and panicle diseases (Rana et al. 1976; Torres-Montalvo et al. 1992). Plant color scored as a qualitative trait (tan vs. purple) mapped to the same region of SBI-06 that contained the linked QTL for foliar diseases (Klein et al. 2001; Murali Mohan et al. 2009).

Rust

Sorghum rust (*Puccinia purpurea* Cooke) is a widespread disease in all sorghum-growing areas of the world. Yield losses from 29 to 50 % were recorded (Hepperly 1990). The disease is important as it reduces the quality and palatability of green fodder and also acts as a predisposing factor to other major diseases like *Fusarium* stalk rots, charcoal rot, and grain mold (Frederiksen and Rosenow 1986). Although only two races of *P. purpurea* have been identified (Bergquist 1974), the presence of more physiological races has been suspected (Indira et al. 1982). The genetics of rust resistance in sorghum appears complicated, with varying numbers of genes at dispersed loci having differing effects and possible modes of action (Miller and Cruzado 1969; Patil-Kulkarni et al. 1972; Rana et al. 1976). The inheritance of resistance was reported to be quantitative with the involvement of minimum of four loci (Tao et al. 1998). Susceptibility was found to be dominant over resistance (Dabholkar et al. 1980). Though strong linkage between tan plant color and rust resistance was observed, sorghum lines with purple pigmentation and rust resistance have also been documented, indicating close linkage between the two loci and not pleiotropy.

Four major QTLs for rust resistance on SBI-01, 2, 3, and 8 explaining 16–42 % of trait variation were identified using a population of 160 RILs (Tao et al. 1998). The major QTL on SBI-08 accounting 42 % of trait variation was found to host the key rust R-gene homologue of *Rp1-D* from maize and sugarcane (McIntyre et al. 2005). This QTL is believed to be the *Pu* gene (Coleman and Dean 1961; Miller and Cruzado 1969). This region of sorghum has been previously shown to be orthologous to maize chromosome 10S, the location of the major rust resistance gene *Rp1* (McIntyre et al. 2004; Ramakrishna et al. 2002), which is located between 2,487,742 and

2,514,226 bp on sorghum physical map (Mace and Jordan 2010). Conservation of gene function at the *Rp1* locus for rust resistance between sorghum, maize, and sugarcane has been established (McIntyre et al. 2005). However, the nature of the conserved function is not known since three different rust pathogens (*P. purpurea* in sorghum, *Puccinia sorghi* in maize, and *P. melanocephala* in sugarcane) cause rust in respective hosts. It is postulated that the products of resistance genes interact with pathogen avirulence proteins or that common avirulence determinants may be recognized in maize, sorghum, and sugarcane rust fungi. Alternatively, R-gene conservation may reflect conserved signaling motifs that activate downstream resistance mechanisms (Childs et al. 1997). Several of the RGAs are found to be associated with rust resistance in sugarcane and were colocated with QTL for rust resistance in sorghum, which are potentially useful as markers for rust resistance breeding in sugarcane and sorghum. In a recent study involving association-mapping panel, Upadhyaya et al. (2013b) identified five SNP loci linked to rust resistance, two of which contained the rust resistance gene homologous to the maize rust resistance gene (*Rp1-D*) and to the wheat rust resistance gene (*Lr1*). The remaining loci contained genes important in defense responses.

Charcoal Rot

Charcoal rot caused by the fungus *Macrophomina phaseolina*, is the most common and destructive root and stalk rot disease of sorghum grown under residual soil moisture. The disease is quite common and is a major factor limiting sorghum production in post-rainy season in India (Rana et al. 1982). The disease is also prevalent in Australia (Trimboli and Burgess 1982) and the United States, particularly in the southern states of Texas, Georgia and Arizona (Edmunds et al. 1964; Edmunds and Zummo 1975). Significant yield reduction in major sorghum-growing areas of Africa due to charcoal rot has been reported (Gebrekidan and Kebede 1979; Hulluka and Esele 1992). Incidence of charcoal rot is highly related with occurrence of moisture stress during post-flowering stage of the crop (Seetharama et al. 1987). Acute moisture stress coinciding with grain-filling stage (terminal drought) of post-rainy sorghum predisposes the plants to root and stalk rots leading to severe crop lodging and loss in grain yield and quality and quantity of stover. Improving tolerance to terminal drought and associated root and stalk rots needs serious consideration (Subudhi et al. 2000).

Epidemics of charcoal rot in several states of India growing post-rainy sorghum was reported (Anahosur and Rao 1977; Kamatar et al. 2000) with a significant loss in grain yield (55–64 %), fodder quality, and quantity (Mughogho 1984). The disease becomes severe when plants are under physiological and environmental stress. Other physiological conditions in plants, such as levels of sugar (Maranville 1974), water in the stalk (Pedgaonkar and Mayee 1990), sink/source balance (Dodd 1980), and non-senescence (Rosenow et al. 1996), are known to influence stalk rotting and level of host resistance.

Genetic control of charcoal rot resistance is complex. Both additive and non-additive gene actions in the genetic control of this trait have also been documented (Rao et al. 1993). Breeding for charcoal rot resistance should be done in a holistic approach (Rosenow et al. 1983), and selection for stiff-stalk and drought tolerance, especially of the non-senescent type (staygreen), combined with high yield, would be more productive than breeding for charcoal rot resistance alone. Screening for resistance to charcoal rot under natural infection has been difficult because of irregularities in disease development. However, considerable progress has been made in the development of techniques for artificial screening for charcoal rot resistance. The toothpick inoculation procedure was successfully developed and used for screening resistance to charcoal rot in post-rainy sorghum (Rao et al. 1980; Seetharama et al. 1987).

Breeding efforts that resulted in the development of improved sorghum varieties and hybrids that transformed sorghum production in India are highly susceptible to charcoal rot (Anahosur and Rao 1977). Until now, there are no high-yielding sorghum cultivars with charcoal rot resistance available for cultivation by farmers. Management of charcoal rot with chemicals is not feasible as it only adds to the cost of cultivation which no farmer is willing to adopt. Hence, the most realistic alternative strategy is to breed for charcoal rot resistance. Though several efforts were made earlier to breed for charcoal rot resistance, the success is very limited due to the quantitative inheritance of resistance and probably also due to the difficulty in the selection for resistance because of strong interaction between host, pathogen and the environment. Resistance was reported to be non-additive (Garud and Borikar 1985; Indira et al. 1983).

Not many efforts have been made in dissecting the resistance to charcoal rot in terms of QTL. Two studies have reported QTL for resistance to charcoal rot using the same RIL population evaluated over three locations and 4 years (Reddy et al. 2008; Patil et al. 2012). The study using 93 RILs of the cross IS22380×E36-1 was able to identify nine consistent QTL over locations and years for three morphological traits (number of internodes crossed by the rot, length of infection, and percent lodging) and three biochemical traits (lignin and total phenols). Candidate genes for each of the QTL influencing both morphological and biochemical traits have been identified.

Ergot

Ergot (sugary disease) is an endemic fungal disease found in major sorghum-growing regions of the world. The disease which was limited to Asia and Africa was also reported from the United States (Isakeit et al. 1998) and subsequently from Australia (Ryley et al. 1999) and Brazil (Casela et al. 2008). Three species of ergot pathogen are prominently prevalent in different parts of the world. *Claviceps africana* is the most predominant throughout the Americas, Australia, Asia, and Africa, while *Claviceps sorghi* is limited to Asia and *Claviceps sorghicola* is confined to Japan (Bandyopadhyay et al. 1998; Pažoutová et al. 2000; Tooley et al. 2000).

The disease develops following the infection of unfertilized ovaries of sorghum panicle during anthesis preventing further pollen fertilization and seed set. Male sterile lines (A-lines) are particularly susceptible to ergot in hybrid seed production plots because of non-availability of viable pollen due to non-synchronous flowering of A-line and restorer lines (R-lines) or due to adverse climatic conditions, especially cooler temperatures (Bandyopadhyay et al. 1998). Loss of total grain yield ranging from 25 to 80 % has been documented from India and Zimbabwe. Therefore, ergot is considered as a major disease causing significant economic loss in seed production plots.

Dissection of genetic factors of ergot resistance in sorghum resulted in the identification of 18 QTL for three component traits (percentage ergot infection, pollen quantity, and pollen viability) and validation of two QTL using different genetic backgrounds (Parh et al. 2008). Both pollen- and non-pollen-based mechanisms were found to operate for ergot resistance. Four major QTL for percent ergot infection (SBI-01, SBI-06, SBI-08, and SBI-09) and one major QTL each for pollen quantity (SBI-06) and pollen viability (SBI-07) were detected besides the colocalization of QTL, signifying the clustering of genes with related function. It was also observed that the major QTL for percent ergot infection on SBI-06 was colocated with QTL for a number of diseases including grain mold, anthracnose, zonate leaf spot, and bacterial leaf spot (Klein et al. 2001; Murali Mohan et al. 2010). Three other regions on SBI-07 and SBI-10 and SBI-08 that are known to contain QTL for grain mold and rust resistance (Klein et al. 2001; Tao et al. 1998) also appear to contain a QTL for ergot resistance (Parh et al. 2008).

4.5.2.3 Weed

Striga

Striga is a devastating parasitic weed in Africa and parts of Asia. A single recessive gene (*lgs*) controls low Striga germination stimulant activity, a well-known resistance mechanism in sorghum. Molecular markers linked to the *lgs* gene can accelerate the development of Striga-resistant cultivars. Using a high-density linkage map constructed with 367 markers (DArT and SSRs) and an in vitro assay for germination stimulant activity towards *Striga asiatica* in 354 RILs derived from SRN39 (low stimulant) and Shanqui Red (high stimulant), Satish et al. (2012a) precisely tagged and mapped the *lgs* gene on SBI-05 between two tightly linked microsatellite markers SB3344 and SB3352 at a distance of 0.5 and 1.5 cm, respectively.

4.5.3 Abiotic Stresses

Staygreen

Among the abiotic stresses, drought is the most important environmental stress in agriculture, which limits crop productivity in the arid and semi-arid regions of the world where many of the world's poorest farmers live. Developing genotypes with improved drought tolerance will enhance the food production and livelihoods of farmers in these regions. With changing climatic conditions and erratic rainfall patterns, the recurrence of drought in these regions is going to increase, and enhancement of drought tolerance and water-use efficiency by crop plants is therefore a global concern. Drought tolerance is a complex phenomenon involving mechanisms like early maturity, avoidance, tolerance, and desiccation tolerance (Blum 1988). A variety of morphological and physiological changes have been identified in response to drought stress in plants, which include root morphology and rooting depth, plant architecture, leaf area, cuticular resistance and thickness, stomatal conductance, osmotic adjustment, antioxidative defense, hormonal regulation, desiccation tolerance (membrane and protein stability), and maintenance of photosynthesis through persistent green leaf area (staygreen). Plants expressing a variety of genes associated with these morphological and physiological traits tolerant to abiotic stresses have been demonstrated (Bohnert et al. 1995; Bray 1997; Nguyen et al. 1997; Shinozaki and Yamaguchi-Shinozaki 1996). In addition, some genes are thought to function not only in stress tolerance but also in the regulation of gene expression and signal transduction in stress response (Shinozaki and Yamaguchi-Shinozaki 1996, 2000).

The term staygreen has been used to describe the post-flowering drought-tolerance response in sorghum (Rosenow et al. 1988). Staygreen is characterized by the plant's ability to maintain functional photosynthetic leaf area during the grain-filling stage even under severe post-flowering drought stress. Staygreen or delayed-senescence is an important trait associated with drought tolerance, and staygreen varieties of sorghum outperform conventional varieties under drought conditions. Sorghum genotypes with this trait continue to fill their grain normally under drought and exhibit increased resistance to charcoal rot and lodging. This is associated with resistance to diseases (Borrell and Hammer 2000), increased cytokinin concentration (McBee 1984), and stem sugars in basal nodes (Duncan 1984).

Due to known association of staygreen trait with drought tolerance and stover quality, its genetic and physiological basis has been studied by many authors using different staygreen sources in sorghum (Crasta et al. 1999; Haussmann et al. 2002; Kebede et al. 2001; Subudhi et al. 2000; Tao et al. 2000; Tuinstra et al. 1997; Xu et al. 2000) and in other crops such as rice (Abdelkhalik et al. 2005; Cha et al. 2002; Jiang et al. 2004), soybean (Luquez and Guiamét 2001), and maize (Zheng et al. 2009). The expression of post-flowering symptoms becomes more prominent when crop growth is favorable prior to flowering, followed by severe moisture stress, particularly during the grain-filling stage. Sorghum genotypes with staygreen trait continue to fill their grain normally under drought and exhibit increased drought tolerance characterized by delaying the onset of leaf senescence and reducing its rate (i.e., two components of the staygreen trait). Therefore, staygreen offers an effective strategy for increasing grain production, fodder quality, and grain crop residues particularly under water-limited conditions. Staygreen in sorghum is genetically and physiologically complex, exhibiting a variety of expression patterns and environmental sensitivities, depending on background genotype (Van Oosterom et al. 1996).

Several sorghum genotypes (B35, SC56, and E36-1) have been identified that exhibit the staygreen trait (Crasta et al. 1999; Haussmann et al. 2002; Rosenow et al. 1983; Tuinstra et al. 1997). The genotype BTx642 (formerly B35) has been a useful source of staygreen for research and the development of commercial hybrids (Henzell et al. 2001). Several QTL associated with staygreen trait have been detected across genetic backgrounds. Comparison of the staygreen QTL from Xu et al. (2000) and Subudhi et al. (2000) along with results obtained by other workers (Tao et al. 2000; Tuinstra et al. 1997) using B35 as a staygreen source resulted in the consistent identification of four major QTL, namely, *Stg1* (SBI-03) and *Stg2* (SBI-03), *Stg3* (SBI-02) and *Stg4* (SBI-05) in different genetic and environment backgrounds which together accounted up to 53.5 % phenotype variance. The QTL *Stg1*, *Stg2*, and *Stg3* overlap with the QTL for chlorophyll content, and molecular markers linked to these QTL are available. The ranking of staygreen QTL based on their contribution to the staygreen phenotype in BTx642 × RTx7000 population is *Stg2* > *Stg1* > *Stg3* > *Stg4* (Xu et al. 2000). Recently, *Stg2*, *Stg3*, and *StgB* were identified to be the three key QTLs for MAS to improve terminal drought tolerance (Nagaraja Reddy et al. 2014). Staygreen QTL individually reduced leaf senescence in introgression lines and may contribute significantly towards breeding drought tolerance (Harris et al. 2007; Kassahun et al. 2010). More recently, the potential use of staygreen QTL in improving transpiration efficiency and water extraction capacity in sorghum for terminal drought tolerance (Vadez et al. 2011) and grain yield particularly under low-yield environments has been demonstrated (Jordan et al. 2012).

Colocation of staygreen and nodal root angle QTL in sorghum (Mace et al. 2012) highlights the probable role of roots in retaining leaves green. Involvement of staygreen in canopy development, leaf anatomy, root growth, water uptake, and grain yield was reported (Borrell et al. 2014a, b). Despite this, the genetic basis of drought tolerance as a complete phenomenon in sorghum has not been well understood.

Cold Tolerance

Sorghum is a tropical crop adapted to warmer environments. Germination and early crop establishment is an important growth stage requiring optimum soil moisture and temperature. Cool temperature (15 °C or less) especially during the early growth stages is one of the major abiotic stress limitations to sorghum cultivation in temperate and higher elevation regions. When grown under low temperatures, sorghum cultivars produced more leaves (Hesketh et al. 1969; Quinby 1973) and had delayed floral initiation (Caddel and Weibel 1971; Quinby 1973), and panicles were found to be male sterile (Downes and Marshall 1971; Singh 1977). Because plant population and spacing finally affect grain yield (Willey and Heath 1969), the improvement in stand establishment and early-season cold survival should ultimately lead to higher and more reliable yields of grain sorghum. Development of cold-tolerant sorghum cultivars would also potentially allow sorghum cultivation in nontraditional areas. Genotypic variations for seedling cold tolerance have been reported from germplasm lines (Singh 1985). Especially the "kaoliang" landraces from China are known to exhibit higher seedling emergence and seedling vigor under cooler conditions (Cisse and Ejeta 2003; Qingshan and Dahlberg 2001) and form an excellent genetic resource for improving cold tolerance in sorghum (Franks et al. 2006). However, these landraces also harbor poor or undesirable agronomic traits as linkage drags (Knoll and Ejeta 2008). Use of sources of cold tolerance lines in genetic studies resulted in the identification of QTL, which pave way for marker-assisted improvement of elite lines for cold tolerance.

Knoll et al. (2008) mapped cold tolerance QTL using a population of 153 RILs from a cross, Shan Qui Red (cold-tolerant) × SRN39 (cold-sensitive) which resulted in the identification of two QTL for germination, one on SBI-03 contributing 12–15 % of variation under both cold and optimal temperatures, while the second QTL on SBI-07 showed greater significance only under cold temperature accounting 10 % trait variation. A major QTL with

8–27 % trait contribution was identified on SBI-01, which showed strong associations with seedling emergence and seedling vigor under early as well as late field plantings. Similarly, one QTL for both early and late seedling emergence was identified on SBI-02, explaining 8–10 % of trait variation. A new source of cold tolerance, PI610727, was used to tag the genomic regions exhibiting significant contributions to traits for early-season cold tolerance (Burow et al. 2011). Fourteen QTL for four component traits of cold tolerance (germination at low and optimum temperature, field emergence, and seedling vigor) on five linkage groups (LG1, LG2, LG4, LG7, and LG9) were detected. Vigorous germination was found to be an important component of cold germinability, which was also reported in rice (Fujino et al. 2004) and in sorghum (Knoll and Ejeta 2008). This strong relation was reinforced by the colocation of QTL for cold and optimal temperature germinability detected on chromosome 2. PI610727 was found to share common loci with other known early-season cold-tolerant sorghum germplasm and harbors novel QTL for enhanced germination and field emergence. Further investigations would be required to validate these results and use of other types of markers for use in MAS.

Even though many QTLs have been identified and mapped to the sorghum genome for important traits, it is very important to select the most effective QTL for transferring into the elite cultivars through marker-assisted breeding by targeting those QTL that contributes substantially to the phenotypic variance of the trait of interest. A list of important QTL for various traits and their contribution to phenotypic variance along with linked markers are given in Table 4.1.

4.6 Major Effect Genes

Integration of previously mapped major effect genes onto a complete genome map, linked to the whole genome sequence, was performed by Mace and Jordan (2010), using common markers across populations, which allows the sorghum breeders and researchers to link this information to QTL studies so that they can be aware of the consequences of selection for major genes. Readily scorable trait-linked morphological traits provide new opportunities for breeders to select the target traits indirectly and develop more efficient breeding strategies. The list of some of the major genes mapped in sorghum is listed in Table 4.2.

4.7 DNA Markers for Seed Purity

Genetic purity of hybrids is one of the crucial aspects of the exploitation of the yield potential of high-yielding hybrids and varieties. Maintenance of genetic purity in the seed supply chain is important to ensure the supply of good quality and genetically pure seeds to the farmers. Characterization and identification of varieties, parental lines, and hybrids are of great significance in modern-day agriculture as they play a major role in the development of diverse parental lines and superior varieties and heterotic hybrids. The success of hybrids or varieties depends mainly on the production and timely supply of genetically pure homogenous seeds to farmers. It is estimated that 1 % impure hybrid seed could lead to a loss of 100 kg ha^{-1} of grain yield in rice (Mao et al. 1998). In seed production plots, the methods adopted by farmers provide opportunities for seed contamination leading to the presence of "off-types" in the hybrid or the varietal seed lot. This is particularly so in a country like India wherein maximum seed production is done through contract farming involving a farmer or farmer groups whose education and technical skill levels are low. Thus, monitoring of seed production in seed production plots becomes necessary at several crop growth stages.

The sorghum hybrid industry depends on three-line system of hybrid development involving a male sterile line (A-line), its cognate maintainer line (B-line), and a restorer line (R-line). The A-line is maintained by crossing with its B-line, whereas the F_1 hybrid seed for commercial exploitation is obtained by crossing A-line with specific R-line. Thus, assessment and maintenance of genetic purity of these lines are crucial for the success of any good hybrid. Assessment

Table 4.1 Important QTL in sorghum with their associated markers

Trait/genes/QTL	LG	LOD	R^2	Reference	Linked markers
I. Agronomic traits					
Plant height (*Dw$_1$*)	9	6	20	Pereira and Lee (1995)	isu140/PIO100016
Plant height (*Dw$_2$*)	6	16	27	Ritter et al. (2008)	AG/CTG9
Plant height (*Dw$_3$*)	7	8	29	Pereira and Lee (1995)	isu123/isu116
Maturity	6	91	86	Lin et al. (1995)	pSB189/pSB580
	1	6	15	Srinivas et al. (2009)	txp58/Dsenhsbm63
	6	11	36	Kebede et al. (2001)	psb521/psb708
Grain yield	2	4	18	Ritter et al. (2008)	AAG/CAA1
	6	5	15	Srinivas et al. (2009)	GlumeT/Xtxp145
	10	3	14	Ritter et al. (2008)	AAG/CTT2
Dry grain yield	10	5	11	Murray et al. (2008)	Xcup67/txa3777
Seed mass	1	13	20		TS138/rio72
	1	–	11	Rami et al. (1998)	bnl6.25/umc84
	1	5	15	Srinivas et al. (2009)	Dsenhsbm64/ Xcup24
	1	3	11	Pereira et al. (1995)	isu027/npi209
	2	–	19	Rami et al. (1998)	umc122/bnl16.06
	3	6	10	Feltus et al. (2006)	psB443a/pSB614
	3	–	12	Rami et al. (1998)	umc152/umc10
	4	4	10	Srinivas et al. (2009)	Xtxp12/ Dsenhsbm39
	4	4	16	Feltus et al. (2006)	txs604/cdo516.1
	4	5	16	Brown et al. (2006)	Xtxp51/txa6257
	6	7	10	Feltus et al. (2006)	pSB521a/pSB428a
	6	8	15	Murray et al. (2008)	txa2873/txa2067
	7	–	31	Rami et al. (1998)	umc23/sscir88
	7	–	35		umc23/sscir88
	8	6	11	Murray et al. (2008)	rio65/rio37
	8	5	12	Brown et al. (2006)	isu145.2/txa558
	9	6	18		txs1703/cdo580
	10	4	14	Feltus et al. (2006)	txs1106/bnl5.04
	10	5	16	Pereira et al. (1995)	isu156/isu034
II. Insect resistance					
Shoot fly (glossiness)	5	3	17	Satish et al. (2012b)	Xtxp65/ XnhsbmSFC61
Shoot fly (deadhearts)	10	7	23		XnhsbmSFC34/ Xnhsbm1039
Shoot fly (trichome density)	10	9	20		XnhsbmSFC34/ Xnhsbm1039
	10	10	24		Xgap1/Xnhsbm1011
Midge	3	3	12	Tao et al. (2003)	rz543/st698
	7	11	34		txs1931/sg37
	9	5	15		ST1017/SG14
Green Bug (biotype I)	1	2	15	Nagaraj et al. (2005)	Xtxp43/Xtxp85
Biotype I	4	4	20		Sb1-10
Biotype I	7	–	10	Katsar et al. (2002)	bdc098/csu61
Biotype K	1	2	16	Nagaraj et al. (2005)	Xtxp335/Xtxp204
Biotype K	4	3	13		Xtxp12/Xcup20
Biotype K	10	–	15	Katsar et al. (2002)	psb0106/rz144

(continued)

Table 4.1 (continued)

Trait/genes/QTL	LG	LOD	R^2	Reference	Linked markers
III. Disease resistance					
Anthracnose	6	13	40	Murali Mohan et al. (2010)	Xtxp95-Plcor
Zonate leaf spot	6	5	14	Murali Mohan et al. (2010)	Xtxp95-Plcor
	6	5	17		Fdnhsbm107-Fdnhsbm24
	3	4	13		Xtxp228-Drenhsbm103
Target leaf spot	6	20	50	Murali Mohan et al. (2010)	Xtxp95-Plcor
Rust	1	3	26	Tao et al. (1998)	bnl5.09/txs1625
	2	3	17		sscir51/txs2042
	3	4	24		rz323/isu102
	8	9	43		psb47/txs422
	6	8	24	Murali Mohan et al. (2010)	Xtxp95-Plcor
Ergot (% infection)	1	5	12	Parh et al. (2008)	sPb-8261
	6	6	14		sPb-1543
	7	4	10		Xtxp168
	8	4	11		AGG+CAG6
	9	3	20		Sb4-32
Ergot (pollen viability)	7	3	13	Parh et al. (2008)	sPb-5594
	8	3	10		Xtxp273
	6	7	20		AAG+CTT6
Charcoal rot (internodes crossed)	2	4	19	Srinivasa Reddy et al. (2008)	Xtxp297
Charcoal rot (% lodging)	4	4	15	Srinivasa Reddy et al.(2008)	Xtxp343
IV. Weed					
Striga (*lgs*)	5			Satish et al. (2012a, b)	SB3344, SB3343, SB3346
V. Abiotic resistance					
Drought-staygreen (*Stg1*)	3	5	20	Xu et al. (2000)	Xtxp442/Xtxp38
Stg2	3	6	30		Xtxp2/Xtxp503
Stg3	2	3	16		Xtxp430/Xtxp1
Stg4	5	2	11		Xtxp225/Xtxp15
Cold tolerance (late emergence %)	1	–	21	Knoll et al. (2008)	PeriCol/OPK18
	2	–	11		Xtxp201/Sb110
Cold tolerance (early vigor)	1	–	20	Knoll et al. (2008)	PeriCol/OPK18
	4	–	12		Xtxp51/Xtxp21
Cold tolerance (late vigor)	1	–	28	Knoll et al. (2008)	PeriCol/OPK18
Cold tolerance (cold germination)	3	–	13	Knoll et al. (2008)	ubc171/SbAGE01
Cold tolerance (optimal germination)	3	–	15	Knoll et al. (2008)	umc60/ubc171
Cold tolerance (germ12-2.1)	2	9	20	Burow et al. (2011)	XXtxp298/Xsbarslbk_2.64
Cold tolerance (fearlygerm-9.2)	9	5	16	Burow et al. (2011)	Xsbarslbk_9.47/Xsbarslbk_9.53
	9	4	13		Xsbarslbk_9.53/Xsbarslbk_9.58

LG linkage group, *LOD* log of the odds, R^2 % contribution to phenotypic variance

Table 4.2 List of major genes mapped in sorghum

Gene	Trait	Linkage group
Tb1	Tillering	1
Sh1	Grain shattering	1
Y	Grain color	1
Ma_3	Maturity	1
Pericarp	Pericarp color	1
Rf_2	Fertility restoration	2
B2	Testa	2
Z	Mesocarp	2
Ma_5	Maturity	2
Pla	Downy mildew resistance	3
R	Pericarp color	3
AltSB	Aluminum tolerance	3
ms_3	Male sterility	3
A	Awn	3
bmr6	Brown midrib	4
PlcorInt	Plant color intensity	4
Opr	Resistance to organophosphate	5
dw_2	Plant height	6
Ma_1	Maturity	6
gc	Glume cover	6
d	Midrib	6
Rs1	Coleoptile color	6
Lg	Ligule	6
P	Plant color	6
Ymrco	Midrib color	6
bmr12	Brown midrib	7
I	Pericarp color	7
dw_3	Plant color	7
Pu	Rust resistance	8
Rf_1	Fertility restoration	8
Shs1	Head smut resistance	8
Sb.Ht9.1	Plant height	9
bm	Bloom	10
rlf	Virus reaction	10
wx	Endosperm	10
Rs2	Coleoptile and leaf axil	10
Ma_4	Maturity	10
Trit	Trichome morphology	10

Adapted from Mace and Jordan (2010)

of genetic purity of the hybrid seed or of the varietal seed lot is an important step in seed production and marketing chain. With increasing number of public and private-bred sorghum hybrids under commercial cultivation, quality control in terms of monitoring seed genetic purity at both parental and hybrid seed production stages is vital for the success of hybrid sorghum technology. The morphological characters have provided the signature of a variety or hybrid and its genetic purity. Later, biochemical markers like proteins and isozymes were used for this purpose since they provided for rapid analysis even at an early stage of crop growth. Conventionally, seed genetic purity assessment is done by conducting a grow out test (GOT) following breeder-defined morphological descriptors. This involves growing plants to maturity and evaluating several morphological characteristics to decide the purity of the seed lot. This, therefore, is time consuming as it takes one full crop season and also demands capital, labor, and space. Since the expression of morphological traits is influenced by environment, GOT assessment can be subjective. It is therefore necessary to adopt a method which is more precise; saves time, labor, capital, and other resources; and can act as an alternative to GOT. This should also help in marketing of the hybrid seed in the following crop season. For this, several workers have advocated the use of DNA markers in different crops such as rice (Nandakumar et al. 2004; Sundaram et al. 2008), maize (Meng et al. 2009), and cotton (Rana et al. 2007). Of the numerous DNA marker systems available at hand for seed purity test, simple sequence repeats (SSRs) have been preferred in many crops due to their ease in use, abundance, ubiquitous, codominance, higher polymorphism levels, amenable for high-throughput genotyping, etc. So far, in sorghum, the application of DNA markers for testing seed purity is very limited. Recently, randomly amplified polymorphic DNA (RAPD) markers were successfully used for rapid verification of hybridity and genetic purity of different hybrid seed lots, allowing the detection of true hybrids and verification of parentage of the hybrids and lines/cultivars in sorghum (Akhare et al. 2008; Tabbasam et al. 2006). A recent study not only proved the utility of SSRs in establishing genetic purity of commercial seed lots of three sorghum hybrids (CSH19, 13, and 14) but also suggested SSRs can safely be employed to replace the conventional GOT (Patil 2011).

4.8 MAS: Challenges

MAS is a great tool in the scheme of modern plant breeding. Compared to conventional phenotype-based selection strategies, MAS has several significant advantages. However, the successful application of MAS in plant breeding so far is limited mostly to simple traits with monogenic or oligogenic inheritance. Improvement of traits of complex inheritance through MAS is still a challenge. Non-availability of robust markers, non-validation of marker-trait associations, imprecise estimates of QTL locations and effects, epistasis, genetic background, g×e interactions, lack of cost-effective marker genotyping systems, lack of wet-lab facilities, knowledge gap in plant breeders, etc., are some of the reasons that can account for the low visible impact of MAS.

Over the past two decades, considerable progress has been made in the development of genomic resources in sorghum. Several DNA marker systems have been developed and are effectively used in the development of linkage maps. Significant efforts have also gone into the integration of various linkage maps and construction of a highly saturated consensus map. Several studies have been undertaken to identify QTL for many traits. Fine mapping of a few traits resulted in the identification of candidate genes involved in trait expression. Some of these QTL have been validated in different genetic backgrounds and are therefore ready for application through MAS in sorghum. Plant breeders need to employ these molecular tools for the improvement of sorghum and integrate MAS into the conventional breeding programs.

References

Abdelkhalik AF, Shishido R, Nomura K, Ikehashi H (2005) QTL-based analysis of leaf senescence in an indica/japonica hybrid in rice (*Oryza sativa* L.). Theor Appl Genet 110:1226–1235

Afanasenko OS, Makarova IG, Zubkovich AA (1999) The number of genes controlling resistance to *Pyrenophora teres* Drechs. strains in barley. Russ J Genet 35:274–283

Agrama H, Widle G, Reese J, Campbell L, Tuinstra M (2002) Genetic mapping of QTLs associated with greenbug resistance and tolerance in *Sorghum bicolor*. Theor Appl Genet 104:1373–1378

Akhare AA, Sakhare SB, Kulwal PL, Dhumale DB, Kharkar A (2008) RAPD profile studies in sorghum for identification of hybrids and their parents. Int J Integr Biol 3:18–24

Anahosur KH, Rao MVR (1977) A note on the epidemic of charcoal rot of sorghum in the regional research station, Dharwar. Sorghum Newsl 20:22

Aruna C, Bhagwat VR, Madhusudhana R, Vittal S, Hussain T, Ghorade RB, Khandalkar HG, Audilakshmi S, Seetharama N (2011) Identification and validation of genomic regions that affect shoot fly resistance in sorghum [*Sorghum bicolor* (L.) Moench]. Theor Appl Genet 122:1617–1630

Audilakshmi S, Stenhouse JW, Reddy TP (2005) Genetic analysis of grain mold resistance in white seed sorghum genotypes. Euphytica 145:95–101

Bandyopadhyay R, Mughogho LK (1988) Evaluation of field screening techniques for resistance to sorghum grain molds. Plant Dis 72:500–503

Bandyopadhyay R, Frederickson DE, McLaren NW, Odvody GN, Ryley MJ (1998) Ergot: a new disease threat to sorghum in the Americas and Australia. Plant Dis 82:356–367

Beil GM, Atkins RE (1967) Estimates of general and specific combining ability in F_1 hybrids for grain yield and its components in grain sorghum, *Sorghum vulgare* Pers. Crop Sci 7:225–228

Bergquist RR (1974) The determination of physiologic races of sorghum rust in Hawaii. Proc Am Phytopathol Soc 1:67–70

Bhat BV, Balakrishna D, Sathish K, Srinivas G, Seetharama N (2004) Molecular marker-aided approaches for sorghum improvement. AgBiotechNet 6:1–13

Blum A (1988) Drought resistance. Plant breeding for stress environments. CRC Press, Boca Raton, pp 43–76

Bohnert HJ, Nelson DE, Jensen RG (1995) Adaptations to environmental stresses. Plant Cell 7:1099–1111

Boora KS, Frederiksen RA, Magill CW (1999) A molecular marker that segregates with sorghum leaf blight resistance in one cross is maternally inherited in another. Mol Gen Genet 261:317–322

Borrell AK, Hammer GL (2000) Nitrogen dynamics and the physiological basis of stay-green in sorghum. Crop Sci 40:1295–1307

Borrell AK, Mullet JE, George-Jaeggli B, van Oosterom EJ, Hammer GL, Klein PE, Jordan DR (2014a) Drought adaptation of stay-green sorghum is associated with canopy development, leaf anatomy, root growth, and water uptake. J Exp Bot 65:6251–6263

Borrell AK, Oosterom EJ, Mullet JE, George-Jaeggli B, Jordan DR, Klein PE, Hammer GL (2014b) Stay-green alleles individually enhance grain yield in sorghum under drought by modifying canopy development and water uptake patterns. New Phytol 203:817–830

Bray EA (1997) Plant responses to water deficit. Trends Plant Sci 2:48–54

Brown PJ, Klein PE, Bortiri E, Acharya CB, Rooney WL, Kresovich S (2006) Inheritance of inflorescence architecture in sorghum. Theor Appl Genet 113:931–942

Brown PJ, Rooney WL, Franks C, Kresovich S (2008) Efficient mapping of plant height quantitative trait loci in a sorghum association population with introgressed dwarfing genes. Genetics 180:629–637

Burow G, Burke JJ, Xin Z, Franks CD (2011) Genetic dissection of early-season cold tolerance in sorghum [*Sorghum bicolor* (L.) Moench]. Mol Breed 28:391–402

Caddel JL, Weibel DE (1971) Effect of photoperiod and temperature on the development of sorghum. Agron J 63:799–803

Casela CR, Ferreira AS, Santos FG, Guimaraes FB (2008) Sorghum diseases in Brazil. In: Leslie JF (ed) Sorghum and millets diseases. Iowa State Press, Ames, p 379

Cha KW, Lee YJ, Koh HJ, Lee BM, Nam YW, Paek NC (2002) Isolation, characterization, and mapping of the stay green mutant in rice. Theor Appl Genet 104:526–532

Childs KL, Miller FR, Cordonnier-Pratt M-M, Pratt LH, Morgan PW, Mullet JE (1997) The sorghum photoperiod sensitivity gene, Ma_3, encodes a phytochrome B. Plant Physiol 113:611–619

Cisse ND, Ejeta G (2003) Genetic variation and relationships among seedling vigor traits in sorghum. Crop Sci 43:824–828

Coleman OH, Dean JL (1961) The inheritance of resistance to rust in sorgo. Crop Sci 1:152–154

Crasta OR, Xu WW, Rosenow DT, Mullet J, Nguyen HT (1999) Mapping of post-flowering drought resistance traits in grain sorghum: association between QTLs influencing premature senescence and maturity. Mol Gen Genet 262:579–588

Dabholkar AR, Sharma HC, Baghel SS (1980) Inheritance of rust resistance in *Sorghum bicolor* (L.) Moench. JNKVV Res J 14:12–14

Deeming JC (1972) A review of the taxonomy of African shoot flies of sorghum. Samaru Res Bull 236:24p

Dhillon MK, Sharma HC, Reddy BVS, Singh R, Naresh JS (2006) Inheritance of resistance to sorghum shoot fly. Crop Sci 46:1377–1383

Dodd JL (1980) The photosynthetic stress-translocation balance concept of sorghum stalk rots. In: Sorghum diseases, a world review. Proceedings of the international workshop on sorghum diseases, sponsored jointly by Texas A&M University (USA) and ICRISAT, Patancheru, pp 300–305

Downes RW, Marshall DR (1971) Low temperature induced male sterility in *Sorghum bicolor*. Aust J Exp Agric Anim Husb 11:352–356

Duncan RR (1984) The association of plant senescence with root and stalk diseases in sorghum. In: Mughogho LK, Rosenberg G (eds) Sorghum root and stalk rots, a critical review. ICRISAT, Patancheru, pp 99–110

Eathington SR (2005) Practical applications of molecular technology in the development of commercial maize hybrids. Proceedings of the 60th annual Corn and Sorghum Seed Research conferences. American Seed Trade Association, Washington, DC

Eddleman BR, Chang CC, McCarl BA (1999) Economic benefits from grain sorghum variety improvement in the United States. In: Wiseman BR, Webster JR (eds) Economic, environmental, and social benefits of resistance in field crops. Thomas Say Publications, Entomological Society of America, Lanham, pp 17–44

Edmunds LK, Zummo N (1975) Sorghum diseases in the United States and their control. Agriculture Handbook No. 468, Agricultural Research Service, United States Department of Agriculture in cooperation with Kansas and Mississippi Agricultural Experimental Stations, Washington, DC

Edmunds LK, Voigt RL, Carasso FM (1964) Use of Arizona climate to induce charcoal rot in grain sorghum. Plant Dis Reptr 48:300–302

Edwards D, Forster JW, Chagné D, Batley J (2007) What are SNPs? In: Association mapping in plants. Springer, New York, pp 41–52

Erpelding JE, Wang ML (2007) Response to anthracnose infection for a random selection of sorghum germplasm. Plant Pathol J 6:127–133

Esele JP, Frederiksen RA, Miller FR (1993) The association of genes controlling caryopsis traits with grain mold resistance in sorghum. Phytopathology 83:490–495

Feltus FA, Hart GE, Schertz KF, Casa AM, Kresovich S, Abraham S, Klein PE, Brown PJ, Paterson AH (2006) Alignment of genetic maps and QTLs between inter- and intra-specific sorghum populations. Theor Appl Genet 112:1295–1305

Folkertsma R, Sajjanar G, Reddy B, Sharma H, Hash CT (2003) Genetic mapping of QTL associated with sorghum shoot fly (*Atherigona soccata*) resistance in sorghum (*Sorghum bicolor*). Final abstracts guide, Plant & Animal Genome XI, San Diego, 11–15 Jan 2003, p 42

Franks CD, Burow GB, Burke JJ (2006) A comparison of US and Chinese sorghum germplasm for early season cold tolerance. Crop Sci 46:1371–1376

Frederiksen RA, Odvody GN (2000) Compendium of sorghum diseases, 2nd edn. American Phytopathological Society, St. Paul

Frederiksen R, Rosenow D (1986) Controlling sorghum diseases. In: Frederiksen RA (ed) Compendium of Sorghum diseases. American Phytopathological Society in cooperation with Dept. of Plant Pathology and Microbiology, Texas A&M University, St. Paul, pp 65–67

Frisch M, Bohn M, Melchinger AE (1999) Minimum sample size and optimal positioning of flanking markers in marker-assisted backcrossing for transfer of a target gene. Crop Sci 39:967–975

Fujino K, Sekiguchi H, Sato T, Kiuchi H, Nonoue Y, Takeuchi Y, Ando T, Lin S, Yano M (2004) Mapping of quantitative trait loci controlling low-temperature germinability in rice (*Oryza sativa* L.). Theor Appl Genet 108:794–799

Garud TB, Borikar ST (1985) Genetics of charcoal rot resistance in sorghum. Sorghum Newsl 28:87

Gebrekidan B, Kebede Y (1979) Highlights of sorghum improvement in Ethiopia. Sorghum Newsl 22:2–5

Glueck JA, Rooney LW (1980) Chemistry and structure of grain in relation to mold resistance. In: Williams RJ, Frederiksen RA, Mughogho LK (eds) Sorghum diseases, a world review. Proceedings of the international workshop on sorghum diseases, 11–15 Dec 1978. ICRISAT, Patancheru, pp 119–140

Gowda PSB, Frederiksen RA, Magill CW, Xu GW (1995) DNA markers for downy mildew resistance genes in sorghum. Genome 38:823–826

Grewal RPS (1988) Genetic basis of resistance to zonate leaf spot disease in forage sorghum. Theor Appl Genet 76:550–554

Halalli M, Gowda B, Kulkarni K, Goud J (1983) Evaluation of advanced generation progenies for resistance to shootfly in sorghum. Indian J Genet 43:291–293

Harris KM (1976) The sorghum midge. Annals Appl Biol 84:114–118

Harris HB, Burns RE (1973) Relationship between tannin content of sorghum grain and preharvest seed molding. Agron J 65:957–959

Harris K, Subudhi PK, Borrell A, Jordan D, Rosenow D, Nguyen H, Klein P, Klein R, Mullet J (2007) Sorghum stay-green QTL individually reduce post-flowering drought-induced leaf senescence. J Exp Bot 58:327–338

Hart G, Schertz K, Peng Y, Syed N (2001) Genetic mapping of *Sorghum bicolor* (L.) Moench QTLs that control variation in tillering and other morphological characters. Theor Appl Genet 103: 1232–1242

Harvey T, Seifers D, Kofoid K (1996) Effect of sorghum hybrid and imidacloprid seed treatment on infestations by corn leaf aphid and greenbug (Homoptera: Aphididae) and the spread of sugarcane mosaic virus strain MDMV-B. J Agric Entomol 13:9–15

Haussmann B, Mahalakshmi V, Reddy B, Seetharama N, Hash CT, Geiger H (2002) QTL mapping of stay-green in two sorghum recombinant inbred populations. Theor Appl Genet 106:133–142

Henzell R, Franzmann B, Brengman R (1994) Sorghum midge resistance research in Australia. Sorghum Improvement Conference of North America, USA; University of Georgia, USA; International Crops Research Institute for the Semi-Arid Tropics, Patacheru

Henzell R, Hare B, Jordan D, Fletcher D, McCosker A, Bunker G, Persley D (2001) Sorghum breeding in Australia: public and private endeavours. Proceedings of the 4th Australian sorghum conference, Kooralbyn

Hepperly PR (1990) Sorghum rust: II. Control and losses. J Agric Univ Puerto Rico 74:37–44

Hesketh J, Chase S, Nanda D (1969) Environmental and genetic modification of leaf number in maize, sorghum, and Hungarian millet. Crop Sci 9:460–463

Holland JB (2004) Implementation of molecular markers for quantitative traits in breeding programs-challenges and opportunities. New directions for a diverse planet. Proceedings of the 4th international crop science congress. The Regional Institute Ltd, Gosford

Hospital F (2003) Marker-assisted breeding. In: Newbury HJ (ed) Plant molecular breeding. Blackwell Publishing/CRC Press, Oxford/Boca Raton, pp 30–59

Hulluka M, Esele J (1992) Sorghum diseases in eastern Africa. In: de-Milliano WJA, Frederiksen RA, Bergston GD (eds) Sorghum and millets diseases: a second world review. ICRISAT, Patancheru, p 21

Ibrahim OE, Nyquist W, Axtell J (1985) Quantitative inheritance and correlations of agronomic and grain quality traits of sorghum. Crop Sci 25:649–654

ICRISAT (1992) Annual Report 1991. International Crop Research Institute for Semi-arid Tropics, Patancheru

Indira S, Rana B, Rao N (1982) Further studies on the incidence and genetics of rust resistance in sorghum. Indian J Genet 42:106–113

Indira S, Rana B, Rao N (1983) Genetics of host plant resistance to charcoal rot in sorghum. Indian J Genet 43:472–477

Isakeit T, Odvody G, Shelby R (1998) First report of sorghum ergot caused by *Claviceps africana* in the United States. Plant Dis 82:592–592

Jiang GL (2013) Molecular markers and marker-assisted breeding in plants. In: Anderson SB (ed) Plant breeding from laboratories to fields. InTech, Croatia, pp 45–83

Jiang GH, He YQ, Xu CG, Li XH, Zhang Q (2004) The genetic basis of stay-green in rice analyzed in a population of doubled haploid lines derived from an *indica* by *japonica* cross. Theor Appl Genet 108:688–698

Jordan D, Hunt C, Cruickshank A, Borrell A, Henzell R (2012) The relationship between the stay-green trait and grain yield in elite sorghum hybrids grown in a range of environments. Crop Sci 52:1153–1161

Jotwani M, Marwaha K, Srivastava K, Young W (1970) Seasonal incidence of shoot fly (*Atherigona varia soccata* Rond) in jowar hybrids at Delhi. Indian J Entomol 32:7–15

Kamatar M, Hiremath R, Palakshappa M (2000) Development of charcoal rot resistant rabi sorghum variety. Karnataka J Agric Sci 13:376–379

Kassahun B, Bidinger F, Hash C, Kuruvinashetti M (2010) Stay-green expression in early generation sorghum [*Sorghum bicolor* (L.) Moench] QTL introgression lines. Euphytica 172:351–362

Katsar CS, Paterson AH, Teetes GL, Peterson GC (2002) Molecular analysis of sorghum resistance to the greenbug (Homoptera: Aphididae). J Econ Entomol 95:448–457

Kebede H, Subudhi P, Rosenow D, Nguyen H (2001) Quantitative trait loci influencing drought tolerance in grain sorghum (*Sorghum bicolor* L. Moench). Theor Appl Genet 103:266–276

Klein R, Rodriguez-Herrera R, Schlueter J, Klein P, Yu Z, Rooney W (2001) Identification of genomic regions that affect grain-mould incidence and other traits of agronomic importance in sorghum. Theor Appl Genet 102:307–319

Klein RR, Mullet JE, Jordan DR, Miller FR, Rooney WL, Menz MA, Franks CD, Klein PE (2008) The effect of tropical sorghum conversion and inbred development on genome diversity as revealed by high-resolution genotyping. Crop Sci 48(Suppl):S12–S26

Knoll J, Ejeta G (2008) Marker-assisted selection for early-season cold tolerance in sorghum: QTL validation across populations and environments. Theor Appl Genet 116:541–553

Knoll J, Gunaratna N, Ejeta G (2008) QTL analysis of early-season cold tolerance in sorghum. Theor Appl Genet 116:577–587

Li J, Thomson M, McCouch SR (2004) Fine mapping of a grain-weight quantitative trait locus in the pericentromeric region of rice chromosome 3. Genetics 168:2187–2195

Lin YR, Schertz KF, Paterson AH (1995) Comparative analysis of QTLs affecting plant height and maturity across the Poaceae, in reference to an interspecific sorghum population. Genetics 141:391–411

Little CR, Perumal R, Tesso T, Prom LK, Odvody GN, Magill CW (2012) Sorghum pathology and biotechnology – a fungal disease perspective: part I. Grain mold, head smut, and ergot. Eur J Plant Sci Biotechnol 6:10–30

Luquez VM, Guiamét JJ (2001) Effects of the 'stay green' genotype GGd1d1d2d2 on leaf gas exchange, dry matter accumulation and seed yield in soybean (*Glycine max* L. Merr.). Annals Bot 87:313–318

Mace ES, Jordan DR (2010) Location of major effect genes in sorghum (*Sorghum bicolor* (L.) Moench). Theor Appl Genet 121:1339–1356

Mace ES, Jordan DR (2011) Integrating sorghum whole genome sequence information with a compendium of sorghum QTL studies reveals uneven distribution of QTL and of gene-rich regions with significant implications for crop improvement. Theor Appl Genet 123:169–191

Mace ES, Singh V, Van Oosterom EJ, Hammer GL, Hunt CH, Jordan DR (2012) QTL for nodal root angle in sorghum (*Sorghum bicolor* L. Moench) co-locate with QTL for traits associated with drought adaptation. Theor Appl Genet 124:97–109

Mao C, Virmani S, Kumar I (1998) Technological innovations to lower the cost of hybrid rice seed production. In: Virmani SS, Siddiq EA, Muralidharan K (eds) Advances in hybrid rice technology. IRRI, Manila, pp 111–128

Maranville JW (1974) What's new in sorghum physiology? In: Wilkinson D (ed) Proceedings of the 29th annual corn and sorghum research conference. American Seed Trade Association, Washington, DC, pp 22–28

Mathur K, Bunker R (2002) Leaf blight of sorghum caused by *Drechslera australiensis* – a new report from India. Int Sorghum Millets Newsl 43:60

McBee G (1984) Relation of senescence, and kernel maturity to carbohydrate metabolism in sorghum. In: Mughogho LK (ed) Sorghum root and stalk diseases, a critical review. Proceedings of the consultative group discussion of research needs and strategies for control of sorghum root and stalk diseases, Bellagio, 27 Nov–2 Dec 1983. ICRISAT, Patancheru, pp 119–129

McIntyre CL, Hermann SM, Casu RE et al (2004) Homologues of the maize rust resistance gene *Rp1-D* are genetically associated with a major rust resistance QTL in sorghum. Theor Appl Genet 109:875–883

McIntyre CL, Casu RE, Drenth J, Knight D, Whan VA, Croft BJ, Jordan DR, Manners JM (2005) Resistance gene analogues in sugarcane and sorghum and their association with quantitative trait loci for rust resistance. Genome 48:391–400

Meng QC, Chen YP, Yang XH, Zhao W, Zhang YB, Yuan JH (2009) Seed purity test for maize hybrid Suyu 20 using SSR markers. Jiangsu J Agric Sci 25:508–512

Menkir A, Ejeta G, Butler L, Melakeberhan A (1996) Physical and chemical kernel properties associated with resistance to grain mold in sorghum. Cereal Chem 73:613–617

Miller FR, Cruzado HJ (1969) Allelic interactions at the *Pu* locus in *Sorghum bicolor* (L.) Moench. Crop Sci 9:336–338

Morris GP, Ramu P, Deshpande SP et al (2013) Population genomic and genome-wide association studies of agroclimatic traits in sorghum. Proc Natl Acad Sci U S A 110:453–458

Mughogho LK (1984) Sorghum root and stalk rots. A critical review. Proceedings of the consultative group discussion of research needs and strategies for control of sorghum root and stalk diseases, Bellagio, 27 Nov–2 Dec 1983. ICRISAT, Patancheru

Multani DS, Briggs SP, Chamberlin MA, Blakeslee JJ, Murphy AS, Johal GS (2003) Loss of an MDR transporter in compact stalks of maize *br2* and sorghum *dw3* mutants. Science 302:81–84

Murali Mohan S, Madhusudhana R, Mathur K, Howarth CJ, Srinivas G, Satish K, Reddy RN, Seetharama N (2009) Co-localization of quantitative trait loci for foliar disease resistance in sorghum. Plant Breed 128:532–535

Murali Mohan S, Madhusudhana R, Mathur K, Chakravarthi DVN, Rathore S, Nagaraja Reddy R, Satish K, Srinivas G, Sarada Mani N, Seetharama N (2010) Identification of quantitative trait loci associated with resistance to foliar diseases in sorghum [*Sorghum bicolor* (L.) Moench]. Euphytica 176:199–211

Murphy RL, Klein RR, Morishige DT, Brady JA, Rooney WL, Miller FR, Dugas DV, Klein PE, Mullet JE (2011) Coincident light and clock regulation of pseudoresponse regulator protein 37 (PRR37) controls photoperiodic flowering in sorghum. Proc Natl Acad Sci U S A 108:16469–16474

Murphy RL, Daryl TM, Brady JA, Rooney WL, Shanshan Y, Klein PE, Mullet JE (2014) *Ghd7* (*Ma*$_6$) represses sorghum flowering in long days: *Ghd7* alleles enhance biomass accumulation and grain production. Plant Gen. doi:10.3835/plantgenome2013.11.0040

Murray SC, Sharma A, Rooney WL, Klein PE, Mullet JE, Mitchell SE, Kresovich S (2008) Genetic improvement of sorghum as a biofuel feedstock: I. QTL for stem sugar and grain non-structural carbohydrates. Crop Sci 48:2165–2179

Nagaraj N, Reese JC, Tuinstra MR, Smith CM, Amand PS, Kirkham MB, Kofoid KD, Campbell LR, Wilde GE (2005) Molecular mapping of sorghum genes expressing tolerance to damage by greenbug (Homoptera: Aphididae). J Econ Entomol 98:595–602

Nagaraja Reddy R, Madhusudhana R, Murali Mohan S, Chakravarthi DVN, Mehtre SP, Seetharama N, Patil JV (2013) Mapping QTL for grain yield and other agronomic traits in post-rainy sorghum [*Sorghum bicolor* (L.) Moench]. Theor Appl Genet 126:1921–1939

Nagaraja Reddy RR, Madhusudhana R, Murali Mohan S, Seetharama N, Patil JV (2014) Detection and validation of stay-green QTL in post-rainy sorghum involving widely adapted cultivar, M35-1 and a popular stay-green genotype B35. BMC Genom 15(1):909

Nandakumar N, Singh AK, Sharma RK, Mohapatra T, Prabhu KV, Zaman FU (2004) Molecular fingerprinting of hybrids and assessment of genetic purity of hybrid seeds in rice using microsatellite markers. Euphytica 136:257–264

Nguyen HT, Chandra Babu R, Blum A (1997) Breeding for drought resistance in rice: physiology and molecular genetics considerations. Crop Sci 37:1426–1434

Nwanze K (1997) Screening for resistance to sorghum shoot fly. In: Sharma HC, Singh F, Nwaze KF (eds) Plant resistance to insect in sorghum. ICRISAT, Patancheru, pp 35–37

Oh BJ, Frederiksen RA, Magill CW (1994) Identification of molecular markers linked to head smut resistance gene (*Shs*) in sorghum by RFLP and RAPD analyses. Phytopathology 84:830–833

Padmaja PG, Madhusudhana R, Seetharama N (2010a) Sorghum shoot fly. Directorate of Sorghum Research, Rajendranagar. ISBN 81-89335-29-4

Padmaja PG, Woodcock CM, Bruce TJ (2010b) Electrophysiological and behavioral responses of sorghum shoot fly, *Atherigona soccata*, to sorghum volatiles. J Chem Ecol 36:1346–1353

Painter RH (1951) Insect resistance in crop plants. Soil Sci 72:481

Parh DK, Jordan DR, Aitken EA, Mace ES, Jun-Ai P, McIntyre CL, Godwin ID (2008) QTL analysis of ergot resistance in sorghum. Theor Appl Genet 117:369–382

Park SJ, Huang Y, Ayoubi P (2006) Identification of expression profiles of sorghum genes in response to greenbug phloem-feeding using cDNA subtraction and microarray analysis. Planta 223:932–947

Paterson AH, Lander ES, Hewitt JD, Peterson S, Lincoln SE, Tanksley SD (1988) Resolution of quantitative traits into Mendelian factors by using a complete linkage map of restriction fragment length polymorphisms. Nature 335:721–726

Patil S (2011) Studies on genetic diversity assessment in sorghum (*Sorghum bicolor* L. Moench) genotypes and hybrid purity test employing SSR markers. MSc thesis submitted to ANGRAU, Rajendranagar

Patil A, Fakrudin B, Salimath PM, Rajkumar (2012) Genome-wide molecular mapping and QTL analysis, validated across locations and years for charcoal rot disease incidence traits in *Sorghum bicolor* (L.) Moench. Indian J Genet 72:296–302

Patil-Kulkarni BG, Puttarudrappa A, Kajjari NB, Goud JV (1972) Breeding for rust resistance in sorghum. Indian Phytopathol 25:166–168

Pažoutová S, Olšovská J, Linka M, Kolínská R, Flieger M (2000) Chemoraces and habitat specialization of *Claviceps purpurea* populations. Appl Environ Microbiol 66:5419–5425

Pedgaonkar SM, Mayee CD (1990) Stalk water potential in relation to charcoal rot of sorghum. Indian Phytopathology 43:192–196

Peleman JD, van der Voort JR (2003) Breeding by design. Trends Plant Sci 8:330–334

Pereira MG, Lee M (1995) Identification of genomic regions affecting plant height in sorghum and maize. Theor Appl Genet 90:380–388

Pereira MG, Ahnert D, Lee M, Klier K (1995) Genetic mapping of quantitative trait loci for panicle characteristics and seed weight in sorghum. Rev Bras Genet 18:249–257

Qingshan L, Dahlberg JA (2001) Chinese sorghum genetic resources. Econ Bot 55:401–425

Quinby JR (1967) The maturity genes of sorghum. Adv Agron 19:267–305

Quinby JR (1973) The genetic control of flowering and growth in sorghum. Adv Agron 25:125–162

Quinby JR, Karper RE (1954) Inheritance of height in sorghum. Agron J 46:211–216

Ramakrishna W, Emberton J, SanMiguel P, Ogden M, Llaca V, Messing J, Bennetzen JL (2002) Comparative sequence analysis of the sorghum Rph region and the maize *Rp1* resistance gene complex. Plant Physiol 130:1728–1738

Ramasamy P, Menz MA, Mehta PJ et al (2009) Molecular mapping of *Cg1*, a gene for resistance to anthracnose (*Colletotrichum sublineolum*) in sorghum. Euphytica 165:597–606

Rami JF, Dufour P, Trouche G, Fliedel G, Mestres C, Davrieux F, Blanchard P, Hamon P (1998) Quantitative trait loci for grain quality, productivity, morphological and agronomical traits in sorghum (*Sorghum bicolor* L. Moench). Theor Appl Genet 97:605–616

Rana BS, Tripathi DP, Rao NGP (1976) Genetic analysis of some exotic × Indian crosses in sorghum. XV. Inheritance of resistance to sorghum rust. Indian J Genet 36:244–249

Rana BS, Anahosur KH, Rao VJM, Parameshwarappa R, Rao NGP (1982) Inheritance of field resistance to sorghum charcoal rot and selection for multiple disease resistance. Indian J Genet 42:302–310

Rana B, Rao M, Indira S, Singh B et al (1999) Technology for increasing sorghum production and value addition. Director and Project Coordinator (AICSIP) National Research Centre for Sorghum, Hyderabad

Rana MK, Singh S, Bhat KV (2007) RAPD, STMS and ISSR markers for genetic diversity and hybrid seed purity testing in cotton. Seed Sci Technol 35:709–721

Rao KN, Reddy VS, Williams RJ, House LR (1980) The ICRISAT charcoal rot resistance program. Proceedings of the international workshop on Sorghum diseases, Hyderabad, 11–15 Dec 1978, pp 315–321

Rao GMR, Patil SJ, Anahosur KH (1993) Genetics of charcoal rot resistance in Rabi sorghum. Karnataka J Agric Sci 6:113–116

Reddy PS, Fakrudin B, Rajkumar, Punnuri SM, Arun SS, Kuruvinashetti MS, Das IK, Seetharama N (2008) Molecular mapping of genomic regions harboring QTLs for stalk rot resistance in sorghum. Euphytica 159:191–198

Ribaut JM, Vicente MC, Delannay X (2010) Molecular breeding in developing countries: challenges and perspectives. Curr Opin Plant Biol 13:1–6

Ritter KB, Jordan DR, Chapman SC, Godwin ID, Mace ES, McIntyre CL (2008) Identification of QTL for sugar-related traits in a sweet×grain sorghum (*Sorghum bicolor* L. Moench) recombinant inbred population. Mol Breed 22:367–384

Rodriguez-Herrera R, Rooney WL, Rosenow DT, Frederiksen RA (2000) Inheritance of grain mold resistance in grain sorghum without a pigmented testa. Crop Sci 40:1573–1578

Rooney WL, Aydin S (1999) Genetic control of a photoperiod-sensitive response in *Sorghum bicolor* (L.) Moench. Crop Sci 39:397–400

Rosenow DT, Frederiksen RA (1982) Breeding for disease resistance in sorghum. In: Sorghum in the eighties. Proceedings of international sorghum workshop, Hyderabad, 2–7 Nov 1981. Oxford & IBH, New Delhi, pp 447–455

Rosenow DT, Quisenberry JE, Wendt CW, Clark LE (1983) Drought tolerant sorghum and cotton germplasm. Agric Water Manag 7:207–222

Rosenow DT, Woodfin CA, Clark LE (1988) Breeding for the stay green trait in sorghum. Agronomy abstracts ASA, Madison

Rosenow DT, Ejeta G, Clark LE, Gilbert M, Henzell R, Borrell A, Muchow R (1996) Breeding for pre-and post-flowering drought stress resistance in sorghum. In: Rosenow DT, Yohe JM (eds) Proceedings of the international conference on genetic improvement of sorghum and pearl millet. INTSORMIL, Lubbock, pp 400–411

Ryley M, Blaney B, Meinke H (1999) Current and future research on sorghum ergot in Australia. Proceedings of the global conference on ergot of sorghum. EMBRAPA-INTSORMIL, Sete Lagoas, pp 59–67

Satish K, Srinivas G, Madhusudhana R, Padmaja PG, Reddy RN, Murali Mohan S, Seetharama N (2009) Identification of quantitative trait loci for resistance to shoot fly in sorghum [*Sorghum bicolor* (L.) Moench]. Theor Appl Genet 119:1425–1439

Satish K, Gutema Z, Grenier C, Rich PJ, Ejeta G (2012a) Molecular tagging and validation of microsatellite markers linked to the low germination stimulant gene (*lgs*) for Striga resistance in sorghum [*Sorghum bicolor* (L.) Moench]. Theor Appl Genet 124:989–1003

Satish K, Madhusudhana R, Padmaja PG, Seetharama N, Patil JV (2012b) Development, genetic mapping of candidate gene-based markers and their significant association with the shoot fly resistance quantitative trait loci in sorghum [*Sorghum bicolor* (L.) Moench]. Mol Breed 30:1573–1591

Sax K (1923) The association of size differences with seed-coat pattern and pigmentation in *Phaseolus vulgaris*. Genetics 8:552–560

Seetharama N, Bidinger FR, Rao KN, Gill KS, Mulgund M (1987) Effect of pattern and severity of moisture deficit stress on stalk rot incidence in Sorghum I. Use of line source irrigation technique, and the effect of time of inoculation. Field Crops Res 15:289–308

Sharma H (1985) Screening for sorghum midge [*Contarinia sorghicola*] resistance and resistance mechanisms. In: Proceedings of international sorghum entomology workshop, College Station, 15–21 Jul 1984. ICRISAT, Patancheru, pp 275–292

Sharma HC, Reddy BVS, Stenhouse JW, Nwanze KF (1994) Host plant resistance to sorghum midge, *Contarinia sorghicola*. Int Sorghum & Millets Newsl 35:30–41

Sharma HC, Abraham CV, Vidyasagar P, Stenhouse JW (1996) Gene action for resistance in sorghum to midge, *Contarinia sorghicola*. Crop Sci 36:259–265

Shinozaki K, Yamaguchi-Shinozaki K (1996) Molecular responses to drought and cold stress. Curr Opin Biotechnol 7:161–167

Shinozaki K, Yamaguchi-Shinozaki K (2000) Molecular responses to dehydration and low temperature: differences and cross-talk between two stress signaling pathways. Curr Opin Plant Biol 3:217–223

Singh S (1977) Breeding grain sorghum for cold tolerance. International Sorghum Workshop held at ICRISAT during, Patancheru, pp 6–12

Singh SP (1985) Sources of cold tolerance in grain sorghum. Can J Plant Sci 65:251–257

Singh S, Bandyopadhyay R (2000) Panicle and seed diseases. In: Frederiksen RA, Odvody GN (eds) Compendium of sorghum diseases, 2nd edn. American Phytopathological Society, St Paul, pp 38–40

Singh M, Chaudhary K, Singal HR, Magill CW, Boora KS (2006) Identification and characterization of RAPD and SCAR markers linked to anthracnose resistance gene in sorghum [*Sorghum bicolor* (L.) Moench]. Euphytica 149:179–187

Smith CM, Khan ZR, Pathak MD (1994) Techniques for evaluating insect resistance in crop plants. CRC Press, Boca Raton

Somani RB, Indira S (1999) Effect of grain molds on grain weight in sorghum. J Myco Plant Pathol 29:22–24

Soto PE (1974) Ovipositional preference and antibiosis in relation to resistance to a sorghum shoot fly. J Econ Entomol 67:265–267

Srinivas G, Satish K, Madhusudhana R, Reddy RN, Murali Mohan S, Seetharama N (2009) Identification of quantitative trait loci for agronomically important traits and their association with genic-microsatellite markers in sorghum. Theor Appl Genet 118:1439–1454

Subudhi PK, Rosenow DT, Nguyen HT (2000) Quantitative trait loci for the stay green trait in sorghum (*Sorghum bicolor* L. Moench): consistency across genetic backgrounds and environments. Theor Appl Genet 101:733–741

Sukhani TR, Jotwani MG (1980) Efficacy of some newer systemic insecticides for the control of sorghum shootfly, *Atherigona soccat*a Rondani. Indian J Entomol 42:76–81

Sundaram RM, Naveenkumar B, Biradar SK, Balachandran SM, Mishra B, IlyasAhmed M, Viraktamath BC, Ramesha MS, Sarma NP (2008) Identification of informative SSR markers capable of distinguishing hybrid rice parental lines and their utilization in seed purity assessment. Euphytica 163:215–224

Tabbasam N, Mehboob-Ur-Rahman, Zafar Y (2006) DNA-based genotyping of sorghum hybrids. Pak J Bot 38:1599–1604

Tanksley SD, Nelson JC (1996) Advanced backcross QTL analysis: a method for the simultaneous discovery and transfer of valuable QTLs from unadapted germplasm into elite breeding lines. Theor Appl Genet 92:191–203

Tao YZ, Jordan DR, Henzell RG, McIntyre CL (1998) Identification of genomic regions for rust resistance in sorghum. Euphytica 103:287–292

Tao YZ, Henzell RG, Jordan DR, Butler DG, Kelly AM, McIntyre CL (2000) Identification of genomic regions associated with stay green in sorghum by testing RILs in multiple environments. Theor Appl Genet 100:1225–1232

Tao YZ, Hardy A, Drenth J, Henzell RG, Franzmann BA, Jordan DR, Butler DG, McIntyre CL (2003) Identifications of two different mechanisms for sorghum midge resistance through QTL mapping. Theor Appl Genet 107:116–122

Tarumoto I (2005) Glossiness of leaf blades in sorghum (*Sorghum bicolor* L. Moench); its visual and ultrastructural studies. Jap Agric Res Quart 39:153–160

Thakur RP, Reddy BVS, Mathur K (2007) Screening techniques for sorghum diseases, Information: Bulletin No. 76. ICRISAT, Patancheru. ISBN 978-92-9066-504-5

Tooley PW, O'Neill NR, Goley ED, Carras MM (2000) Assessment of diversity in Claviceps africana and other *Claviceps* species by RAM and AFLP analyses. Phytopathology 90:1126–1130

Torres-Montalvo H, Mendoza-Onofre LE, Gonzalez-Hernandez VA, Williams-Alanis H (1992) Reaction of tan and non-tan isogenic genotypes to head blight. Sorghum Newsl 33:36

Trimboli DS, Burgess LW (1982) The fungi associated with stalk and root rot of grain sorghum in New South Wales [Australia]. Sorghum Newsl 25

Tuinstra MR, Grote EM, Goldsbrough PB, Ejeta G (1997) Genetic analysis of post-flowering drought tolerance and components of grain development in *Sorghum bicolor* (L.) Moench. Mol Breed 3:439–448

Tuinstra MR, Ejeta G, Goldsbrough P (1998) Evaluation of near-isogenic sorghum lines contrasting for QTL markers associated with drought tolerance. Crop Sci 38:835–842

Upadhyaya HD, Wang YH, Sharma R, Sharma S (2013a) Identification of genetic markers linked to anthracnose resistance in sorghum using association analysis. Theor Appl Genet 126:1649–1657

Upadhyaya HD, Wang YH, Sharma R, Sharma S (2013b) SNP markers linked to leaf rust and grain mold resistance in sorghum. Mol Breed 32:451–462

Vadez V, Deshpande SP, Kholova J, Hammer GL, Borrell AK, Talwar HS, Hash CT (2011) Stay-green quantitative trait loci's effects on water extraction, transpiration efficiency and seed yield depend on recipient parent background. Funct Plant Biol 38:553–566

Van Oosterom EJ, Jayachandran R, Bidinger FR (1996) Diallel analysis of the stay-green trait and its components in sorghum. Crop Sci 36:549–555

Wang YH, Bible P, Loganantharaj R, Upadhyaya HD (2012) Identification of SSR markers associated with height using pool-based genome-wide association mapping in sorghum. Mol Breed 30:281–292

Waniska RD, Poe JH, Bandyopadhyay R (1989) Effects of growth conditions on grain molding and phenols in sorghum caryopsis. J Cereal Sci 10:217–225

Willey RW, Heath SB (1969) The quantitative relationships between plant population and crop yield. Adv Agron 21:281–321

Wu Y, Huang Y (2008) Molecular mapping of QTLs for resistance to the greenbug *Schizaphis graminum* (Rondani) in *Sorghum bicolor* (Moench). Theor Appl Genet 117:117–124

Xu W, Subudhi PK, Crasta OR, Rosenow DT, Mullet JE, Nguyen HT (2000) Molecular mapping of QTLs conferring stay-green in grain sorghum (*Sorghum bicolor* L. Moench). Genome 43:461–469

Yoshimura S, Yoshimura A, Iwata N, McCouch SR, Abenes ML, Baraoidan MR, Mew TW, Nelson RJ (1995) Tagging and combining bacterial blight resistance genes in rice using RAPD and RFLP markers. Mol Breed 1:375–387

Zheng HJ, Wu AZ, Zheng CC, Wang YF, Cai R, Shen XF, Xu RR, Liu P, Kong LJ, Dong ST (2009) QTL mapping of maize (*Zea mays*) stay-green traits and their relationship to yield. Plant Breed 128:54–62

Zhu-Salzman K, Salzman RA, Ahn JE, Koiwa H (2004) Transcriptional regulation of sorghum defense determinants against a phloem-feeding aphid. Plant Physiol 134:420–431

Heterosis Prediction Using DNA Markers

5

P. Rajendrakumar

Contents

5.1 **Introduction** 102

5.2 **Classical and Molecular Hypotheses on Heterosis** 102

5.3 **Methods for Prediction of Heterosis** 103

5.3.1 Heterosis Prediction Using Agro-morphological Data 103

5.3.2 Heterosis Prediction Using Molecular Marker Data 105

5.4 **Recent Trends in Heterosis Prediction** 107

5.5 **Factors Affecting the Effectiveness of Heterosis Prediction** 109

5.6 **Impact of Heterosis Prediction on the Development of Hybrids** 110

5.7 **Future Prospects** 110

References 110

Abstract

Hybrids are commercially successful in many crops, including sorghum. Development of hybrids through the exploitation of heterosis involves evaluation of hundreds of test crosses in the field, making it input and resource intensive. Therefore, plant breeders are interested in methods that can forecast the potential parental combinations so that only limited test crosses can be evaluated for heterosis. The availability of genomic tools such as DNA markers and gene expression platforms has encouraged research groups globally to work toward the prediction of heterosis. Once DNA markers for the prediction of heterosis are identified, potential parental combinations can be predicted by DNA marker-based analysis, thereby increasing the speed of hybrid development without large-scale field evaluation. Whole transcriptome and metabolome analyses will help in dissecting the genes or metabolic networks involved in heterosis. Advanced genomic tools along with mathematical modeling offer excellent opportunities for the development of simple and reliable methods for the prediction of heterosis. This chapter critically discusses the different methods of heterosis prediction, recent trends, factors affecting heterosis prediction and the impact of prediction in the development of hybrids.

P. Rajendrakumar (✉)
Biotechnology, ICAR-Indian Institute of Millets Research, Rajendranagar, Hyderabad, Telangana 500 030, India
e-mail: rajendra@millets.res.in

R. Madhusudhana et al. (eds.), *Sorghum Molecular Breeding*, DOI 10.1007/978-81-322-2422-8_5

Keywords
Heterosis • Prediction of heterosis • Molecular markers • Prediction methods

5.1 Introduction

Exploitation of hybrid vigor (heterosis) has contributed significantly for increased productivity of food crops in the last several decades. Charles Darwin was the first to describe heterosis in crop plants where he observed that cross-pollinated maize progeny was 25 % taller than inbred progeny (Darwin 1876). Since its rediscovery (Shull 1908; East 1908), the area planted to hybrids in many crops has steadily increased, especially in those crops where cytoplasmic male sterility (CMS) system is available. Exploitation of heterosis in sorghum has resulted in the development and release of first commercial hybrids, CSH 1 and CSH 2, as early as 1965 in India. Till now, more than 30 hybrids have been released centrally for all India cultivation. Sorghum hybrids are grown predominantly in rainy season though very few hybrids were released for post-rainy season. Though the phenomenon of heterosis is being exploited globally in several crop species successfully, its genetic and molecular basis remains unexplained. The advent of molecular markers has helped to dissect the complex traits like yield into its genetic components, enabling a better understanding of the phenomenon of heterosis. As a first step in understanding the molecular basis of heterosis, several studies have been undertaken to dissect and predict heterosis using molecular markers. This will help in the identification of markers and genomic regions associated with heterosis expression resulting in its better understanding at the molecular level. Increasing area under hybrids in many important crops necessitates the prediction of heterosis using phenotypic as well as molecular data, which in turn will help in the identification of superior hybrids in an efficient way with limited resources and inputs by the evaluation of limited experimental hybrids developed from predicted parental combinations. This chapter gives a glimpse of the progress made during the last two decades in the prediction of heterosis using various molecular tools with critical analysis and its future prospects.

5.2 Classical and Molecular Hypotheses on Heterosis

Heterosis refers to the superior performance of heterozygous F_1 hybrids in terms of increased biomass, size, yield, vigor, fertility, and resistance to biotic and abiotic stresses compared to the average of their homozygous inbred parents (Shull 1952; Falconer and Mackay 1996). The genetic basis of heterosis has been discussed by several researchers for almost a century, but little consensus has emerged. Two classical hypotheses, namely, dominance and over-dominance, have been proposed to explain the genetic basis of heterosis. The dominance hypothesis attributes heterosis in the hybrid to the accumulation of superior/dominant alleles and masking effects of deleterious recessive alleles at multiple loci (Davenport 1908; Bruce 1910; Keeble and Pellow 1910). Over-dominance hypothesis suggests that the heterozygous combination of alleles at one or multiple loci in a hybrid is superior to either of the homozygous combinations in the parental inbreds (Shull 1908; East 1908). Nevertheless, these two hypotheses may not be adequate to explain the molecular basis of heterosis since they are based on single-locus theory (Birchler et al. 2003). Moreover, most of the quantitative traits are governed by multiple loci, and as such, each gene replacement may have effects on several characters due to interaction between genes (Wright 1968). This led to the proposal of epistasis as a third hypothesis which suggests that heterosis results from interactions of superior alleles at different loci from two parents, and they may show additive, dominance, or over-dominance effects (Yu et al. 1997; Li et al. 2001).

Two models have been put forth to explain heterosis at the molecular level (Birchler et al. 2003). According to the first model, heterosis is due to the combined allelic expression of various genes that are brought together in the hybrid. The second model explains heterosis as an interaction of different alleles in the hybrid that causes gene expression leading to deviation from the mid-parent values. This condition might be due to the result of allelic interaction between regulatory genes. In addition to these two models, modified gene regulation resulting from differences in the expression of specific transcription factors (Osborn et al. 2003), altered regulation (Song and Messing 2003), and diverse transcriptional and post-transcriptional regulatory processes as well as epigenetic changes such as DNA methylation, histone acetylation, and chromatin remodulation (Yao et al. 2005) were suggested to be involved in the heterosis-associated gene expression. With the advancements in molecular biology, it is possible to study the phenomenon of heterosis in a more refined way.

5.3 Methods for Prediction of Heterosis

5.3.1 Heterosis Prediction Using Agro-morphological Data

Development of a superior hybrid involves synthesis and field evaluation of hundreds of test cross hybrids for their combining ability and yield heterosis. This process is time consuming and resource intensive. Moreover, the limitations in the land availability lead to the evaluation of a limited number of test crosses, and the results are subjective. Therefore, plant breeders have been exploring other methods for the selection of potential parental combinations without field evaluation that are expected to show high levels of heterosis. Methods employed to predict heterosis can be grouped into (1) *per se* performance of parental lines, (2) combining ability, (3) genetic diversity, and (4) mitochondrial complementation. Even though different methods are available for the prediction of heterosis, the crop breeders generally predict heterosis based on the level of genetic diversity of the parental lines. Conventionally, genetic diversity among the parental lines has been determined through multivariate analysis using phenotypic data. Later, biochemical markers were used to estimate the genetic diversity among genotypes. However, due to their inherent limitations, they were not popular. This helped DNA markers to gain popularity, and several studies in different crops employed them in the estimation of genetic diversity in relation to heterosis. The present chapter discusses mainly on the prediction of heterosis based on genetic diversity as estimated from phenotypic, pedigree, and molecular marker data. In addition, gene expression data, such as transcript abundance and expression levels/patterns that are becoming popular in the prediction of heterosis, will also be discussed.

5.3.1.1 *Per se* Performance and Combining Ability

Choice of suitable parents through careful and critical evaluation of the available parental gene pool is of paramount importance for the identification of superior hybrid combinations. In general, it is assumed that the inbred parents with superior *per se* performance will give highly heterotic hybrids. However, this assumption may not be true always. For example, in a study of heterosis in wheat, the hybrids derived from parents with the largest phenotypic differences showed greater mid-parent advantage while the better parent heterosis was found to be more when the parents had the least phenotypic difference. This suggests that the beneficial effects of hybridization due to the dispersion of dominant genes between parents were not sufficient to mask the detrimental effects of other genes (Morgan 1998). *Per se* performance of the parents along with their combining ability may be considered as important criteria in the selection of parents for hybridization in sorghum (Harer and Bapat 1982).

A close association between *per se* performance of hybrids and heterosis was observed for days to 50 % flowering, plant height, leaf area index, brix, panicle length, number of grains per panicle, 100 grain weight, and grain yield per plant, suggesting the utility of *per se* performance in the selection of the crosses (Premalatha et al. 2006). Several studies in different crops reported contrasting conclusions regarding the effectiveness of *per se* performance in the prediction of heterosis. Average *per se* performance of the inbred parent could not predict hybrid performance in maize (Lee et al. 2007) while no association was observed between *per se* performance and heterotic response in sugarcane (Verma and Singh 2004). In contrast, a study in tropical maize suggested that the performance of hybrid progenies under excessive moisture stress can be predicted, to some extent, based on *per se* performance of their inbred parents that were selected and improved for excessive moisture stress (Zaidi et al. 2007). Therefore, it is clear that *per se* performance of parents may not be reliable for the prediction of heterosis since its success depends on crop species, traits of interest, and the season or environment.

Combining ability analysis is a powerful tool to test the value of parental lines to produce superior hybrids. Tests on combining ability such as topcross, polycross, single cross, diallel mating, and line × tester analysis were generally employed by plant breeders to identify parental lines with good combining ability for developing superior heterotic hybrids, though with different levels of success (Virmani 1994). In general, selection of parental lines based on general combining ability (GCA) increases the probability of developing heterotic hybrids in any crop. In contrast, sometimes, the parental lines with low GCA effects will result in the identification of heterotic combinations as noticed in rice (Kumar and Saini 1981), and such combinations could not be obtained from parents exhibiting high GCA (Shrivastava and Seshu 1983). Selection of heterotic parents that are good general and specific combiners is important for substantial yield improvement in any crop, including sorghum.

Estimates of the GCA of parental lines offer a simple and widely used approach for the prediction of heterosis (Cockerham 1967; Melchinger et al. 1987). A study of bioenergy traits in sweet sorghum by Sandeep et al. (2010) revealed that crosses with higher specific combining ability (SCA) and heterotic potential can be produced from parents with contrasting GCA effects indicating the predictive power of combining ability. Studies in different crops have shown moderate to strong correlation between combining ability and *per se* performance (Betran et al. 2003; Dreisigacker et al. 2005; Qian et al. 2007) while others reported no significant relation between F_1 performance and parental GCA or SCA (Verma and Singh 2004; Yadav et al. 2007). If pedigree and *per se* performance data were used, a best linear unbiased prediction of general and specific combining ability increased the efficiency of the prediction of hybrid performance for grain yield and grain dry matter content in maize (Schrag et al. 2010). Moreover, the approach based on GCA accounts for additive genetic variance while the non-additive genetic variance or specific combining ability (SCA) is ignored, which will affect this correlation. Therefore, weightage should be given to both GCA and SCA during the selection of the parental lines. Even though this approach is extensively used for the prediction of heterosis, it is speculative and relies heavily on field evaluation.

5.3.1.2 Genetic Diversity

Conventionally, genetic diversity is estimated based on the pedigree data and phenotypic trait data. It has been suggested that the genetic distance between parents is positively correlated with heterosis of F_1 hybrids, and therefore, the extent of genetic diversity between the two parents has been proposed as a possible measure for the prediction of heterosis (Zhang et al. 1994; Falconer and Mackay 1996). However, a strong correlation between heterosis and genetic distance between parents has been observed rarely (Melchinger 1999; Dixit and Swain 2000; Singh and Singh 2004). Even though several reports have been published in maize (Smith et al. 1990) and rice (Liu and Wu 1998; Zhang et al. 1994, 1996), there is wide variation in the correlation depending on the trait of interest and the dataset. In sorghum, Rani and Rao (2009) observed lack of perfect correspondence between level of heterosis and genetic divergence between parents. Though a positive relationship existed between genetic distance and heterobeltiosis for some characters in sunflower, it was not sufficient to predict the level of heterosis (Rao et al. 2004). Several reports concluded that the prediction of heterosis could not be possible based on genetic divergence, as there was a lack of direct relationship between genetic distance and heterosis (Dave and Joshi 1995; Singh and Singh 2004; Shukla and Singh 2006).

Even though the genetic distance based on phenotypic data and combining ability among parental lines can be used to predict heterosis a priori, these methods do not exclude extensive field tests and crossing. Therefore, a need was felt to establish heterotic groups based on diversity observed at the molecular level, which is environmentally neutral, and testing only those crosses which are most likely to exhibit higher heterosis, thereby accelerating the development of superior hybrids.

5.3.2 Heterosis Prediction Using Molecular Marker Data

Heterosis is presumed to be related to the genetic divergence of the parental lines. Therefore, genetic diversity among the parental lines assessed by biochemical and molecular markers may serve as a potential parameter for prediction of heterosis. Before the advent of DNA markers, prediction of heterosis was usually performed by isozymes, a class of biochemical markers. Studies in different crops with isozyme loci and parental lines resulted in low and non-significant correlations between isozymes diversity and specific heterosis (Smith and Smith 1989; Peng et al. 1988; Frei et al. 1986). Generally, isozymes are not considered to be suitable for the prediction of heterosis due to their limited number, and these loci may not have a direct effect on the target phenotype (Stuber et al. 1999).

Heterosis prediction gained momentum with the advent of DNA markers. Several studies were conducted in different crops using multi-locus as well as single-locus markers, but the conclusions were not consistent. Such inconsistency may be due to the type of markers, number of parental lines, and molecular markers used since these factors influence the correlation between genetic diversity among parental lines and heterosis. Significant correlations between genetic distance and heterosis using DNA markers have been reported in rice (Liu and Wu 1998; Zhang et al. 2010), maize (Smith et al. 1990), wheat (Corbellini et al. 2002), sunflower (Cheres et al. 2000), and rapeseed (Diers et al. 1996). However, weak or non-significant correlations have also been reported (Xiao et al. 1996; Jordan et al. 2003; Liu et al. 1999). Such low or weak prediction ability of markers can be due to (1) a poor association between marker-based estimate of heterozygosity and heterozygosity at quantitative trait loci (QTLs) affecting the trait of interest, (2) a poor association between heterozygosity and heterosis at quantitative trait loci in the crosses (Charcosset et al. 1991), and (3) epistasis (Moll et al. 1965). Some of the studies on the prediction of heterosis using DNA markers in different crops and their conclusions are summarized in Table 5.1.

Earlier studies, mostly used multi-locus markers such as random amplified polymorphic DNA (RAPD), Inter-simple sequence repeat (ISSR) and Amplified fragment length polymorphism (AFLP) for predicting heterosis in different crop species (Shieh and Thseng 2002; Joshi et al. 2001; Schrag et al. 2009). Even hybridization-based markers such as restriction fragment length polymorphism (RFLP) were used for the prediction of heterosis in sorghum (Jordan et al. 2003). With the prominence of single-locus simple sequence repeat (SSR) markers during the late 1990s, several efforts were made to assess the genetic diversity in sorghum (Agrama and Tuinstra 2003; Anas and Yoshida 2004; Mutegi et al. 2011; Rakshit et al. 2012), but little information is available on classification of parental lines into heterotic groups using DNA markers (Menz et al. 2004). Recently, a classification of parental lines based on fertility groups was demonstrated in sorghum using SSR markers (Ganapathy et al. 2012), which has significant implications in classifying genotypes into heterotic groups.

Molecular markers associated with QTL for yield and its contributing traits have been reported by various research groups in sorghum (Srinivas et al. 2009; Nagaraja Reddy et al. 2013). Unfortunately, studies on the prediction of heterosis based on markers associated with yield QTL are limited. In rice, the QTL for 1,000-grain weight highly contributed to heterosis; moreover, the heterosis over mid-parents of single QTL, pair of epistatic QTL, and overall QTL could also be predicted (Gao and Zhu 2007). Thirteen heterotic loci were detected in maize (Tang et al. 2010), genome regions containing heterosis-related QTLs were identified in oilseed rape (Basunanda

Table 5.1 Heterosis prediction studies in important food crops

Crop	Molecular marker	Conclusions	Reference
Rice	RFLP	Relationship between marker heterozygosity and hybrid performance is complex	Zhao et al. (1999)
	RAPD, ISSR, RFLP, and STMS	Genome-wide/QTL-linked markers may not be always useful for parental selection	Joshi et al. (2001)
	SSR and RAPD	GD based on marker data may not be useful for the prediction of heterosis	Kwon et al. (2002)
	AFLP	Only few loci contributed to GD among the parental lines were related to heterosis	Liu et al. (2002)
	Pedigree record, quantitative traits, and SSR	Prediction using SSR and pedigree-based diversity for complex traits is difficult	Xu et al. (2002)
	RAPD and SSR	Proposed concept of "key" DNA markers comprising of markers associated with hybrid vigor and weakness	Cho et al. (2004)
	SSR	Effect-increasing loci were identified for a more effective prediction of heterosis	Renming et al. (2008)
	SSR and EST-SSR	EST-SSRs predict heterosis better than genomic SSR markers	Jaikishan et al. (2010)
Maize	RAPD	Useful in aiding the choice of superior crosses to be made	Lanza and de Souza Jr (1997)
	RFLP	RFLP-based GDs are not suitable for the prediction of heterosis	Benchimol et al. (2000)
	AFLP	Some QTLs of grain yield were located near the loci possessing quantitative effects on grain yield	Vuylsteke et al. (2000)
	Morphological data and RAPD	Correlation between RAPD-based GD and SCA for yield was low and positive	Parentoni et al. (2001)
	RAPD	Coefficient based on RAPDs is not suitable for the prediction of yield performance of hybrids	Shieh and Thseng (2002)
	AFLP and SSR	AFLP-based GD is suitable for predicting the single cross performance	Barbosa et al. (2003)
	SSR	Prediction of yield heterosis is difficult	Xu et al. (2004)
	AFLP	Estimates of prediction efficiency (R^2) varied for grain yield (0.46–0.86) and grain dry matter content (0.59–0.96)	Schrag et al. (2006)
	AFLP	Variable haplotype block length improved hybrid prediction compared with single AFLP markers	Schrag et al. (2007)
	AFLP	Integration of molecular and phenotypic data can provide better predictions	Kiula et al. (2008)
	AFLP	Correlations of GDs with mid-parent heterosis were significantly positive but with low magnitude to be of predictive value	Legesse et al. (2008)
	AFLP	Hybrid performance was predicted efficiently	Schrag et al. (2009)

(continued)

Table 5.1 (continued)

Crop	Molecular marker	Conclusions	Reference
Wheat	RFLP and RAPD	Weak correlation between parental diversity and hybrid performance	Perenzin et al. (1998)
	RAPD	Marker-based GD was not significantly correlated with hybrid performance and heterosis	Liu et al. (1999)
	RAPD	GDs between parents can be used to predict performance of hybrids for selected traits	Krystkowiak et al. (2009)
Sorghum	RFLP	Association of GD with yield is too weak to be of value for identifying superior hybrids	Jordan et al. (2003)

et al. 2010), and ten genomic positions associated with biomass heterosis were detected in *Arabidopsis* (Meyer et al. 2010). Similar studies in sorghum will help in the identification of QTL/genomic regions/genes associated with heterosis. Three heterotic trait loci (HTL) with synergistic intra-locus effects on over-dominant grain yield heterosis were reported recently in sorghum (Ben-Israel et al. 2012). More recently, a set of 30 SSR markers useful for the prediction of grain yield heterosis in sorghum was identified by Rajendrakumar et al. (2013) by correlating the coefficient of marker polymorphism with grain yield heterosis, which when validated in a set of 210 new experimental hybrids and their parental lines revealed a significant, positive, and moderate correlation of marker polymorphism among parental lines with mid-parent ($r=0.48^*$) and better parent heterosis ($r=0.65^*$) for grain yield. Recently, yield-related QTLs were selected by Lu et al. (2014) based on the phenotypic values of eight traits, and these loci were used to construct prediction models of the traits for the hybrids by stepwise regression. This analysis revealed an average correlation ($r=0.65$) between marker value and yield heterosis by accounting dominance and additive effects separately. This information will help molecular breeders to exploit the markers associated with such regions for a reliable and better prediction of heterosis. If successful, such prediction methods may help in the forecasting of potential parental combinations that may exhibit high level of heterosis, which may be used for synthesizing experimental hybrids for field evaluation leading to the identification of better hybrid.

Due to inconsistencies in the prediction of heterosis in earlier studies, concerted efforts were made to improve the power of prediction. Based on computer simulations, Bernardo (1992) suggested that a minimum of 30–50 % of the molecular markers should be linked to QTL while about 20–30 % of them should be randomly positioned. Trait marker-best linear unbiased prediction (TM-BLUP) developed by (Bernardo 1998, 1999), which considers both the molecular marker and trait data, brought about a slight improvement in the quality of the prediction. A promising predictive model based on linear regression, namely, TCSM (total sum of selected markers), was developed by Vuylsteke et al. (2000), which considers the sum of joint effect of genetic loci for the calculation of genotypic value of hybrids. Later, this approach was improved by Schrag et al. (2006, 2007, 2009) by including multiple regression and haplo-block marker organization.

5.4 Recent Trends in Heterosis Prediction

The concept of heterosis prediction has evolved from phenotypic data, isozymes, and multi-locus markers such as RAPDs, ISSRs, and AFLPs to the present single-locus markers such as SSRs and SNPs. The use of multi-locus markers was met with inconsistent results and limited success. The dominance of SSR markers during the late 1990s led to several studies on the prediction of heterosis. Initially, methods involving random/genomic molecular markers resulted in inconsistencies in the prediction of heterosis in different crops. This may be due to the use of random genomic SSR markers distributed across the genome, which may not be associated with the genomic regions related to heterosis expression. Later, the focus was shifted to the selection of specific markers in the genomic regions associated with heterosis or yield QTL. Hence, prediction methods employing EST-SSR markers and QTL-linked markers were developed since such markers can reveal the locus or loci that could be targeted for improvement of a particular trait. Since EST-SSRs detect polymorphism in the expressed regions of the genome that are more likely associated with the traits of interest, they may be more dependable for the prediction of heterosis than markers like RFLPs, RAPDs, AFLPs, and genomic SSRs.

This led to a series of studies in rice by different research groups resulting in the shift in focus from the random markers to more specific markers. Such studies classified the markers as favorable and unfavorable SSR alleles (Liu and Wu 1998), loci with positive or negative effects (He et al. 2002), and markers associated with hybrid vigor and hybrid weakness (Cho et al. 2004). The concept of "effect-increasing" loci was proposed for the prediction of hybrid performances in *indica* rice (Renming et al. 2008). EST-SSR markers were used successfully for the prediction of heterosis in rice, and about 10 informative EST-SSR markers for the prediction of heterosis were identified, which were found to be related to the genes such as putative nitrate transporter protein and MADS-box transcription factors that were mostly associated with the expression of heterosis (Jaikishan et al. 2010). Such markers should be validated in diverse pools of parental lines to achieve consistency in the prediction of heterosis. In general, it may be assumed that the molecular markers can be classified into informative markers (effective and defective) and neutral markers based on their effect on the prediction of heterosis.

Moderate level of success in the prediction of heterosis using DNA markers in some studies encouraged researchers to explore different ways and means to further improve the accuracy of heterosis prediction. In this endeavor, it was strongly felt that the transcriptomic and metabolomic studies are very important since heterosis is a genome-wide phenomenon involving a network of genes associated with plant metabolism. In general, the techniques such as differential display analysis, cDNA-AFLP, and microarrays have been employed for the analysis of gene expression in relation to heterosis. This has led to the identification and characterization of a number of heterosis-related differentially expressed genes (DEGs). Patterns of differential gene expression were correlated with heterosis (Xiong et al. 1998; Wu et al. 2001); the genes exhibiting additive expression were found to be positively correlated to heterosis for yield (Guo et al. 2006). Parental expression levels of certain differentially expressed genes were highly correlated with heterosis for grain yield and also with grain dry matter content (Thiemann et al. 2010). In maize, it was reported that the prediction of hybrid performance with transcriptome-based distances of selected markers was more precise than those involving DNA markers or general combining ability (Frisch et al. 2010).

A combination of two or more prediction methods is gaining popularity in recent times due to the use of diverse as well as robust data. Heterosis prediction involving transcriptome technology and mathematical modeling was developed by Stokes et al. (2006) to identify genes related to heterosis using linear regression algorithms, and the allelic combinations of inbred lines can be used as markers for the prediction of heterosis in hybrids. This approach, successfully demonstrated in *Arabidopsis thaliana* and maize, has great potential to reliably predict heterosis in other crops also. Molecular markers developed from the transcriptome data were found to be feasible for the prediction of heterosis even with the limited number of hybrids as well as limited performance data (Stokes et al. 2010), which need to be validated further before being applied in hybrid breeding programs. A combination of genetic markers, morphological characters, isozymes, and proteins was employed for the prediction of heterosis in oilseed rape (Yu et al. 2005). Significant improvement in the prediction of biomass heterosis was achieved in *Arabidopsis*

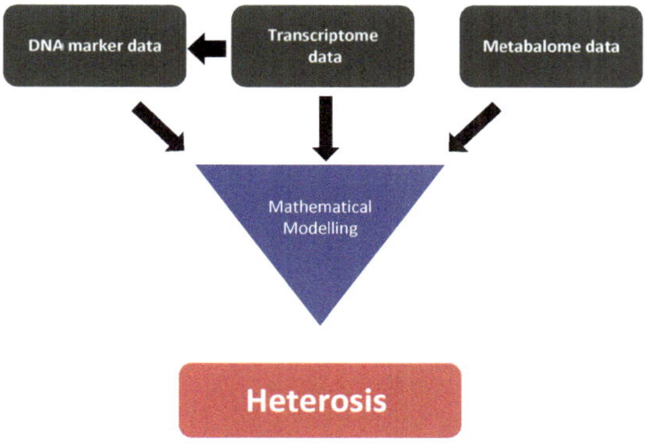

Fig. 5.1 Ideal model for the prediction of heterosis

thaliana with the use of parental genetic markers in combination with metabolic markers identified via feature selection (Gartner et al. 2009). Inclusion of plant metabolites along with genetic markers will improve the heterosis prediction since they represent the combined effect of many genes. A new complementary approach for the prediction of hybrid biomass in *Arabidopsis thaliana* was developed involving metabolite profiles, SNP markers, and feature selection instead of the complete set of SNPs and metabolites (Steinfath et al. 2010). An ideal model for the prediction of heterosis includes molecular markers, transcriptome, metabolome, and mathematical modeling (Fig. 5.1).

5.5 Factors Affecting the Effectiveness of Heterosis Prediction

Though exploited successfully in many agricultural and horticultural crops, heterosis is a highly complex biological phenomenon. During the discussion in the earlier sections, it was clear that molecular marker-assisted prediction of heterosis can be improved when "informative" or "specific" markers were used. In addition, Renming et al. (2008) suggested that the power of prediction was affected by factors such as parental lines used, traits studied, prediction methods, and environments. Therefore, the important factors affecting the prediction of heterosis can be summarized as follows:

- The pollination behavior of the crop, genetic background or nature of plant material, and the sample size of the parental lines and progenies also affect the heterosis prediction.
- The prediction of heterosis can be affected by the genetics of the trait. For example, in quantitative traits, the correlation between genetic diversity and heterosis was affected by the environmental effect (Renming et al. 2008). Moreover, the prediction was better for the traits with high heritability than those with low heritability (Manjarrez-Sandoval et al. 1997).
- Selection of appropriate molecular markers is essential for a reliable prediction of heterosis. Initially, it was thought that inadequate genome coverage and use of anonymous markers may be the reason for the poor prediction of heterosis. Later, several studies suggested that it is important to identify specific marker loci, which are tightly linked to those genomic regions that determine the expression of heterosis. In confirmation to this, Jordan et al. (2003) suggested that important QTLs contributing to grain yield heterosis in sorghum were located in specific chromosome regions and not distributed evenly over the genome. Similarly, in maize, McMullen et al. (2009) concluded that pericentromeric regions may contribute disproportionately to heterosis.
- The approach including the prediction methodology and parameters used to determine the genetic diversity among the parental lines affects the heterosis prediction. For instance, "positive" markers gave a better prediction than total markers (Zhang et al. 1994, 1995), and prediction based on the effect-increasing loci was more effective than total and positive loci (Renming et al. 2008). The parameters used to determine the genetic distance, such as coefficient of parentage, trait variation among the parents and progenies, and biochemical and molecular markers' polymorphism among the parental lines, also influence the prediction of heterosis (Renming et al. 2008). With the current trend of using a combination of prediction approaches, it is important to use the appropriate mathematical modeling procedures.

5.6 Impact of Heterosis Prediction on the Development of Hybrids

Like in many other crops, heterosis has been exploited successfully in sorghum for the past five decades, even without a clear understanding of its genetic and molecular basis. In this situation, the knowledge of molecular markers, transcripts, and metabolites associated with the heterosis will help the plant breeders to undertake a targeted development of heterotic hybrids with the use of molecular tools. This will lead to achieving a higher rate of success in a short time with limited resources. The prediction of heterosis will have a positive impact on the development of hybrids in the following ways:

- Marker-assisted selection of potential parents from a large pool of parental lines for early hybrid performance testing and its advancement to yield trials greatly increases the speed of hybrid development.
- Dissection of regulatory pathways associated with heterosis through identification of molecular markers, candidate genes, and their expression in relation to heterosis.
- Improvement of parental lines for their heterotic potential by marker-assisted introgression of effective or favorable alleles.

5.7 Future Prospects

Heterosis is a complex biological phenomenon that is governed by many genes with small effects that contribute to a particular trait. A reliable and accurate prediction of heterosis will depend on exploitation of various metabolic pathways involving multiple genes or genomic regions. The studies discussed in this chapter highlighted the fact that a major proportion of the markers used for the prediction of heterosis should be linked to QTL for the trait of interest. With the advances in genomics, a large number of SNPs have been detected in sorghum, which has great potential for the prediction of heterosis. Metabolome and protein profiling is important in heterosis prediction as they represent the reliable targets for measuring the expression of heterosis. A lot of genomic, transcriptomic as well as metabolomic data are generated from different crops from the projects associated with heterosis. The data from rice, maize, and wheat have been organized into a Heterosis-related Gene Database (HRGD, http://hrgd.big.ac.cn/index.html) (Song et al. 2009), which contains data on differentially expressed genes and transcription factors associated with heterosis at different growth stages of the crop. This database should include data from other major crops so that it offers to perform comparative analysis that could lead to the identification of a common set of genes involved in the expression of heterosis. With the help of mathematical modeling, the future prediction strategies should integrate the valuable information in molecular marker, transcriptome, metabolome, and proteome data related to heterosis to develop a suitable model for a reliable and efficient prediction of heterosis.

References

Agrama HA, Tuinstra MR (2003) Phylogenetic diversity and relationships among sorghum accessions using SSRs and RAPDs. Afr J Biotechnol 2:334–340

Anas, Yoshida T (2004) Sorghum diversity evaluated by simple sequence repeat (SSR) markers and phenotypic performance. Plant Prod Sci 7:301–308

Barbosa AMM, Geraldi IQ, Benchimol LL, Garcia AAF, Souza CL Jr, Souza AP (2003) Relationship of intra- and inter-population tropical maize single cross hybrid performance and genetic distances computed from AFLP and SSR markers. Euphytica 130:87–99

Basunanda P, Radoev M, Ecke W, Friedt W, Becker HC, Snowdon RJ (2010) Comparative mapping of quantitative trait loci involved in heterosis for seedling and yield traits in oilseed rape (*Brassica napus* L.). Theor Appl Genet 120:271–281

Benchimol LL, de Souza Jr CL, Garcia AAF, Kono PMS, Mangolin CA, Barbosa AMM, Coelho ASG, de Souza AP (2000) Genetic diversity in tropical maize inbred lines: heterotic group assignment and hybrid performance determined by RFLP markers. Plant Breed 119:491–496

Ben-Israel I, Kilian B, Nida H, Fridman E (2012) Heterotic trait locus (HTL) mapping identifies intra-locus interactions that underlie reproductive hybrid vigor in *Sorghum bicolor*. PLoS One 7:e38993

Bernardo R (1992) Relationship between single cross performance and molecular marker heterozygosity. Theor Appl Genet 83:628–634

Bernardo R (1998) Predicting the performance of untested single crosses: trait and marker data. In: Lamkey KR, Staub JE (eds) Concepts and breeding of heterosis in crop plants. Crop Science Society of America, Madison, pp 117–127

Bernardo R (1999) Marker-assisted best linear unbiased prediction of single-cross performance. Crop Sci 39:1277–1282

Betran FJ, Ribaut JM, Beck D, Gonzalez de Leon D (2003) Diversity, specific combining ability, and heterosis in tropical maize under stress and non-stress environments. Crop Sci 43:797–806

Birchler JA, Auger DL, Riddle NC (2003) In search of the molecular basis of heterosis. Plant Cell 15:2236–2239

Bruce AB (1910) The Mendelian theory of heredity and the augmentation of vigor. Science 32:627–628

Charcosset A, Lefort-Buson M, Gallais A (1991) Relationship between heterosis and heterozygosity at marker loci: a theoretical computation. Theor Appl Genet 81:571–575

Cheres MT, Miller JF, Crane JM, Knapp SJ (2000) Genetic distance as a predictor of heterosis and hybrid performance within and between heterotic groups in sunflower. Theor Appl Genet 100:889–894

Cho YI, Park CW, Kwon SW, Chin JH, Ji HS, Park KJ, McCouch SR, Koh HJ (2004) Key DNA markers for predicting heterosis in F_1 hybrids of japonica rice. Breed Sci 54:389–397

Cockerham CC (1967) Prediction of double crosses from single crosses. Der Zuchter 37:160–169

Corbellini M, Perenzin M, Accerbi M, Vaccino P, Borghi B (2002) Genetic diversity in bread wheat, as revealed by coefficient of parentage and molecular markers, and its relationship to hybrid performance. Euphytica 123:273–285

Darwin CR (1876) The effects of cross- and self-fertilization in the vegetable kingdom. John Murray, London

Dave RV, Joshi P (1995) Divergence and heterosis for fodder attributes in pearl millet. Indian J Genet 55:392–397

Davenport CB (1908) Degeneration, albinism and inbreeding. Science 28:454–455

Diers BW, McKetty PBE, Osborn TC (1996) Relationship between heterosis and genetic distance based on restriction fragment length polymorphism markers in oilseed rape (*Brassica napus* L). Crop Sci 36:79–83

Dixit UN, Swain D (2000) Genetic divergence and heterosis in sesame. Indian J Genet 60:213–219

Dreisigacker S, Melchinger AE, Zhang P, Ammar K, Flachenecker C, Hoisington D, Warburton ML (2005) Hybrid performance and heterosis in spring bread wheat, and their relations to SSR-based genetic distances and coefficients of parentage. Euphytica 144:51–59

East EM (1908) Inbreeding in corn. Conn Agric Exp Stn Rep 1907:419–428

Falconer DS, Mackay TFC (1996) Introduction to quantitative genetics, 4th edn. Longman, Harlow

Frei OM, Stuber CW, Goodman MM (1986) Use of allozymes as genetic markers for predicting performance in maize single cross hybrids. Crop Sci 26:37–42

Frisch M, Thiemann A, Fu J, Schrag TA, Scholten S, Melchinger AE (2010) Transcriptome-based distance measures for grouping of germplasm and prediction of hybrid performance in maize. Theor Appl Genet 120:441–450

Ganapathy KN, Gomashe SS, Rakshit S, Prabhakar B, Ambekar SS, Ghorade RB, Biradar BD, Saxena U, Patil JV (2012) Genetic diversity revealed utility of SSR markers in classifying parental lines and elite genotypes of sorghum (*Sorghum bicolor* L. Moench). Aust J Crop Sci 6:1486–1493

Gao YM, Zhu J (2007) Mapping QTLs with digenic epistasis under multiple environments and predicting heterosis based on QTL effects. Theor Appl Genet 115:325–333

Gartner T, Steinfath M, Andorf S, Lisec J, Meyer RC, Altmann T, Willmitzer L, Selbig J (2009) Improved heterosis prediction by combining information on DNA and metabolic markers. PLoS One 4:e5220

Guo M, Rupe MA, Yang X, Crasta O, Zinselmeier C, Smith OS, Bowen B (2006) Genome-wide transcript analysis of maize hybrids: allelic additive gene expression and yield heterosis. Theor Appl Genet 113:831–845

Harer PN, Bapat DR (1982) Line × tester analysis of combining ability in grain sorghum. J Maharashtra Agric Univ 7:230–232

He GH, Hou L, Li DM, Luo XY, Niu GQ, Tang M, Pei Y (2002) Prediction of yield and yield components in hybrid rice by using molecular markers. Acta Genet Sin 29:438–444

Jaikishan I, Rajendrakumar P, Ramesha MS et al (2010) Prediction of heterosis for grain yield in rice using 'key' informative EST-SSR markers. Plant Breed 129:108–111

Jordan DR, Tao Y, Godwin ID, Henzell RG, Cooper M, McIntyre CL (2003) Prediction of hybrid performance in grain sorghum using RFLP markers. Theor Appl Genet 106:559–567

Joshi SP, Bhave SG, Chowdari KV, Apte GS, Dhonukshe BL, Lalitha K, Ranjekar PK, Gupta VS (2001) Use of DNA markers in prediction of hybrid performance and heterosis for a three-line hybrid system in rice. Biochem Genet 39:179–200

Keeble F, Pellow C (1910) The mode of inheritance of stature and time of flowering in peas (*Pisum sativum*). J Genet 1:47–56

Kiula BA, Lyimo NG, Botha AM (2008) Association between AFLP-based genetic distance and hybrid performance in tropical maize. Plant Breed 127:140–144

Krystkowiak K, Adamski T, Surma M, Kaczmarek Z (2009) Relationship between phenotypic and genetic diversity of parental genotypes and the specific combining ability and heterosis effects in wheat (*Triticum aestivum* L.). Euphytica 165:419–434

Kumar I, Saini SS (1981) Diallel analysis in rice. Combining ability for various quantitative characters. Genet Agric 35:243–252

Kwon SJ, Ha WG, Hwang HG, Yang SJ, Choi HC, Moon HP, Ahn SN (2002) Relationship between heterosis and genetic divergence in 'Tongil'-type rice. Plant Breed 121:487–492

Lanza LLB, de Souza Jr CL (1997) Genetic distance of inbred lines and prediction of maize single-cross performance using RAPD markers. Theor Appl Genet 94:1023–1030

Lee EA, Ash MJ, Good B (2007) Re-examining the relationship between degree of relatedness, genetic effects, and heterosis in maize. Crop Sci 47:629–635

Legesse BW, Myburg AA, Pixley KV, Afriyie ST, Botha AM (2008) Relationship between hybrid performance and AFLP based genetic distance in highland maize inbred lines. Euphytica 162:313–323

Li ZK, Luo LJ, Mei HW et al (2001) Overdominant epistatic loci are the primary genetic basis of inbreeding depression and heterosis in rice. I. Biomass and grain yield. Genetics 158:1737–1753

Liu XC, Wu JL (1998) SSR heterogenic patterns of parents for marking and predicting heterosis in rice breeding. Mol Breed 3:263–268

Liu ZQ, Pei Y, Pu ZJ (1999) Relationship between hybrid performance and genetic diversity based on RAPD markers in wheat, *Triticum aestivum* L. Plant Breed 118:119–123

Liu X, Ishiki K, Wang W (2002) Identification of AFLP markers favorable to heterosis in hybrid rice. Breed Sci 52:201–206

Lu XP, Liu DD, Wang SY, Mi FG, Han PA, Lu ES (2014) Genetic effects and heterosis prediction model of *Sorghum bicolor*×*S. sudanense* Grass. Acta Agron Sin 40(3):466–475

Manjarrez-Sandoval P, Carter TE, Webb DM, Burton JW (1997) RFLP genetic similarity estimates and coefficient of parentage as genetic variance predictors for soybean yield. Crop Sci 37:698–703

McMullen MD, Kresovich S, Villeda HS et al (2009) Genetic properties of the maize nested association mapping population. Science 325:737–740

Melchinger AE (1999) Genetic diversity and heterosis. In: Coors JG, Pandey S (eds) The genetics and exploitation of heterosis in crops. CSSA, Madison, pp 99–118

Melchinger AE, Geiger HH, Seitz G, Schmidt GA (1987) Optimum prediction of three-way crosses from single crosses in forage maize (*Zea mays* L.). Theor Appl Genet 74:339–345

Menz MA, Klein RR, Unruh NC, Rooney WL, Klein PE, Mullet JE (2004) Genetic diversity of public inbreds of sorghum determined by mapped AFLP and SSR markers. Crop Sci 44:1236–1244

Meyer RC, Kusterer B, Lisec J et al (2010) QTL analysis of early stage heterosis for biomass in *Arabidopsis*. Theor Appl Genet 120:227–237

Moll RH, Longquist JH, Fortuna JV, Johnson EC (1965) The relation of heterosis and genetic divergence in maize. Genetics 52:139–144

Morgan CL (1998) Mid-parent advantage and heterosis in F_1 hybrids of wheat from crosses among old and modern varieties. J Agric Sci 130:287–295

Mutegi E, Sagnard F, Semagn K, Deu M, Muraya M, Kanyenji B, deVilliers S, Kiambi D, Herselman L, Labuschagne M (2011) Genetic structure and relationships within and between cultivated and wild sorghum [*Sorghum bicolor* (L.) Moench] in Kenya as revealed by microsatellite markers. Theor Appl Genet 122:989–1004

Nagaraja Reddy R, Madhusudhana R, Murali Mohan S, Chakravarthi DV, Mehtre SP, Seetharama N, Patil JV (2013) Mapping QTL for grain yield and other agronomic traits in post-rainy sorghum [*Sorghum bicolor* (L.) Moench]. Theor Appl Genet 126:1921–1939

Osborn TC, Pires JC, Birchler JA et al (2003) Understanding mechanisms of novel gene expression in polyploids. Trends Genet 19:141–147

Parentoni SN, Magalhaes JV, Pacheco CAP et al (2001) Heterotic groups based on yield-specific combining ability data and phylogenetic relationship determined by RAPD markers for 28 tropical maize open pollinated varieties. Euphytica 121:197–208

Peng JY, Glaszmann JC, Virmani SS (1988) Heterosis and isozyme diversions in *indica* rice. Crop Sci 28:561–563

Perenzin M, Corbellini M, Accerbi M, Vaccino P, Borghi B (1998) Bread wheat: F_1 hybrid performance and parental diversity estimates using molecular markers. Euphytica 100:273–279

Premalatha N, Kumaravadivel N, Veerabadhiran P (2006) Heterosis and combining ability for grain yield and its components in sorghum [*Sorghum bicolor* (L.) Moench]. Indian J Genet 66:123–126

Qian W, Sass O, Meng J, Li M, Frauen M, Jung C (2007) Heterotic patterns in rapeseed (*Brassica napus* L.): I. Crosses between spring and Chinese semi-winter lines. Theor Appl Genet 115:27–34

Rajendrakumar P, Hariprasanna K, Jaikishan I, Madhusudhana R, Patil JV (2013) Potential of microsatellite marker polymorphism in the prediction of grain yield heterosis in sorghum [*Sorghum bicolor* (L.) Moench]. In: Rakshit S, Das IK, Shyamprasad G, Mishra JS, Ratnavathi CV, Chapke RR, Tonapi VA, Dayakar Rao B and Patil JV (eds) Compendium of papers and abstracts: Global consultation on millets promotion for health and nutritional security, 18–20 Dec 2013. Society of Millets Research, 11–127, Directorate of Sorghum Research, Rajendranagar, p 329

Rakshit S, Gomashe SS, Ganapathy KN, Elangovan M, Ratnavathi CV, Seetharama N, Patil JV (2012) Morphological and molecular diversity reveal wide variability among sorghum Maldandi landraces from India. J Plant Biochem Biotechnol 21:145–156

Rani KJ, Rao SS (2009) Relationship between heterosis and genetic divergence in rabi sorghum [*Sorghum bicolor* (L.) Moench]. Res Crops 10:319–322

Rao M, Reddy GL, Kulkarni RS, Ramesh S, Lalitha Reddy SS (2004) Prediction of heterosis based on genetic divergence of parents through regression analysis in sunflower (*Helianthus annuus* L.). Helia 27:51–58

Renming Z, Yinghua L, Zhenglin Y, Fangming Z, Bingqiang Z, Rong X, Xianchun S, Guanghua H (2008) Prediction of hybrid grain yield performances in *indica* rice (*Oryza sativa* L.) with effect-increasing loci. Mol Breed 22:467–476

Sandeep RG, Gururaja Rao MR, Ramesh S, Chikkalingaiah, Shivanna H (2010) Parental combining ability as a good predictor of productive crosses in sweet sorghum [*Sorghum bicolor* (L.) Moench]. J Appl Nat Sci 2:245–250

Schrag TA, Melchinger AE, Sørensen AP, Frisch M (2006) Prediction of single-cross hybrid performance for grain yield and grain dry matter content in maize using AFLP markers associated with QTL. Theor Appl Genet 113:1037–1047

Schrag TA, Maurer HP, Melchinger AE, Piepho HP, Peleman J, Frisch M (2007) Prediction of single-cross hybrid performance in maize using haplotype blocks associated with QTL for grain yield. Theor Appl Genet 114:1345–1355

Schrag TA, Mohring J, Maurer HP, Dhillon BS, Melchinger AE, Piepho HP, Sørensen AP, Frisch M (2009) Molecular marker-based prediction of hybrid performance in maize using unbalanced data from multiple experiments with factorial crosses. Theor Appl Genet 118:741–751

Schrag TA, Mohring J, Melchinger AE, Kusterer B, Dhillon BS, Piepho HP, Frisch M (2010) Prediction of hybrid performance in maize using molecular markers and joint analyses of hybrids and parental inbreds. Theor Appl Genet 120:451–461

Shieh GJ, Thseng FS (2002) Genetic diversity of Tainan-white maize inbred lines and prediction of single cross hybrid performance using RAPD markers. Euphytica 124:307–313

Shrivastava MN, Seshu DV (1983) Combining ability for yield and associated characters in rice. Crop Sci 23:741–744

Shukla S, Singh SP (2006) Genetic divergence in relation to heterosis in opium poppy (*P. somniferum* L.). J Med Arom Plant Sci 28:4–8

Shull GH (1908) The composition of a field of maize. Am Breeders Assoc Rep 4:296–301

Shull GH (1952) Beginnings of the heterosis concept. In: Gowen JW (ed) Heterosis: a record of researches directed toward explaining and utilizing the vigor of hybrids. Iowa State College Press, Ames, pp 14–48

Singh SP, Singh M (2004) Multivariate analysis in relation to genetic improvement in *Cuphea procumbens*. J Genet Breed 58:105–112

Smith JSC, Smith OS (1989) Comparison of heterosis among hybrids as a measure of relatedness with that to be expected on the basis of pedigree. Maize Genet Coop Newsl 63:86–87

Smith OS, Smith JSC, Bowen SL, Tenborg RA, Wall SJ (1990) Similarities among a group of elite maize inbreds as measured by pedigree, F1 grain yield, grain yield heterosis and RFLPs. Theor Appl Genet 80:833–840

Song R, Messing J (2003) Gene expression of a gene family in maize based on non-collinear haplotypes. Proc Natl Acad Sci U S A 100:9055–9060

Song S, Huang Y, Wang X et al (2009) HRGD: a database for mining potential heterosis-related genes in plants. Plant Mol Biol 69:255–260

Srinivas G, Satish K, Madhusudhana R, Nagaraja Reddy R, Murali Mohan S, Seetharama N (2009) Identification of quantitative trait loci for agronomically important traits and their association with genic-microsatellite markers in sorghum. Theor Appl Genet 118:1439–1454

Steinfath M, Gartner T, Lisec J, Meyer RC, Altmann T, Willmitzer L, Selbig J (2010) Prediction of hybrid biomass in Arabidopsis thaliana by selected parental SNP and metabolite markers. Theor Appl Genet 120:239–247

Stokes D, Morgan C, O'Neill CM, Fraser F, Bancroft I (2006) Transcriptome-based predictive modeling of heterosis. Abstract international symposium on heterosis in plants, Potsdam-Golm, 18–20 May 2006

Stokes D, Fraser F, Morgan C, O'Neill CM, Dreos R, Magusin A, Szalma S, Bancroft I (2010) An association transcriptomics approach to the prediction of hybrid performance. Mol Breed 26:91–106

Stuber CW, Polacco M, Senior ML (1999) Synergy of empirical breeding, marker-assisted selection, and genomics to increase crop yield potential. Crop Sci 39:1571–1583

Tang J, Yan J, Ma X, Teng W, Wu W, Dai J, Dhillon BS, Melchinger AE, Li J (2010) Dissection of the genetic basis of heterosis in an elite maize hybrid by QTL mapping in an immortalized F_2 population. Theor Appl Genet 120:333–340

Thiemann A, Fu J, Schrag TA, Melchinger AE, Frisch M, Scholten S (2010) Correlation between parental transcriptome and field data for the characterization of heterosis in *Zea mays* L. Theor Appl Genet 120:401–413

Verma PS, Singh SB (2004) Heterosis in relation to *per se* performance and effects of general combining ability in sugarcane. Sugar Tech 6:181–185

Virmani SS (1994) Heterosis and hybrid rice breeding. Monographs on Theor Appl Genet 22. Springer, Berlin/Heidelberg; IRRI, Manila

Vuylsteke M, Kuiper M, Stam P (2000) Chromosomal regions involved in hybrid performance and heterosis: their AFLP® -based identification and practical use in prediction models. Heredity 85:208–218

Wright S (1968) Evolution and genetics of population. University of Chicago Press, Chicago

Wu LM, Ni ZF, Wang ZK, Lin Z, Sun QX (2001) Relationship between differential gene expression patterns of multigene families and heterosis in a wheat diallel crosses. Acta Genet Sin 28:256–266

Xiao J, Li J, Yuan L, McCouch SR, Tanksley SD (1996) Genetic diversity and its relationships to hybrid performance and heterosis in rice as revealed by PCR-based markers. Theor Appl Genet 92:637–643

Xiong LZ, Yang GP, Xu CG, Zhang Q, Saghai-Maroof MA (1998) Relationships of differential gene expression in leaves with heterosis and heterozygosity in a rice diallel cross. Mol Breed 4:129–136

Xu W, Virmani SS, Hernandez JE, Sebastian LS, Redoña ED, Li Z (2002) Genetic diversity in the parental lines and heterosis of the tropical rice hybrids. Euphytica 127:139–148

Xu SX, Liu J, Liu GS (2004) The use of SSRs for predicting the hybrid yield and yield heterosis in 15 key inbred lines of Chinese maize. Hereditas 141:207–215

Yadav HK, Shukla S, Singh SP (2007) Genetic divergence in parental genotypes and its relation with heterosis, F_1 performance and general combining ability (GCA) in opium poppy (*Papaver somniferum* L.). Euphytica 157:123–130

Yao Y, Ni Z, Zhang Y, Chen Y, Ding Y, Han Z, Liu Z, Sun Q (2005) Identification of differentially expressed genes in leaf and root between wheat hybrid and its parental inbreds using PCR-based cDNA subtraction. Plant Mol Biol 58:367–384

Yu SB, Li JX, Xu CG, Tan YF, Gao YJ, Li XH, Zhang Q, Saghai Maroof MA (1997) Importance of epistasis as the genetic basis of heterosis in an elite rice hybrid. Proc Natl Acad Sci U S A 94:9226–9231

Yu CY, Hu SW, Zhao HX, Guo AG, Sun GL (2005) Genetic distances revealed by morphological characters, isozymes, proteins and RAPD markers and their relationships with hybrid performance in oilseed rape (*Brassica napus* L.). Theor Appl Genet 110:511–518

Zaidi PH, Mani Selvan P, Sultana R, Srivastava A, Singh AK, Srinivasan G, Singh RP, Singh PP (2007) Association between line *per se* and hybrid performance under excessive soil moisture stress in tropical maize (*Zea mays* L.). Field Crops Res 101:117–126

Zhang QF, Gao YJ, Yang SH, Ragab RA, Saghai Maroof MA, Li ZB (1994) A half-diallel analysis of heterosis in elite hybrid rice based on RFLP and microsatellites. Theor Appl Genet 89:185–192

Zhang QF, Gao YJ, Yang SH, Saghai-Maroof MA, Li JN (1995) Molecular divergence and hybrid performance in rice. Mol Breed 1:133–142

Zhang QF, Zhou ZQ, Yang GP, Xu CG, Liu KD, Saghai Maroof MA (1996) Molecular marker heterozygosity and hybrid performance in *indica* and *japonica* rice. Theor Appl Genet 93:1218–1224

Zhang T, Ni XL, Jiang KF, Deng HF, He Q, Yang QH, Yang L, Wan XQ, Cao YJ, Zheng JK (2010) Relationship between heterosis and parental genetic distance based on molecular markers for functional genes related to yield traits in rice. Rice Sci 17:288–295

Zhao MF, Li XH, Yang JB, Xu CG, Hu RY, Liu DJ, Zhang Q (1999) Relationship between molecular marker heterozygosity and hybrid performance in intra- and inter-sub-specific crosses of rice. Plant Breed 118:139–144

Part III

Advances in Genomics Research

Genomics and Bioinformatics Resources

6

P. Rajendrakumar and Sujay Rakshit

Contents

6.1 **Introduction** ... 118

6.2 **Sequence Databases and Comparative Genomics Resources** ... 118
6.2.1 Gramene ... 119
6.2.2 PlantGDB ... 121
6.2.3 Phytozome ... 123
6.2.4 GreenPhylDB ... 125
6.2.5 CoGe ... 127
6.2.6 PLAZA ... 127
6.2.7 CSGRqtl ... 128
6.2.8 SorGSD ... 130

6.3 **Genomics Resources for Analyzing DNA Sequence Variation** ... 132

6.4 **Bioinformatics Resources for DNA Sequence Analysis** ... 134

6.5 **Resources for Analyzing Transcriptomes** ... 137

6.6 **Genetic Resources for Mapping Agronomically Important Traits** ... 140

6.7 **Mutant Resources for Analyzing Gene Function** ... 145

6.8 **Future Prospects** ... 146

References ... 146

P. Rajendrakumar (✉)
Biotechnology, ICAR-Indian Institute of Millets Research, Rajendranagar, Hyderabad, Telangana 500 030, India
e-mail: rajendra@millets.res.in

S. Rakshit
Plant Breeding, ICAR-Indian Institute of Millets Research, Rajendranagar, Hyderabad, Telangana 500 030, India
e-mail: sujay@millets.res.in

Abstract

Rapid generation of genome and transcriptome data from various research projects across the globe resulted in the accumulation of enormous amounts of data of various crop species or groups of crops. These data are organized and stored in different databases, which offers user-friendly search and retrieval of the desired information for further analyses and use. The information in these databases is used to develop DNA-based markers such as SSRs and SNPs, which are the most popular genomics resources applied for QTL mapping and marker-assisted selection. Several bioinformatics resources such as algorithms, stand-alone software, as well as web-based tools are developed by several research groups and made available in the public domain or sold commercially for the rapid and systematic analysis of DNA sequence or gene expression data. Genetic resources such as biparental, multi-parental, natural, as well as mutant populations for various target traits were developed by researchers, which are utilized for the mapping of QTLs as well as identification of candidate genes associated with the traits of interest by the application of genomics tools developed using bioinformatics resources in these mapping populations. This review discusses the most relevant databases useful for sorghum, development of genomics resources such as DNA markers using various bioinformatics tools, and the genetic resources available in sorghum.

R. Madhusudhana et al. (eds.), *Sorghum Molecular Breeding*, DOI 10.1007/978-81-322-2422-8_6

Keywords
Genetic resources • Genomics • Databases • Transcriptomics • Mutant resources

6.1 Introduction

In the current era of genomics and next-generation sequencing, a plethora of information is generated on genome and gene sequences, which is of paramount importance to understand gene function and the regulatory networks involved in plant growth, development, as well as stress tolerance at the molecular level. Plant genomics took a giant leap with the sequencing of the whole genome of *Arabidopsis thaliana* in 2000 (The Arabidopsis Genome Initiative 2000). This was followed by genome sequencing of important food crops like rice (Yu et al. 2002; Goff et al. 2002), sorghum (Paterson et al. 2009), maize (Schnable et al. 2009), barley (The International Barley Genome Sequencing Consortium 2012), pigeon pea (Varshney et al. 2012), chickpea (Varshney et al. 2013), and others. Rapid progress in plant genomics led to the discovery and isolation of important genes that regulate economically important traits and tolerance to biotic and abiotic stresses. With the availability of genome sequences in sorghum, the biggest challenge is to determine the functions of over 30,000 genes and deploy them in a practical genetic improvement program.

Being a C_4 cereal and a drought-tolerant crop, sorghum always remained in the focus of genomics researchers. With the publication of sorghum genome sequence (Paterson et al. 2009), sorghum genomics has taken a paradigm shift and generated enormous information that has been integrated with other related crop species through comparative genomic studies. The completion of sequencing of sorghum genome and the creation of related genomics resources together with advances in the development of mapping populations and molecular marker resources have allowed researchers to accelerate the identification of agronomically important quantitative trait loci (QTLs) (Satish et al. 2009; Aruna et al. 2011; Mace et al. 2012; Nagaraja Reddy et al. 2013; Madhusudhana and Patil 2013). Advances in sequencing technologies such as next-generation sequencing (NGS) have resulted in sequence-based resources and related resource platforms for specific organisms including sorghum. Availability of whole-transcriptome profiling methods, advances in plant proteomics, simultaneous profiling of many metabolites, mutant populations, and biological databases have brought significant change in the approach in dealing with the biological processes. This chapter deals with the available genomics and bioinformatics resources in sorghum, which will help the sorghum researchers to develop useful information for the genetic improvement of sorghum.

6.2 Sequence Databases and Comparative Genomics Resources

Sorghum genome is comparatively small (~730 Mb), making it an attractive model for functional genomics of Saccharinae and other C_4 grasses. It is the first C_4 plant to be sequenced in 2009 by Andrew Paterson and his group involving 20 laboratories across the USA, Germany, China, Switzerland, and India. They observed that sorghum has ~75 % larger heterochromatin DNA as compared to rice. However, sorghum and rice have similar quantities of euchromatin. The net size expansion of the sorghum genome relative to rice predominantly involved long terminal repeat (LTR) retrotransposons. It was found that the sorghum genome contains 55 % retrotransposons, which is intermediate between rice (26 %) and maize (79 %) genomes. Paterson et al. (2009) modeled 34,496 sorghum genes, out of which ~27,640 were bona fide protein-coding genes. The accumulated information/data related to genome, transcriptome, proteome, and metabolome as a result of various high-throughput sequencing projects and proteomic and metabolomic studies are stored in different databases, which are summarized in Table 6.1. The comparative genomic databases like Gramene, PlantGDB, Phytozome, GreenPhylDB, CoGE,

Table 6.1 Database resources for sorghum

Group	Resources	Databases
Genome	Genome sequence, gene annotation	PlantGDB, Phytozome, CoGE, PLAZA
	Molecular markers, DNA variation, quantitative trait locus	Gramene, Phytozome, PIP database, NCBI dbSNP, CSGRqtl, SorGSD
	Genome re-sequencing	$(GIGA)^n$ DB
	Focused gene family database	GRASIUS, Phytozome
Transcriptome	Full-length cDNAs, ESTs	PlantGDB, Phytozome, NCBI dbEST
	Non-coding RNA	NRDR
	microRNA	PMRD, miRBase
Proteome	Proteome/modificome profile	GreenPhylDB, Phytozome
	Sub-cellular localization	Gramene, Phytozome, PlantGDB
Metabolome	Metabolic map	SorghumCyc

Modified from Mochida and Shinozaki (2010)

PLAZA, OrthologID, PlantTribes, SynBrowse, etc., along with associated web portals provide a uniform set of tools and automated analyses across a wider range of plant genomes. Among them, the first six resources deal with sorghum sequences along with other plant species. The salient features of these databases and their utility in comparative genomics are discussed below.

6.2.1 Gramene

Gramene (http://www.gramene.org/) is a curated, open-source, data resource for comparative genome analysis in the grasses developed in recognition of the importance of the grass family and put on public domain in 2002 from Cold Spring Harbor Laboratory (Ware et al. 2002). Initially, the rice sequence was used as base information to facilitate genomics research in other grass families like maize, sorghum, millet, sugarcane, wheat, oats, and barley. Subsequently, Ensembl Genomes at the European Bioinformatics Institute joined the group at the Cold Spring Harbor Laboratory, to further make this database as a resource for plant comparative genomics based on Ensembl technology. Besides rice, the database has information on barley, *Brachypodium*, foxtail millet, maize, oats, pearl millet, rye, wheat, and sorghum. In the recent version, information from other plant species like *Glycine*, *Musa*, *Solanum*, *Brassica*, *Arabidopsis*, *Vitis*, *Populus*, etc., have also been included. The goal of this database is to facilitate the study of cross-species homology relationships using information derived from public projects involved in genomic and EST sequencing, protein structure and function analysis, genetic and physical mapping, interpretation of biochemical pathways, gene and QTL localization, and descriptions of phenotypic characters and mutations. Even though a new version has been launched very recently, the information can be accessed through the old version also, which is organized in a better way. Information in the database organized in different modules (Fig. 6.1) are detailed as follows:

"Genome" Module This contains detailed information about the species, assembly, annotation, structural variation, besides references, and link to other species-related sites on the above-mentioned plant species.

"Genetic Diversity" Module This stores information on genotypes, phenotypes and their environments, germplasm, and association data. It

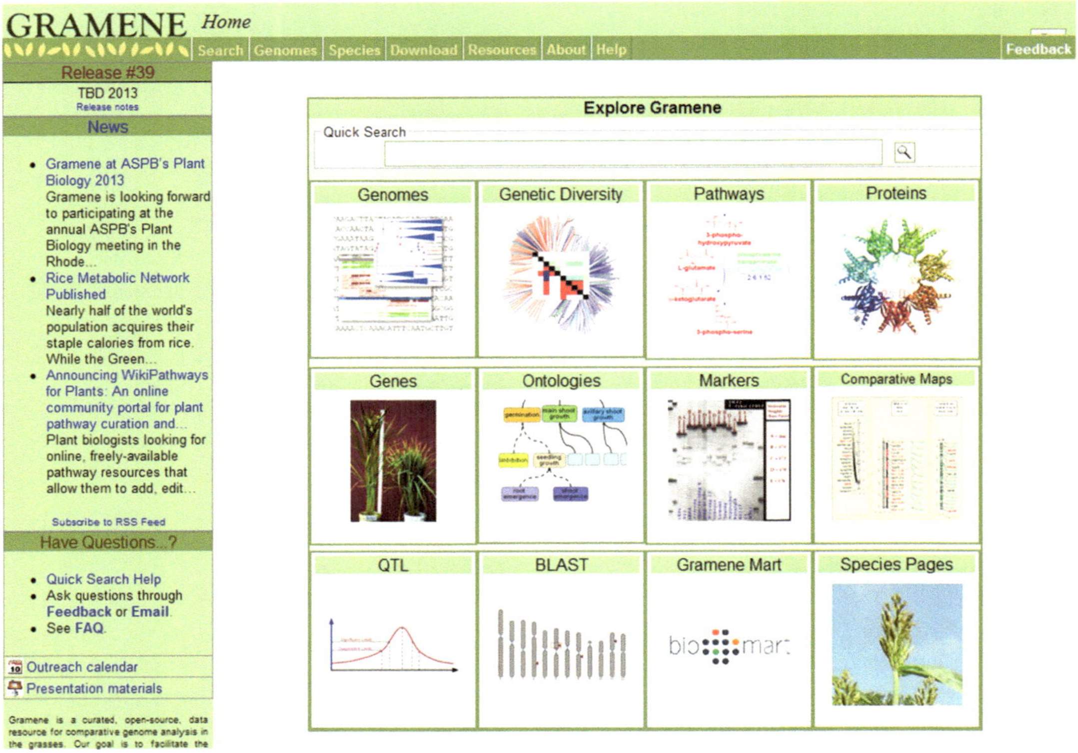

Fig. 6.1 Organization of Gramene database (http://www.gramene.org/)

also contains information from small-scale SSR diversity studies to large-scale SNP/InDel-based genotype-phenotype studies conducted in the mandated crops. With respect to sorghum, the database contains six datasets [Hamblin et al. (2004, 2006, 2007), White et al. (2004) and Casa et al. (2006)] and the results of the Sorghum Diversity Project.

"Pathways" Module This contains four submodules, viz., RiceCyc, MaizeCyc, BrachyCyc, and SorghumCyc, which deals with information on pathway databases of the respective crops. It also provides mirrors of pathway databases from *Arabidopsis*, tomato, potato, pepper, coffee, *Medicago*, *E. coli*, and the MetaCyc and PlantCyc reference databases, thereby enabling comparative genome analysis. In this database, 297 pathways, 1,838 enzymatic reactions, and 9 transport reactions have been described. Known and/or predicted biochemical pathways and genes from sorghum are catalogued in SorghumCyc, which is primarily based on the genome annotations of *Sorghum bicolor* cv. BTx623. Many of the pathways might be incomplete or may contain errors since the functions of many of the sorghum genes are either provided by homology or HMM-based predictions.

"Protein" Module This contains information on Swiss-Prot-TrEMBL protein entries from family Poaceae, which are annotated by the following three concepts of Gene Ontology (GO): (1) molecular function, (2) biological process in which it is involved, and (3) cellular component where it is localized. The associations assigned are based on annotations in the published literatures or generated through in silico approaches. Each association is supported with evidence (reference) and the evidence code (experiment type). On sorghum and related species, information on 35,817 proteins are available.

"Genes" or "Gene and Allele" Module This contains detailed information on publicly available genes in cereal crops. Genes and their alleles associated with morphological, developmental, and agronomically important phenotypes, variants of physiological characters, biochemical functions, and isozymes are described here. Species-wise search for different gene types like "CDS, rRNA, tRNA, miRNA, siRNA, pseudogenes, not classified, sequenced gene loci or all gene types" is possible using wild cards.

"Ontologies" Module This contains a collective information on controlled internationally accepted vocabularies and their associations to various objects such as QTL, phenotype, gene, proteins, and Ensembl rice genes for the following knowledge domains: Plant Ontology (PO), Trait Ontology (TO), Gene Ontology (GO), Environment Ontology (EO), and Gramene's Taxonomy Ontology.

"Markers" Module This contains the basic/primary information on the marker name, synonyms, source species, and a list of map positions of various markers used for mapping. This module has the link to "SSRIT tool," which is useful for the identification of microsatellites.

"Maps" Module This is primarily a visualization tool useful for visualizing the genetic, physical, sequence, and QTL maps for species dealt in the database. Comparative Map Viewer, referred as *CMap*, allows users to construct and compare different maps. All the data including the map sets, maps, features, and correspondences in this module are built from the "Markers" module. In *CMap*, the genomes of rice, sorghum, and *Brachypodium* are compared using syntenic blocks.

"QTL" Module This contains quantitative trait loci (QTL) identified for numerous agronomic traits in the crops dealt in the database. Information on QTL along with associated traits and the mapped locus on the genetic map are available. With respect to sorghum, information on 136 QTL along with details of associated markers, linkage group, trait symbol, etc., are available.

"BLASTView" Module This module provides an integrated platform for homology search against Ensembl plant databases, offering access to both BLAST (Basic Local Alignment Search Tool) and BLAT (BLAST-like Alignment Tool) programs. Species-wise search is possible in both DNA and protein databases using BLASTN (aligns the nucleotide sequences) and BLASTX (aligns translated sequences of any nucleotide sequence in all six reading frames), respectively.

"Gramene Mart" Module This module has four databases, viz., Plant Gene 37, Plant variation 37, Gramene mapping, and Gramene QTL 37. Each database can be searched in 10 datasets of which sorghum is one.

"Species Page" This contains detailed information on all the 11 cereal species dealt in the database with full phylogenetic information.

6.2.2 PlantGDB

PlantGDB was first reported by Dong et al. (2005), in which EST sequences were assembled into contigs that represent tentative unique genes. The functional annotation of these contigs was performed with the information derived from known protein sequences that were highly similar to the putative translation products. Initially, the database started with the data from only two plant species, viz., *Arabidopsis* and rice. Subsequently, PlantGDB (http://www.plantgdb.org) was published as a resource for comparative genomics across 14 plant species by Duvick et al. (2007). The aim of this web resource is to develop robust genome annotation methods, tools, and standard training sets for a number of sequenced or soon to be sequenced plant genomes. PlantGDB has four modules, viz., Sequence module, Genome module, Tools module, and Datasets module. Organization of the PlantGDB is shown in Fig. 6.2.

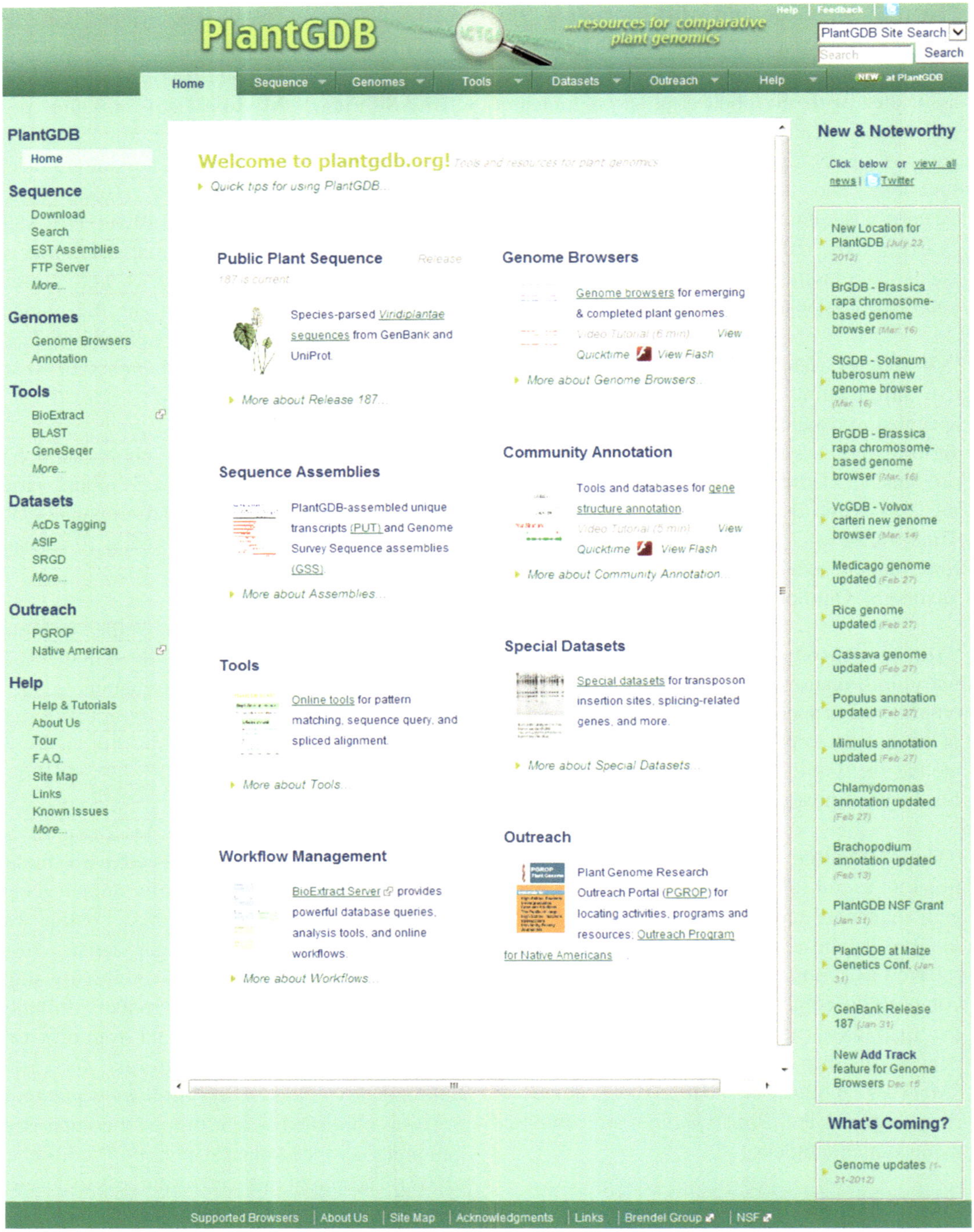

Fig. 6.2 Organization of PlantGDB (http://www.plantgdb.org/)

"Sequence" Module This module can be used to BLAST search or to download nucleotide or protein sequences as well as access custom transcript assemblies. This module contains EST assemblies comprising PlantGDB-derived unique transcripts (PUT) assembled from plant mRNA sequences available at GenBank. Genome survey sequence (GSS) assemblies for maize and sorghum are also available.

"Genome" Module This module contains genome sequence information on 16 dicots and seven monocots including sorghum (SbGDB). It has genome browsers to display current gene structure models and transcript evidence from spliced alignments of EST and cDNA sequences. The browsers also link community annotation tools to refine the gene annotations or to identify novel annotations. Each genome assembly is splice-aligned to transcripts as well as proteins from similar species and presented in a simple graphical interface (the *xGDB platform*).

"Tools" Module This module provides a variety of tools for sequence analysis as follows:

- *BioExtract* – a web interface to automate bioinformatics workflows. It is useful to query sequence databases, analyze data with bioinformatics tools, save results, and create and manage workflows.
- *Standard NCBI BLAST* – useful to search against single or multiple BLAST databases simultaneously.
- *Distributed Annotation System (DAS)* – useful to access PlantGDB annotations from the remote genome browsers.
- *GeneSeqer* and *GenomeThreader* – useful to develop gene structure models based on spliced alignment to genomic sequences of both native and homologous ESTs, cDNAs, and protein sequences.
- *MuSeqBox* – useful to examine multi-query sequence BLAST output, filter the BLAST hits based on user-defined criteria, and extract the informative parameters in tabular form.
- *PatternSearch* – useful to search the specific patterns in genome sequence, i.e., short matches interspersed with mismatches and InDels.
- *ProbeMatch* – allows the user to query his sequence against PLEXdb Probe Sequences.
- *TableMaker* – an online search tool to access GenBank tables at PlantGDB using MySQL queries.
- *yrGATE* – useful to create gene annotations in an xGDB genome browser itself. It shows all splice junctions revealed by EST/cDNA evidence and helps to create gene models and validate them.

"Datasets" Module This module has datasets on *AcDs* Tagging Project, Alternative Splicing in Plants (ASIP) database, Plant Expression database (PLEXdb), Rescue-Mu tagged maize sequences, Maize-RFLP Full-Length Insert Sequencing Project, Splicing-Related Gene database (SRGD), and Uniform-Mu tagged maize sequences.

6.2.3 Phytozome

Phytozome (http://phytozome.jgi.doe.gov/pz/) is another online resource that was first released in 2008 to facilitate comparative genomic studies among green plants. It enables users with different computational abilities to access annotated plant gene families, navigate their evolutionary history, examine them in genomic context, assign putative function, and provide uniform access to complete genomes, gene and related sequences and alignments, gene functional information, and gene families, either as bulk information or as the result of user-defined queries (Goodstein et al. 2012). A number of commonly used open-source tools like Lucene, GBrowse (Stein et al. 2002), Jalview (Waterhouse et al. 2009), BioMart (Smedley et al. 2009), mView (Brown et al. 1998), and pygr are integrated in this portal which help in the gene family search, inspection, and evaluation. The Phytozome v7.0 contains data and analyses for 25 plant genomes, 18 of which are sequenced, assembled, and partially or completely annotated at the Joint Genome

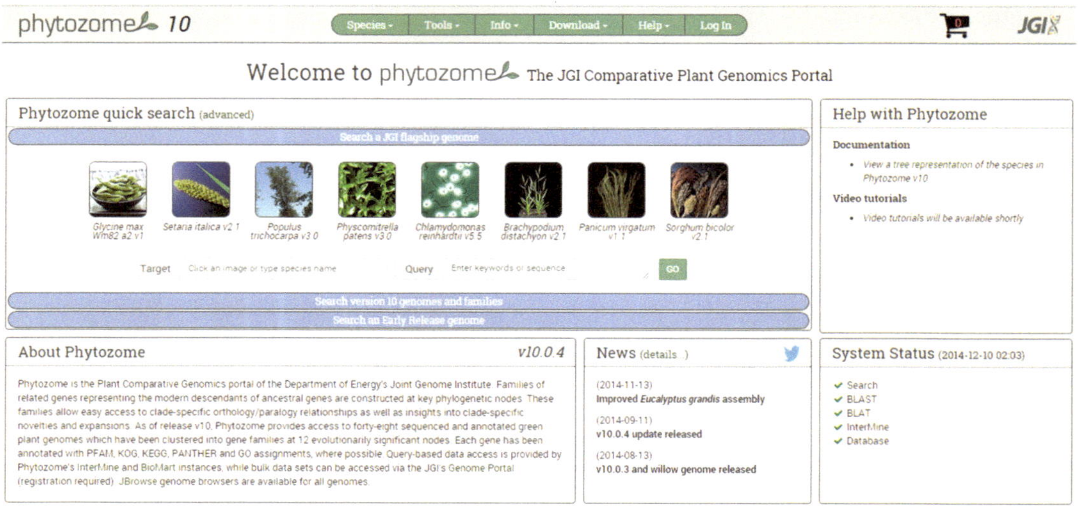

Fig. 6.3 Organization of Phytozome database (http://phytozome.jgi.doe.gov/pz/)

Institute (JGI). However, the recently released Phytozome v10 provides access to 47 sequenced and annotated green plant genomes including early-release genomes. With respect to sorghum, the initial release comprises the Sbi1 assembly and a Sbi1.4 gene set, which are the assembly and annotation reported by Paterson et al. (2009). Now, v2.1 is available as an early release as a result of modern annotation with additional RNA-seq data, comprising v2.0 assembly and v2.1 gene set. The genome is in 10 chromosomes with many short unmapped fragments; some may contain annotated genes (http://www.phytozome.net/sorghum_er.php).

Phytozome contains three important modules, viz., Species, Tools, and Info. "Species" and "Tools" modules have the options for keyword search, BLAST search, BLAT search, JBrowse, and bulk data. In addition to this, the "Tools" module has the options for the InterMine and BioMart, which is useful for data warehousing and construction of customized datasets with information on gene or gene families and annotation. With respect to poplar, *Brachypodium*, eucalyptus, and cassava genomes, both expression and diversity data can be viewed in JBrowse, searched, and downloaded from InterMine, as well as in bulk from the JGI Genome Portal. Screenshot of Phytozome database is depicted in Fig. 6.3.

Keyword and Sequence Similarity Search Information on relevant attributes of gene and gene family like names, symbols, synonyms, external database identifiers, definitions, and functional annotation IDs can be retrieved by keyword search. BLAST and BLAT can be used to identify the genomic regions, gene transcripts, peptides, and gene families most similar to the query sequence. Gene families at a particular evolutionary node and families matching particular phylogenetic profiles can be searched. The database can also be searched for functional annotations to retrieve all matching functional identifiers and gene families.

Information on Gene and Gene Families The *Gene Family view* gives information on each gene family and its constituent members. The default *Genes in this family* tab displays members of a particular gene family along with their source identifier, aliases, synonyms, and gene symbols. The *family page* has a set of lower tabs (Functional Annotation, MSA, and Family History) and upper tabs (Find related families, Align family members, Get Data, and Display options). The lower tab helps in the exploration of the evolutionary history of the gene family, while the upper tab is useful for analyzing the similarity among the related sub-families of genes. Besides depicting single gene functional annotations and evolutionary history, the *Gene Page* has links to alternatively spliced transcripts, if any; access to genomic, transcript, and peptide sequences associated with the gene; and a graphical view of other Phytozome peptides aligned against the peptide of the gene of interest.

Genome Browser In *GBrowse* module, genome-centric views are provided for all the 41 genomes currently available in the Phytozome. This module can be accessed directly from the Phytozome home page, from individual member gene links available on the Gene Family or Gene Page, and from the BLAST/BLAT results page. Each browser shows a gene prediction track, where homologous peptides from related species, supporting ESTs, and one or more syntenic VISTA tracks identifying regions of this genome are depicted. The gene features and VISTA tracks are hyperlinked to the Gene Page and corresponding regions in the VISTA browser, respectively. On sorghum, 697,578,683 base pairs are arranged, which correspond to 34,496 loci and 36,338 protein-coding transcripts.

Data Retrieval Bulk data files containing the genome assembly sequence, gene structure, transcript, coding, and peptide sequence in FASTA format and general annotation information are available for the genomes hosted in Phytozome database. Repeat-masked genome assemblies as well as supporting annotation data are also available for download. By using BioMart module, customized datasets can be constructed consisting of information on gene or gene family sequences and annotations based on user-defined data filters, attributes, and output formats. This module can be accessed from the "Get Data" tab on the Gene Family page or directly from the main menu of Phytozome.

6.2.4 GreenPhylDB

GreenPhylDB is a database specifically designed for comparative and functional genomics based on completely sequenced genomes. The development of GreenPhylDB v1.0 (5) by Conte et al. (2008) was inspired by the availability of whole-genome sequences of *Arabidopsis thaliana* and *Oryza sativa* genomes that offered opportunity for comparative genomics in plants. Since then, the most popular and reliable approach for the functional annotation of genes is by analyzing genes between species to identify orthologous genes (Kuzniar et al. 2008; Gabaldon et al. 2009). GreenPhylDB v2.0 was published during 2011 by Rouard et al. (2011) by adding 14 new genomes belonging to a major phylum of the plant kingdom including rodophytes, chlorophytes, mosses, lycophytes, and flowering plants with monocotyledons and dicotyledons. Six genomes were added in version 3, while 14 genomes were added in the current version (v4). Currently, the database has the genome information on 37 species. GreenPhylDB is accessible at http://www.greenphyl.org/cgi-bin/index.cgi.

The database represents a catalogue of gene families based on complete genomes. GreenPhylDB comprises complete proteome sequences from the major plant phylum, which are clustered to define a consistent and extensive set of homeomorphic plant families. Lists of plant- or species-specific gene families and several tools are provided to facilitate comparative genomics within plant genomes. The analyses include clustering of gene family followed by a phylogenomic analysis of the generated gene families. Upon validation of a cluster, phylogenetic analyses are performed to predict orthologs and ultraparalogs. Results of clustering are first manually annotated and then analyzed by a phylogenetic-based approach to predict orthologs, which is particularly useful for functional genomics and candidate gene identification of genes affecting agronomic traits of interests. This resource has 2,915 annotated gene families, of which 53 are specific to sorghum. Schematic diagram of the web resource is given in Fig. 6.4.

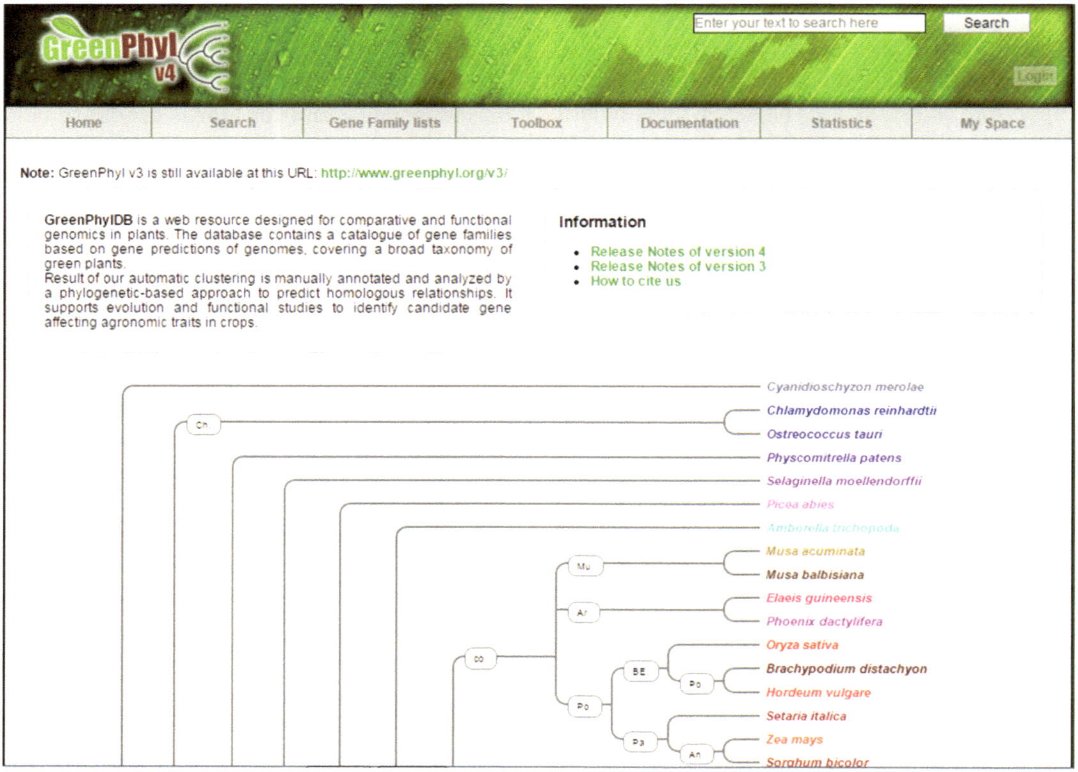

Fig. 6.4 Schematic diagram of GreenPhylDB (http://www.greenphyl.org/cgi-bin/index.cgi)

GreenPhylDB has three important modules that can be used for comparative genomics analysis, viz., "Search" module, "Gene Family Lists" module, and "Tool Box" module.

"Search" Module This module has the options of Quick Search and Advanced Family Search. The former searches based on text/keywords, while the latter on InterPro domains.

"Gene Family Lists" Module This module contains the list of annotated gene families comprising 2,915 clusters and transcription factor families comprising 33 clusters. It has the option to list the gene families specific to a particular species/phylum. A GO Browser was developed as a web interface, which displays a list of terms defined in the Plant GO. By selecting a specific GO entry, the users can access a list of gene families potentially involved in plant growth and development along with the sub-classification of each identified gene family.

"Tool Box" Module This contains the option for BLAST (sequence search with BLASTP or BLASTX), sequences to families (family classification of given sequence), Homolog sequences (get homologs and/or similar sequences with sequence ID inferred from phylogeny), InterPro domain (domain distribution is displayed by sequence and by species), Export sequences (provides a list of sequence ID used in database that can be exported in the selected format), TreePattern (to explore phylogenetic trees), and Create family (to create gene families). It has 32,796 sorghum-specific proteome datasets in it. The biggest advantage of this portal is to construct gene family tree and identify homologs.

6.2.5 CoGe

CoGE stands for Accelerating Comparative Genomics (https://genomevolution.org/CoGe/). It is a unique web resource having many interconnected tools to create open-ended analysis networks. The features of CoGe were published by Lyons and Freeling (2008) and Lyons et al. (2008). CoGe is designed to address four issues: (1) single platform to store multiple versions of multiple genomes from multiple organisms, (2) rapid identification of sequences of interest in genomes of interest (with associated information), (3) comparison of multiple genomic regions using any algorithms, and (4) visualization of the results for easy and quick identification of "interesting" patterns. Organization of CoGe database is given in Fig. 6.5. CoGe database has a set of tools for comparative genome analysis. They are as follows:

- *OrganismView* – searches and gives an overview of an organism and its genomic information
- *CoGeBlast* – BLAST sequences against any number of organisms of user's choice
- *FeatView* – searches for genomic features by name or description
- *SynMap* – generates syntenic dotplots of any two genomes
- *SynFind* – identifies syntenic regions across many genomes
- *GEvo* – compares multiple genomic regions using a variety of sequence comparison algorithms for high-resolution analysis to quickly identify patterns of genome evolution

An Integrative *Orthology Viewer* combines information from different orthology prediction methodologies. Central tools and access points of CoGe allow to find sequences of interest, and "*hub*" points direct from one part of the system to another. For example, if a region with an inversion is identified during the comparison of sorghum with the maize genome using *SynMap*, breakpoints of that region may be compared using *GEvo* in high detail, and the maize sequence can be extracted out using *SeqView*. Subsequently, *FeatView* can be used to identify all the protein-coding regions, and the information generated can be used to find homologs in other plant genomes using *CoGeBlast*. *GEvo* can be used to validate putative syntenic regions. If, say, a gene with extra copy number is identified in a syntenic region, its sequence may be obtained using *FeatView* once again. Putative intra- and interspecific homologs of it may be obtained using *CoGeBlast*, which will generate a FASTA file using *FastaView*. This can be aligned using *CoGeAlign* and used to build a phylogenetic tree through *TreeView* or exported to more expansive phylogenetic platform such as *CIPRES*. Simultaneously, the codon and protein usage variation of the genes may be checked using *FeatList*. If some interesting variation is observed in some genes, their overall GC content and wobble-position GC content may be checked in *FeatView*. Horizontal transfer of DNA fragments/genes from the mitochondria can be identified using *CoGeBlast* or *GEvo*.

6.2.6 PLAZA

A centralized plant genomics platform is essential for performing evolutionary and comparative analyses of gene families and genome organization, which integrates all the information generated by various sequencing projects along with advanced tools for data mining. PLAZA is a versatile plant comparative genomics resource centralizing genomic data from different genome sequencing initiatives (http://bioinformatics.psb.ugent.be/plaza/) published by Proost et al. (2009). Plant sequence data and comparative genomics methodologies are integrated in an online platform with interactive tools to study gene function and gene and genome evolution within the green plant lineage. It has integrated structural and functional annotation of 25 green plant species, which includes 909,850 genes. Out of these genes, 85.8 % are protein coding, which are clustered in 32,294 multigene families, resulting in 18,547 phylogenic trees. In addition to the basic

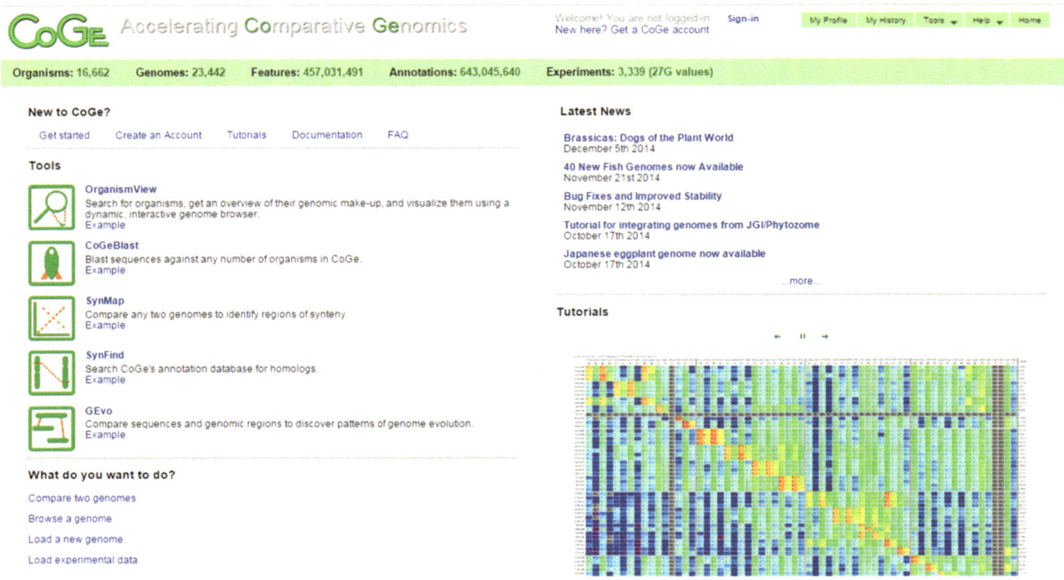

Fig. 6.5 Organization of CoGe database (https://genomevolution.org/CoGe/)

information related to gene structure and function such as genome coordinates, mRNA and protein sequences, and gene description, PLAZA offers various tools to browse genomic data for homology, ranging from local synteny to gene-based colinearity views useful for comparative genomics plant genome evolution. Organization of PLAZA database is given in Fig. 6.6.

- *Synteny plot* – a basic tool that shows all genes from the specified gene family along with their neighboring genes, thereby helping to study genomic homology in comparison to colinearity.
- *WGDotplot* – useful for analyzing genome-wide colinearity leading to the identification of large-scale duplications or to study genomic rearrangements within or between species.
- *Skyline plot* – useful for browsing multiple homologous genomic segments and provides a comprehensive view of the regions that are colinear in the species selected by the user.
- *Workbench* – useful to analyze multiple genes in batch that are uploaded through gene identifiers or based on similarity search and calculate different genome statistics for user-defined gene sets.
- *Whole Genome Mapping tool* – useful to display a selection of genes on the chromosomes and to view the distribution of different classes of genes such as protein coding, pseudogene, or transposable element. It also provides information about the gene duplication.
- *Advanced query system* – useful for the rapid retrieval of relevant information using different data types and research tools.

6.2.7 CSGRqtl

CSGRqtl is a comparative genomic database (http://helos.pgml.uga.edu/qtl/) developed by Zhang et al. (2013a) as a data mining resource specific to crops, weeds, and models of Saccharinae clade. This database complements and supplements the database Gramene, which contains mapping data from a wide spectrum of grasses. CSGRqtl uses sorghum genome sequence as a reference with an aim of anchor-

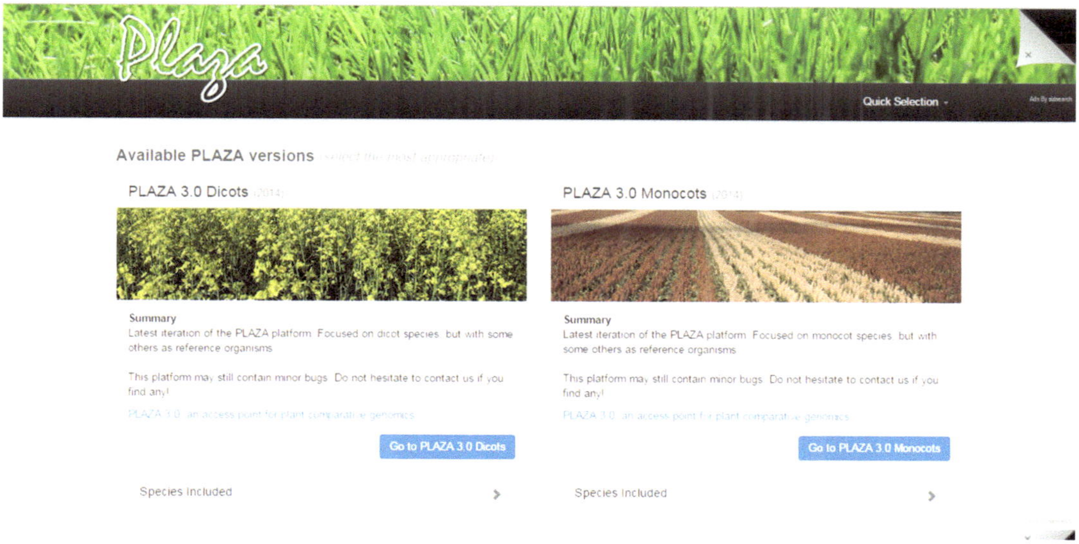

Fig. 6.6 Organization of PLAZA database (http://bioinformatics.psb.ugent.be/plaza/)

ing published QTLs of the Saccharinae clade to the sorghum genome. This database uses the Plant Trait Ontology defined by Gramene and facilitates the data comparisons among the grasses of Saccharinae clade and between Saccharinae and other taxa to categorize quantitative trait loci (QTLs) with their approximate physical positions. CSGRqtl also integrates gene annotations, genetic markers, and paleoduplicated regions, which facilitate QTL mapping and the study of candidate gene underlying the QTLs. Organization of CSGRqtl database is given in Fig. 6.7.

CSGRqtl is equipped with a number of tools for analysis, such as text-based search, trait ontology browser, QTL correspondence, and CMap database, to allow a user to query and visualize the background database.

Text-Based Search QTL search based on the trait of interest gives a set of QTLs underlying the trait. Genome-wide overview of QTL distribution is depicted by a circular plot created by the bioinformatics resource Circos. It also identifies the potential QTL hot spots in the genome. QTL search based on a sorghum gene identifier or annotation gives a list of QTLs containing the queried gene, and the approximate positions of genes and QTLs are depicted by a plot.

Trait Ontology Browser The trait ontology browser displays the hierarchy of trait ontology and lists QTLs associated with each trait accession since each QTL is allotted a trait accession defined by Gramene Plant Trait Ontology.

QTL Correspondence The associations between paleoduplicated regions and QTLs in rice and sorghum are depicted by circular plots to indicate non-overlapping QTLs divided by inter-genomic synteny or narrowed by intra-genomic synteny. The users can download orthologs/paralogs for genes associated with non-overlapping QTLs for the trait of interest.

CMap Database Alignments between genetic maps and the sorghum genome sequence can be viewed by the user. It also gives information on RFLP probe and SSR primer sequences for each anchored marker that is amenable for alignment.

Genome Browser This is implemented using Generic Genome Browser version 2.39 (Stein et al. 2002) and used to associate sorghum QTL

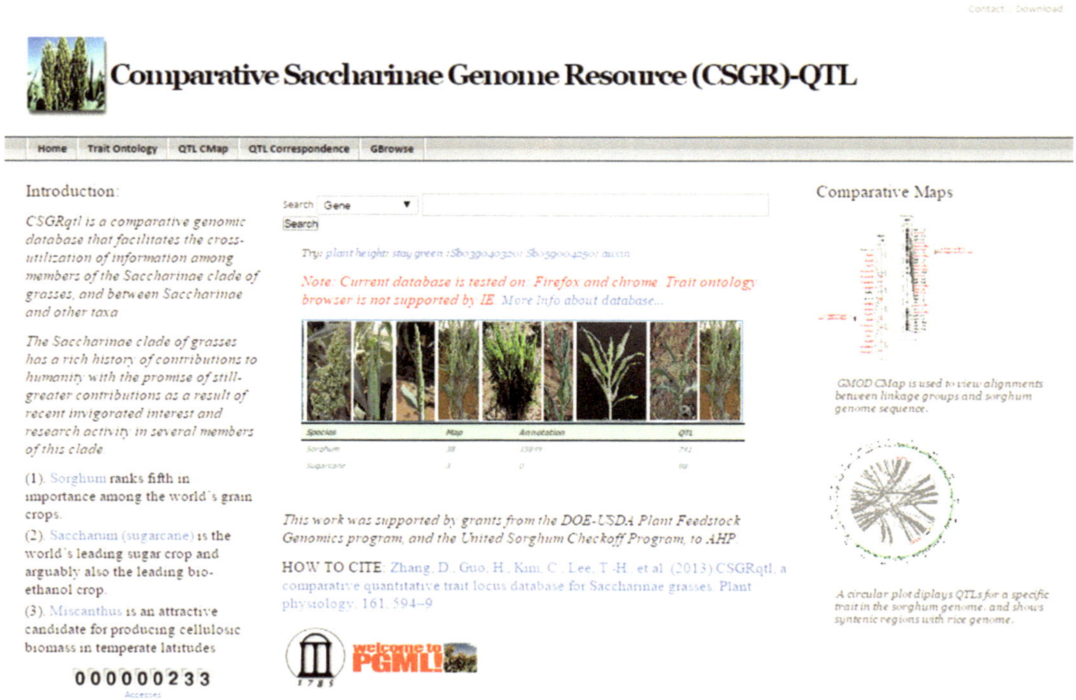

Fig. 6.7 Organization of CSGRqtl database (http://helos.pgml.uga.edu/qtl/)

data with gene annotations. This browser contains gene models from standard sorghum genome annotation version 1.4 and about 209,828 sorghum ESTs from the NCBI. It also contains GC content, six-frame translation, and restriction sites for each genomic region. The user can get information of all annotated genes for a particular QTL region and also can access all QTLs in any genomic region.

6.2.8 SorGSD

DNA sequence variations between diverse sorghum lines are an important pre-requisite for the genetic improvement of sorghum for agronomic traits as well as tolerance to biotic and abiotic stresses through breeding by design and high-efficiency genomic selection. Advances in the next-generation sequencing (NGS) technologies have brought about a surge in the re-sequencing of the diverse sorghum accessions belonging to different categories such as improved inbreds, landraces, wild/weedy sorghums, and wild relatives. Recently, a diverse panel of 48 sorghum accessions which were divided into four groups, including improved inbreds, landraces, wild/weedy sorghums, and a wild relative *Sorghum propinquum*, has been re-sequenced leading to the generation of enormous amount of SNP data (Mace et al. 2013). Proper organization of this SNP data will offer excellent opportunity for researchers to identify variation in their genes of interest, explore evolutionary relationships among cultivated and wild types, develop DNA markers for future genetic studies, and utilize this data for genome-wide association studies (GWAS). With this premise, SorGSD, a web-based large-scale genome variation database, was developed during August 2014 and maintained by the Data Management Center, Beijing Institute of Genomics (BIG), Chinese Academy of Sciences, and the Laboratory for Conservation and Utilization of Bio-resources, Institute of

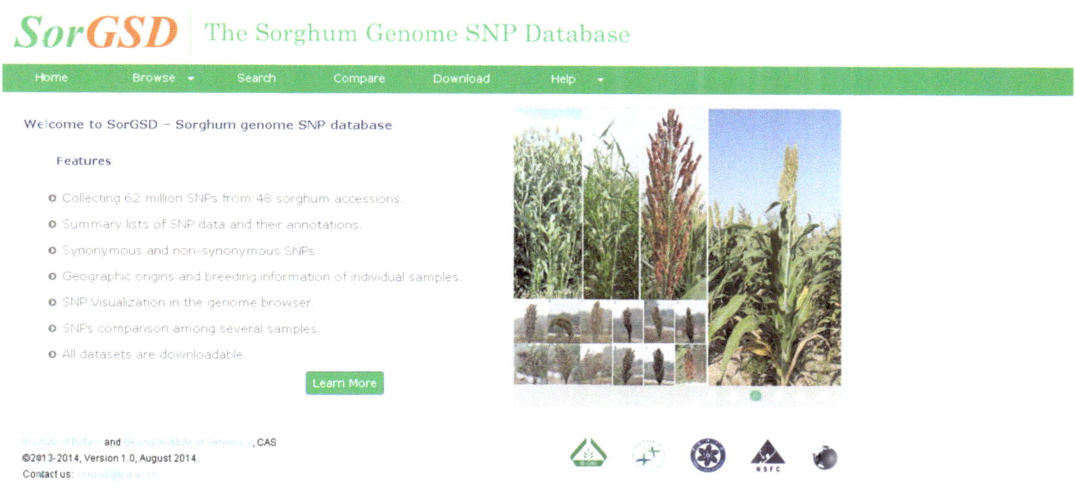

Fig. 6.8 Organization of SorGSD database (http://sorgsd.big.ac.cn/snp/)

Botany, Chinese Academy of Sciences. The database contains 62 million SNPs with annotations assisted by an easy-to-use web interface for users for efficient browsing, searching, and analysis of the SNPs. The pipeline for SNP calling included trimming of adapter and filtering of all low-quality reads, the use of BWA (version 0.6.2-r126) to map clean read to sorghum reference sequence (V1.4), SAMtools package to convert mapping results to BAM format, Picard (version 1.87) program to eliminate duplicated reads generated during the process of library construction, SNP calling by GATK (version 2.5-2-gf57256b) toolkit, SNP identification based on the quality estimation scores generated by GATK (quality value ≥30 and depth of coverage ≥5), and SnpEff program for the annotation of SNPs. This database is a rich repository to molecular breeders for the identification of biomarker, genetic analysis, and marker-assisted breeding of sorghum and other crops. The SorGSD can be accessed from http://sorgsd.big.ac.cn/snp/. The SorGSD has four modules, viz., Browse, Search, Compare, and Download as shown in Fig. 6.8.

Browse Module This module can be used to browse total SNPs as well as gene-wise and chromosome-wise SNPs. SNPs in Gene lists SNPs located in gene and coding regions. The "Coding," "Synonymous," and "Non-synonymous" lists are used to view SNP located in coding region, annotated as synonymous and non-synonymous. Query can be given based on the chromosome number also. The users can browse SNP information and their relevant annotations for each sorghum line. The database uses GBrowse to visualize InDel, SNP, gene, transcript, density information of SNP/300 kb, and allele frequency.

Search Module This module helps in searching SNPs in a single individual by setting parameters such as chromosome location, SNP class, and SNP location in gene region and genotype. Options are provided for the selection of SNP annotation and SNP genotype. The search results can be either visualized graphically in a genome browser or displayed in formatted tables.

Compare Module This module helps in comparing SNPs in two or more individuals by setting parameters such as chromosome location, SNP class, and SNP location in gene region and genotype. The SNPs in selected individuals can be compared with that of a single reference genotype or more genotypes. Options are provided for the selection of SNP annotation and SNP genotype.

Download Module This module contains the datasets, viz., SNP files, InDel files, SRA files, and Fastq files, which can be directly downloaded for further analysis.

6.3 Genomics Resources for Analyzing DNA Sequence Variation

In the current era of genomics, large-scale EST as well as genome sequencing projects resulted in the generation of enormous amount of DNA sequence data that are organized and stored in various databases. Such data available in the public domain are the main targets for the development of molecular markers such as simple sequence repeats (SSRs), insertion-deletions (InDels), single nucleotide polymorphisms (SNPs), etc., through various computational approaches and bioinformatics tools available. These molecular markers are the best tools available for the plant geneticists for analyzing the DNA sequence variations through an easy and rapid PCR assay and are useful for assessing the genetic diversity and population structure of germplasm lines, varietal identification, genetic purity testing of hybrids and parental lines, mapping of genetic loci through QTL mapping, and marker-assisted selection.

Assessment of genetic diversity in the germplasm accessions or the parental line gene pool is the primary step in any plant breeding program. Earlier, morphological as well as quantitative traits were used for the assessment of genetic diversity. Due to their inherent limitations in the number as well as environmental influence, molecular markers have become the choice of such genetic diversity assessments since these markers are environmentally neutral. Several studies were undertaken over the years in the assessment of genetic diversity using molecular markers such as RAPD (Ayana et al. 2000; Uptmoor et al. 2003), RFLP (Tao et al. 1993; Ahnert et al. 1996), AFLP (Geleta et al. 2006; Ritter et al. 2007), and ISSR (Aruna et al. 2012). These markers are also used in the mapping of major genes (Knoll et al. 2008; McIntyre et al. 2008) as well as QTL (Srinivas et al. 2009b; Satish et al. 2009).

SSR markers are the widely used PCR-based markers for various genetics and mapping studies in sorghum due to their abundance in the genome, highly polymorphic nature, and easy assay. Prior to the completion of genome sequencing of sorghum in 2009, several research groups have developed a large number of SSR markers which were subsequently used for various genetic studies in sorghum. These studies helped in the development of genomic SSR markers [Xtxp series (Kong et al. 2000; Bhattramakki et al. 2000, http://sorgblast3.tamu.edu/search/marker.htm), XSb series (Taramino et al. 1997), Xgap series (Brown et al. 1996)], SSR markers derived from cDNAs [Xcup series (Schloss et al. 2002)], whole-genome sequence-based SSR markers [SB series (Yonemaru et al. 2009)], expressed sequence tag (EST)-based SSR markers [Xisep series (Ramu et al. 2009), Xiabt series (Arun 2006; Reddy et al. 2008), Stgnhsbm and Dsenhsbm series (Srinivas et al. 2008, 2009a)], SSR markers derived from unigenes [Ungnhsbm series (Srinivas et al. 2009b; Nagaraja Reddy et al. 2012)], $(GATA)_n$ motif-based SSR markers [SbGM series (Jaikishan et al. 2013)], and other SSR markers of an unknown type [gpsb and mSbCIR series (developed at CIRAD, France, and partially published in Mace et al. 2009)]. The details of the SSR markers developed by different sorghum groups are given in Table 6.2.

Even though several studies on the assessment of genetic diversity were reported over the years, very few studies have done a comprehensive analysis and resulted in a set of robust SSR markers that can be used universally across laboratories for this purpose. A set of 38 SSR markers distributed across 10 chromosomes of sorghum selected based on three different linkage maps were used to establish the diversity research set comprising 107 sorghum accessions (Shehzad et al. 2009) from a set of 320 sorghum germplasm

Table 6.2 SSR markers developed by different sorghum research groups

Type of SSR markers	Marker series	No. of markers developed	No. of markers experimentally tested	Reference
Genomic SSR	Xtxp	206	165	Bhattramakki et al. (2000), Kong et al. (2000)
		38	38	
	XSb	15	13	Taramino et al. (1997)
	Xgap	149	149	Brown et al. (1996)
cDNA-derived SSR	Xcup	74	60	Schloss et al. (2002)
Whole-genome SSR	SB	5,599	970	Yonemaru et al. (2009)
EST-derived SSR	Xisep	600	386	Ramu et al. (2009)
	Xiabt	520		Arun (2006)
	Stgnhsbm	50	50	Srinivas et al. (2008, 2009a)
		116	109	
Unigene-derived SSR	Dsenhsbm	50	50	Srinivas et al. (2009b), Nagaraja Reddy et al. (2012)
	Ungnhsbm	1,519	302	
$(GATA)_n$ motif-based SSR	SbGM	110	50	Jaikishan et al. (2013)
Other SSRs	gpsb and mSbCIR	30	24	Mutegi et al. (2011); Billot et al. (2012)

accessions. A diversity analysis kit was developed (Billot et al. 2012), which contains information on 48 robust sorghum SSR markers that can be used to calibrate SSR genotyping data acquired with different technologies and compare them to genetic diversity references. A reference set comprising a wide range of sorghum genetic diversity was screened with 40 EST-SSR markers by Ramu et al. (2013), and the analysis highlighted the greater discriminating power of these markers as compared to the genomic SSR markers.

In the current era of genome sequencing, the discovery of SNPs and insertion and (or) deletions (InDels) through high-throughput methods has led to a revolution in their use as DNA markers (Batley and Edwards 2007; Batley et al. 2007; Edwards et al. 2007). Advancements in sequencing technologies, execution of re-sequencing projects, and availability of the enormous amount of ESTs along with the development of efficient computational platforms have helped in the rapid discovery of SNPs and InDels in sorghum. SNPs may be considered the ultimate genetic marker as they represent the finest resolution of a DNA sequence, generally are abundant in populations, and have a low mutation rate (Syvanen 2001). The mining of readily available sequence data for SNPs through in silico approaches significantly reduces the costs (Taillon-Miller et al. 1998), and several SNP mining tools have been developed (Barker et al. 2003; Batley et al. 2003; Savage et al. 2005; Chagne et al. 2007). Sorghum researchers across the globe have utilized different types of data, such as ESTs, whole-genome re-sequencing data, and genotyping-by-sequencing data for the discovery of SNPs with the help of various computational tools. The detail of the SNPs developed by different sorghum groups is given in Table 6.3.

InDels are next only to SNPs in terms of their abundance in the genome. However, InDels can be converted into PCR-based markers and can be resolved through routine gel electrophoresis systems. InDels exhibit length polymorphisms that have been successfully exploited in sorghum in the mapping of important loci such as waxy (McIntyre et al. 2008) and tannin (Wu et al. 2012). In sorghum, about 99,948 InDels of 1 to 10 bp in length were detected by Zheng et al. (2011) through a genome-wide analysis in sweet and grain sorghum. Potential intron polymorphism (PIP) markers developed by Yang et al.

Table 6.3 SNPs developed by different sorghum research groups

Target data	Computational tool used	No. of SNPs identified	Reference
ESTs	CodonCode Aligner	12,421 SNPs	Girma (2009)
ESTs	HaploSNPer	77,094 potential and 40,589 reliable SNPs	Singhal et al. (2011)
Re-sequencing data	SOAPsnp software	1,057,018 SNPs	Zheng et al. (2011)
Eight genome equivalents to reference genome	SOAP v2 and NovoAlign	283,000 SNPs	Nelson et al. (2011)
GbS data of 971 diverse sorghum accessions	TASSEL 3.0 GBS pipeline	265,487 SNPs	Morris et al. (2013)
Re-sequencing data	realSFS and SOAPsnp	4,946,038 SNPs	Mace et al. (2013)

(2007) are a unique type of markers that targets both the SNPs and InDels. Among the two types of polymorphisms, intron length polymorphism (ILP) can be easily detected by exon-primed intron-crossing PCR (EPIC-PCR) (Palumbi 1995), where primers are designed in exonic regions flanking the target introns. Potential intron polymorphism (PIP) database for plants was developed by Yang et al. (2007) comprising a total of 57,658 PIP markers for 59 plant species, of which 4314 are of sorghum. These markers can be exploited for genetic diversity assessment, cultivar identification, mapping, and marker-assisted selection.

A new high-throughput hybridization-based marker technology that could serve as an efficient alternative to low-throughput gel-based marker systems was reported in sorghum by Mace et al. (2008). This system does not require sequence information and is amenable for high multiplexing. A genotyping array was developed with ~12,000 genomic clones using PstI + BanII complexity with a subset of clones obtained through the suppression subtractive hybridization (SSH) method. About 508 markers were polymorphic and were used for the genetic diversity analysis of 90 diverse sorghum genotypes and for the construction of a genetic linkage map for a cross between R931945-2-2 and IS 8525. These markers are useful for whole-genome profiling and can be used for diversity analyses and construction of medium-density genetic linkage maps.

A robust SNP array platform was developed recently by Bekele et al. (2013) using 2,124 selected Infinium Type II SNPs from a total of over one million high-quality SNPs identified by the alignment of whole-genome sequences (6–12× coverage) of genetically diverse genotypes comprising two grain and three sweet sorghum genotypes of *S. bicolor* and an additional 876 SNPs selected based on their phenotypic association with early-stage chilling tolerance identified by phenotype-based pool sequencing. Testing this array with selected SNPs using 564 genotypes comprising four unrelated RIL and F_2 populations and a genetic diversity collection resulted in the validation of 2,620 robust and polymorphic SNPs. This SNP array platform is very useful for genetic mapping, genome-wide association, and genomic selection.

6.4 Bioinformatics Resources for DNA Sequence Analysis

DNA sequence variations arise either due to point or gross mutations. Point mutations are mostly due to base substitution (transition or transversion). Gross mutation may be insertion or deletion (InDels) of few to large sequences. It also may arise by duplication, inversion, and translocation. Gross mutations involving large sequences can easily be detected cytologically, while it is difficult to detect gross mutations of smaller dimension (say < 500 bases). Sequencing helps us to detect this variation very precisely. For this purpose, the mutant sequence is aligned with the wild-type sequence, and sequence variation is detected.

Sequence alignment is a way of arranging the nucleic acid (DNA, RNA) or protein sequences to identify regions of similarity that may be a result of functional, structural, or evolutionary relationships between the sequences (Mount 2004). Aligned nucleotide or amino acid sequences are generally represented as a row matrix by inserting gaps between the residues such that identical characters are aligned in successive columns. In the case of two sequences sharing a common ancestor, mismatches are interpreted as point mutations and gaps as InDels. Conservation of base pairs indicates a similar functional or structural role of the target sequence. Very short or very similar sequences can be aligned manually. However, with sequencing projects, large sequences are available, and these need to be aligned in large number, which is not possible manually. Different algorithms have been developed to facilitate high-quality sequence alignments. These include dynamic programming, heuristic algorithms, or probabilistic methods. Computational approaches to sequence alignment are of two categories: global alignments and local alignments.

Global Alignment With this approach, both sequences are aligned along their entire lengths, including every nucleotide or amino acid, and the best alignment is found. This approach is most useful if the query sequences are similar and of roughly equal size. Dynamic programming called Needleman-Wunsch algorithm is very popularly used for this purpose.

Local Alignment In this approach, the best subsequence alignment, which includes only the most similar sequence, is found. This approach is useful, particularly for dissimilar sequences that are expected to contain similar sequence motifs within their larger sequence. The dynamic programming method, namely, Smith-Waterman algorithm, is commonly used for this purpose.

Both the global and local alignments lead to erroneous conclusion when the downstream part of one sequence overlaps with the upstream part of the other sequence. In such situations, hybrid methods such as "semi-global" or "glocal" (short for *global-local*) methods are employed. These methods help in finding the best possible alignment that includes the start and end of one or the other sequence. The sequence alignment may be of two types based on the number of sequences used for alignment, viz., pair-wise sequence alignment (PSA) and multiple sequence alignment (MSA).

PSA This is the comparison of two biological sequences (nucleic acid or protein) at a time to reveal the similarity or homology between them. This helps in finding the best-matching local or global alignments of two query sequences. This method is most useful in situations where extreme precision like searching of database for sequences with high similarity is not required. Dot matrix methods, dynamic programming, and word methods are mostly used for pair-wise alignments. Several PSA tools have been developed by several workers (Table 6.4), many of them are free to use, and some of them are available as commercial software. BLAST is the best example for pair-wise sequence alignment. BLAST searches a query sequence (discovered sequence) against a database of known sequences in order to find similarities. The program compares nucleotide or protein sequences to sequence databases and calculates the statistical significance of matches. BLAST searching can be performed using the web-based applications (Web BLAST) or can be run locally in the PC provided it has an existing database to aid searching. There are five types of BLAST:

- BLASTN: Compares a nucleotide query sequence against a nucleotide sequence database
- BLASTP: Compares an amino acid query sequence with a protein database
- BLASTX: Translates a DNA sequence into six protein sequences using all six possible reading frames and then compares each of these proteins to protein database
- TBLASTN: Translates every DNA sequence in a database into six potential proteins and then compares the protein query against each of those translated proteins
- TBLASTX: Translates DNA from both a query and a database into six potential proteins and then performs 36 protein-protein database searches

Table 6.4 Important resources for pair-wise alignment

Name	Sequence type	Alignment type	Availability
AlignMe	P	G, L	http://www.bioinfo.mpg.de/AlignMe
BLAST	P, N	L	http://blast.ncbi.nlm.nih.gov/Blast.cgi
JAligner	P, N	L	http://jaligner.sourceforge.net/
LALIGN	P, N	L (non-overlapping)	http://www.ch.embnet.org/software/LALIGN_form.html
mAlign	N	G, L	ftp://ftp.csse.monash.edu.au/software/m-align/
MCALIGN2	N	G	http://homepages.ed.ac.uk/eang33/mcalign/mcinstructions.html
MUMmer	N	G	http://mummer.sourceforge.net/
needle	P, N	SG	http://www.ebi.ac.uk/Tools/emboss/align/
PatternHunter	N	L	http://www.bioinformaticssolutions.com/products/ph/download_academ.php
PyMOL	P	G (by selection)	www.pymol.org
REPuter	N	L	http://bibiserv.techfak.uni-bielefeld.de/reputer/
SEQALN	P, N	L or G	http://www-hto.usc.edu/software/seqaln/seqaln-query.html
tranalign	N	NA	http://mobyle.pasteur.fr/?form=tranalign
UGENE	P, N	G, L	http://ugene.unipro.ru/

P protein, *N* nucleic acid, *G* global, *L* local, *SG* semi-global

MSA This is the comparison of more number of sequences (three or more) simultaneously. It helps in predicting the structure and function of a given protein sequence due to its ability to detect common features and conserved domains across the sequences analyzed. It also helps to discover novel and related sequences and to construct and search for sequence patterns. Detection of conserved regions among homologous sequences is useful in designing PCR primers. There are a number of MSA programs available in the public domain (Table 6.5).

Clustal is a widely used multiple sequence alignment tool (Chenna et al. 2003). There are three main variations: ClustalW (command line interface, Larkin et al. 2007), ClustalX (this version has a graphical user interface, Thompson et al. 1997), and Clustal Omega [(allows hundreds of thousands of sequences to be aligned in only a few hours). It will also make use of multiple processors, where present. In addition, the quality of alignments is superior to previous versions (Sievers et al. 2011)]. A wide range of input formats, including NBRF/PIR, FASTA, EMBL/Swiss-Prot, Clustal, GCC/MSF, GCG9 RSF, and GDE, are acceptable in this program. The output format can be one or many of the following: Clustal, NBRF/PIR, GCG/MSF, PHYLIP, GDE, or NEXUS. There are three main steps in the alignment process, i.e., pair-wise alignment, followed by the creation of a guide tree (or use a user-defined tree) and finally the use of the guide tree to carry out a multiple alignment. If "Do Complete Alignment" option is selected, all these steps are done automatically, or else the task may be carried out following options, viz., "Do Alignment from guide tree" and "Produce guide tree only." There is an option for default setting or customized settings.

Alignments may be represented graphically and in text format. An asterisk or pipe symbol is commonly used to show identity between two columns. Colors are also used by several programs to display identity and dissimilarity. The sequence alignment results are stored in a variety of text-based file formats. Most common input and output formats are FASTA format and GenBank format.

6.5 Resources for Analyzing Transcriptomes

Transcriptome analysis involves the screening of candidate genes, predicting its function, and discovery of regulatory elements through high-throughput gene expression analysis. Initially, large-scale sequencing of ESTs was used as the main approach for transcriptome analysis. Later, the hybridization-based methods such as microarrays/GeneChips were developed and popularly used for large-scale gene expression analysis. Sequencing-based methods such as serial analysis of gene expression (SAGE) and massively parallel signature sequencing (MPSS) have been successfully employed for the identification of a large number of transcripts along with quantitative comparison of transcriptomes (Velculescu et al. 1995; Brenner et al. 2000). Furthermore, as a next-generation DNA sequencing application, deep sequencing of short fragments of expressed RNAs, including sRNAs, is quickly becoming an efficient tool for use with genome-sequenced species (Harbers and Carninci 2005; de Hoon and Hayashizaki 2008).

A sorghum cDNA microarray providing data on 12,982 unique gene clusters was used by Buchanan et al. (2005) to examine genome-wide changes in gene expression in sorghum seedlings under high salinity (150 mM NaCl), osmotic stress (20 % polyethylene glycol), or abscisic acid (125 μM ABA). A total of 3,508 cDNAs selected from the two cDNA libraries constructed from a strong greenbug resistance sorghum line (M627) and a susceptible line (Tx7000) with or without infestation were used to develop a cDNA microarray for the identification of sorghum genes responsive to greenbugs (Park et al. 2006).

An Agilent rice gene expression microarray (product number: G2519F, 44 K) was used to study tissue-specific gene expression profiles of *S. propinquum* with special emphasis on rhizome development by Zhang et al. (Zhang et al. 2013b) that contained 45,220 independent probes (60-mer) corresponding to 21,495 *O. sativa* mRNA sequences available in GenBank due to the non-availability of microarray platform in sorghum. Only recently, the first whole-transcriptome

Table 6.5 Important resources for multiple sequence alignment

Name	Sequence type	Alignment type	Availability
AMAP	P, N	G	http://baboon.math.berkeley.edu/mavid/
Base-By-Base	P, N	L or G	http://athena.bioc.uvic.ca/virology-ca-tools/base-by-base/
ClustalW	P, N	L or G	http://www.ebi.ac.uk/Tools/msa/clustalw2/
CodonCode Aligner	N	L or G	http://www.codoncode.com/aligner/
Compass	P	G	http://prodata.swmed.edu/compass/compass_advanced.php
DNA Baser Sequence Assembler	N	L or G	www.DnaBaser.com
FSA	P, N	G	http://orangutan.math.berkeley.edu/fsa/
MSA	P, N	L/G	http://www.ncbi.nlm.nih.gov/CBBresearch/Schaffer/msa.html
MULTALIN	P, N	L/G	http://multalin.toulouse.inra.fr/multalin/
MUSCLE	P, N	L/G	http://www.ebi.ac.uk/Tools/msa/muscle/
ePROBALIGN	P	G	http://probalign.njit.edu/index.html
PSAlign	P, N	L/G	http://faculty.cs.tamu.edu/shsze/psalign/
RevTrans	N/P (special)	L/G	http://www.cbs.dtu.dk/services/RevTrans/
SAM-T08	P	L/G	http://compbio.soe.ucsc.edu/SAM_T08/T08-query.html
T-Coffee	P, N	L/G	http://tcoffee.crg.cat/

P protein, *N* nucleic acid, *G* global, *L* local, *SG* semi-global microarray was developed in sorghum by Shakoor et al. (2013) comprising a gene chip containing 1,026,373 probes covering 149,182 exons (27,577 genes) across the nuclear, chloroplast, and mitochondrial genome along with putative non-coding RNAs to identify tissue-specific genes and novel regulatory sequences. Toward identification and functional characterization of genes in sorghum genome, Shakoor et al. (2014) used the first commercial whole-transcriptome sorghum microarray chip (Sorgh-WTa520972F) to identify tissue- and genotype-specific expression patterns using grain, sweet, and bioenergy sorghums. Microarray dataset was generated using 78 samples involving different tissue types (shoot, root, leaf, and stem) and dissected stem tissues (pith and rind) of six diverse genotypes (R159, Atlas, Fremont, PI152611, AR2400, and PI455230), which revealed tissue- and genotype-specific expression patterns of different metabolic pathways indicating the importance of intraspecies variations in sorghum.

Next-generation sequencing technologies have enabled the researchers to characterize small RNA component of the transcriptomes in many plant species. According to the recent release of the miRBase database (http://www.mirbase.org, release 20: June 2013), about 205 miRNAs are described for sorghum, whereas 592 miRNAs are described for rice. The sorghum genome sequencing consortium identified 149 predicted miRNAs belonging to 27 miRNA families (Paterson et al. 2009). The identification of miRNAs from different target tissues, developmental stages, and stress treatments offer an excellent opportunity to understand the role of miRNAs in the regulation of expression of genes influencing traits of agronomic importance.

Prior to whole-transcriptome microRNA (miRNA) sequencing projects, computational approaches based on homology search were used for the identification of miRNAs in different plant species. Using this approach, Du et al. (2010) identified a total of 17 new miRNAs based on the GSS and the miRNA secondary structure that were distributed unevenly among 11 miRNA families. Analysis of these miRNAs via online software miRU revealed that they might regulate 64 target genes, most of which are involved in RNA processing, metabolism, cell cycle, protein degradation, stress response, and transportation. The small RNA component of the transcriptome of grain and sweet sorghum stems was characterized by Calviño et al. (2011) using F_2 population derived from the cross between BTx623 (grain sorghum) and Rio (sweet sorghum) that segregated for sugar content and flowering time. They reported that the variation in miR172 and miR395 expression correlated with flowering time, while that of miR169 correlated with sugar content in stems.

With the increasing number of whole genomes, large-scale cDNA sequencing, and microarray projects in the recent years, an enormous amount of transcriptome data is generated and stored in public databases. This data serves as a valuable resource for many secondary uses, such as co-expression and comparative transcriptome analyses. NCBI's Gene Expression Omnibus (GEO) (http://www.ncbi.nlm.nih.gov/geo/) and the European Bioinformatics Institute (EBI)'s ArrayExpress (http://www.ebi.ac.uk/arrayexpress/) are similar such databases, which serve as the primary archives of transcriptome data in the public domain (Parkinson et al. 2007; Barrett et al. 2009). Transcriptome data of *S. propinquum* in relation to rhizome development and the data of grain (BTx623) and sweet (Keller) sorghum are stored in GEO. ArrayExpress database has the transcriptome datasets of tissue-specific transcriptomic profiling of *S. propinquum* using a rice genome array, sorghum gene expression using Agilent custom 4x44K microarray, RNA-Seq of *S. bicolor* 9d seedlings in response to osmotic stress and abscisic acid, and high-throughput sequencing of small RNAs in *S. bicolor*. Another database, namely, EGENES (http://www.genome.jp/kegg-bin/create_kegg_menu?category=plants_egenes), is a multi-species resource integrating the genomic, chemical, and network information comprising genes, molecules, and biological pathways representing the cellular functions. This resource is useful for the comparison and mutual validation of genome-based pathway annotation and EST-based annotation. The EGENES consists of data on 25 eukaryotic species including sorghum. The sorghum datasets include 190,946 ESTs, 19,597 contigs, 23,171 singletons, 122 pathway maps, and 1,189 mapped contigs.

6.6 Genetic Resources for Mapping Agronomically Important Traits

The majority of the agriculturally important traits such as yield, biotic and abiotic stress tolerance is complex and is governed by quantitative trait loci (QTLs). These QTLs are influenced by the environment and the interaction between QTL and environment. Linkage mapping (biparental mapping) and linkage disequilibrium (LD) mapping (association mapping) are the two most commonly used approaches for the dissection of such complex traits. The first step in QTL mapping is the development of mapping populations by crossing parental lines that are contrasting for the trait of interest (biparental population and multiparental population) or assembling of diverse germplasm lines (natural population), which serve as an important genetic resource.

Biparental Population This approach involves the mating between two parental lines that are contrasting for the trait of interest and advancing them to develop F_2, backcross populations, recombinant inbred lines (RILs), backcross inbred lines (BILs), and double haploid (DH). In addition to this, introgression lines (ILs) and near isogenic lines (NILs) are also developed. These mapping populations are known as a first-generation mapping resource. Even though several preliminary studies used F_2 populations, the use of advanced generations, particularly RILs derived by single-seed descent from F_2 individuals from a cross between two distinct homozygotes, is most commonly used for QTL mapping purposes (Keurentjes et al. 2011) because they are immortal and can be multiplied any number of times (Huang et al. 2011) for phenotyping in different environments/seasons. To tackle the problem of epistasis due to the interaction between multiple loci, the biparental populations such as ILs and NILs are also used (Rakshit et al. 2012). The advantage of employing NIL over RIL is mainly the detection of minor QTL that are missed while using RILs (Keurentjes et al. 2007). Globally, many sorghum research groups have developed and used several RIL populations for various traits (Table 6.6).

Multi-parental Mapping Populations Biparental populations, though popularly used for QTL mapping, have two major limitations such as relying on the recombination events happening in the F_1 and subsequent generation and mapping only the allelic pairs that are present in the two contrasting parents (Rakshit et al. 2012). This affects the map resolution of the QTL since QTL will be placed on a large chromosomal region (Li et al. 2010). In the recent past, the second-generation mapping resources such as association mapping, nested association mapping (NAM), and multiparent advanced generation inter-cross (MAGIC) were developed to overcome the limitations of biparental populations. NAM populations are developed by crossing a central parent with other diverse parents in a star design (Huang et al. 2011), and such populations have been established in maize (Yu et al. 2008; Buckler et al. 2009; McMullen et al. 2009) and *Arabidopsis* (Bentsink et al. 2010; Brachi et al. 2010). Development of a large NAM population in sorghum was reported recently by Jordan et al. (2012) comprising more than 4,000 lines from 100 sub-populations derived from a large BC_1F_1 population using a single elite line as the recurrent parent resulting in the sampling of the diversity of sorghum including wild relatives. Such populations help in fine mapping of QTL; however, the interaction of QTL with genetic background cannot be analyzed since one parent is common in all sub-populations. Consequently, the concept of MAGIC population was proposed by Cavanagh et al. (2008) to address the major limitations of biparental mapping populations. This concept was used as an additional resource for dissecting the genetics of natural varieties in *Arabidopsis* multi-parent recombinant inbred line (AMPRIL) population (Huang et al. 2011). Recently, another multi-parent mapping population

Table 6.6 Biparental mapping populations in sorghum

Trait	Mapping population	Population size	Reference
Plant phenology			
Plant height			
dw_2	*S. bicolor* × *S. propinquum*	F_2: 320	Lin et al. (1995)
dw_2	Shan Qui Red × SRN39	RIL: 153	Klein et al. (2008)
	296B × IS18551	RIL: 168	Madhusudhana and Patil (2013)
Photoperiod sensitivity	BTx623 × IS3620C	RIL: 127	Childs et al. (1997)
	IS2807 × IS7680	RIL: 85	Chantereau et al. (2001)
Flowering time	Kikuchi Zairai × SC112	F_2: 144	El Mannai et al. (2012)
	BTx642 and Tx7000	RIL: 90	Yang et al. (2014)
Maturity			
Ma_3	100 M × 58 M	RIL: 137	Childs et al. (1997)
Ma_1	*S. bicolor* × *S. propinquum*	F_2: 320	Lin et al. (1995), Klein et al. (2008)
Ma_4	BTx623 × IS3620C	RIL: 137	Hart et al. (2001)
Plant color	IS2807 × 379	RIL: 110	Rami et al. (1998)
	IS2807 × 249	RIL: 91	
	RTx430 × Sureno	RIL: 125	Klein et al. (2001)
	296B × IS18551	RIL: 168	Srinivas et al. (2009b)
Stem	B35 × Tx7000	RIL: 98	Xu et al. (2000)
	BTx623 × IS3620C	RIL: 137	Hart et al. (2001)
	296B × IS18551	RIL: 168	Srinivas et al. (2009b)
Tillering	BTx623 × IS3620C	RIL: 137	Hart et al. (2001)
			Kebrom et al. (2006)
Inflorescence architecture	BTx623 × IS3620C	RIL: 119	Brown et al. (2006)
Nodal root angle	B923296 × SC170-6-8	RIL: 141	Mace et al. (2012)
Agronomic traits			
Agronomically important traits	296B × IS18551	RIL: 168	Srinivas et al. (2009b)
	Sorghum bicolor × *S. sudanense*	$F_{2:3}$: 248	Fang et al. (2011)
	654 × LTR108	RIL: 244	Zou et al. (2012)
	M35-1 × B35	RIL: 245	Nagaraja Reddy et al. (2013)
Agronomic traits and yield components	IS2449 × IS1488	RIL: 100	Phuong et al. (2013)
Grain quality, productivity, morphological and agronomical traits	IS2807 × 379	RIL: 110	Rami et al. (1998)
	IS2807 × 249	RIL: 90	
Hybrid-related traits			
Fertility restoration			
Rf_1	ATx623 × RTx432	F_2: 373	Klein et al. (2001)
Rf_2	R931945-2-2 × IS8525	RIL: 285	Jordan et al. (2010)
	B923296 × SC170-6-8	RIL: 233	
rf_4	(A3Tx398*4/IS1112C//B3Tx398)	BC_3F_1: 378	Wen et al. (2002)
Rf_5	BTx642 × QL12	RIL: 218	Jordan et al. (2011)
Seed-/grain-related traits			
Seed weight	Tx7078 × B35	HILs	Tuinstra et al. (1997)

(continued)

Table 6.6 (continued)

Trait	Mapping population	Population size	Reference
Grain shattering	*S. bicolor* × *S. propinquum*	F_2: 370	Wise et al. (2002)
Grain color	Shan Qui Red × SRN39	RIL: 153	Knoll et al. (2008)
	B35 × Tx7000	RIL: 98	Xu et al. (2000)
Grain testa	IS2807 × 379	RIL: 110	Dufour et al. (1997),
	IS2807 × 249	RIL: 91	Rami et al. (1998)
Grain texture	IS2807 × 379	RIL: 110	Boivin et al. (1999)
	IS2807 × 249	RIL: 91	
	QL39 × QL41	RIL: 160	Tao et al. (2000)
Awns	IS2807 × 379	RIL: 110	Boivin et al. (1999)
	IS2807 × 249	RIL: 91	Tao et al. (2000)
	BTx623 × IS3620C	RIL: 137	Hart et al. (2001)
Glumes	296B × IS18551	RIL: 168	Srinivas et al. (2009b)
Grain pericarp color	QL39 × QL41	RIL: 160	Tao et al. (2000)
Grain protein digestibility	Sureno × P850029	RIL: 277	Winn et al. (2009)
Endosperm texture (waxy)	BTxARG1 x QL39	F_2	McIntyre et al. (2008)
	RTx2907 x QL39		
Biofuel-related traits			
Sugar-related traits	Early Folger × N32B	$F_{2:3}$: 207	Yun-long et al. (2006)
	R9188 × R9403463-2-1	RIL: 184	Ritter et al. (2008)
Brown midrib			
bmr6	Brown County × *bmr6-ref* line	RIL: 218	Saballos et al. (2009)
bmr12	bmr12 × N12	NIL	Bout and Vermerris (2003)
bmr2	AMP11 × Theis	F_2: 200	Saballos et al. (2012)
Biotic stress tolerance			
Head smut resistance	HC325 × RTx7078	F_2: 52	Oh et al. (1994)
Rust resistance	QL39 × QL41	RIL: 160	Tao et al. (1998) McIntyre et al. (2004)
Grain mold tolerance	Sureño × RTx430	RIL: 125	Klein et al. (2001)
Greenbug tolerance	GBIK × Redlan	RIL: 93	Agrama et al. (2002)
	Westland A line × PI550610	F_2: 217	Wu and Huang (2008)
	BTx623 × PI 607900	F_2: 371	Punnuri et al. (2013)
Midge resistance	ICSV745 × 90562	RIL: 120	Tao et al. (2003)
Sorghum aphid resistance	BTx623 × Henong 16	F_3: 64 (homozygous susceptible) and 571 F_4 seedlings	Wang et al. (2013)
Striga resistance	IS9830 × E36-1	RIL: 226	Haussmann et al. (2004)
	N13 × E36-1		
Head bug resistance	Malisor 84–7 × S 34	F_2: 217	Deu et al. (2005)
Anthracnose resistance	HC136 × G73	F_2: 110	Monika Singh et al. (2006)
	BTx623 × SC748-5	$F_{2:3}$	Perumal et al. (2009)
Ergot resistance	R931945-2-2 × IS8525	RIL: 146	Parh et al. (2008)
Stalk rot resistance	IS22380 × E36-1	RIL: 93	Srinivasa Reddy et al. (2008)

(continued)

Table 6.6 (continued)

Trait	Mapping population	Population size	Reference
Shoot fly resistance	296B × IS18551	RIL: 168	Satish et al. (2009)
	27B × IS2122	RIL: 210	Aruna et al. (2011)
Foliar diseases	296B × IS18551	RIL: 168	Murali Mohan et al. (2010)
Abiotic stress tolerance			
Drought tolerance	Tx7078 × B35	$F_{5:8}$ HIFs: 98	Tuinstra et al. (1998)
	B35 × Tx430	RIL: 96	Crasta et al. (1999)
	SC56 × Tx7000	RIL: 125	Kebede et al. (2001)
	B35 × Tx7000	RIL: 98	Sanchez et al. (2002)
	E36-1 × SPV570	RIL: 184	Rajkumar et al. (2013)
Stay green (drought)	B35 × Tx7000	RIL: 98	Xu et al. (2000)
	QL39 × QL41	RIL: 152	Tao et al. (2000)
	IS9830 × E36-1	RIL: 226	Haussmann et al. (2002)
	N13 × E36-1	RIL: 226	
	BTx642 × RTx7000	NIL	Harris et al. (2007)
	296B × IS18551	RIL: 168	Srinivas et al. (2008)
Early-season cold tolerance	Shan Qui Red × SRN39	RIL: 153	Knoll et al. (2008)
Aluminum tolerance	BR007 × SC283	RIL: 354	Magalhaes et al. (2007)
Bloom (drought)	BTx623 × KFS2021	F_2: 220	Burow et al. (2009)
Other traits			
Stem and leaf structural carbohydrates	BTx623 × Rio	RIL: 176	Murray et al. (2008)
Fiber-related traits	SS79 × M71	RIL: 188	Shiringani and Friedt (2011)
Preharvest sprouting resistance	Redland B2 × IS9530	F_2	Lijavetzky et al. (2000)

RIL recombinant inbred line, *NIL* near isogenic line, *HIL* heterogeneous inbred lines of NILs

known as wide diallel population derived from 19 founder lines of sorghum selected from a wide gene pool was used to map the heterotic trait locus and to identify intra-locus interactions underlying hybrid vigor (Ben-Israel et al. 2012).

Natural Populations Linkage analysis involves analyzing a limited number of recombination events that occur during the construction of mapping populations, which results in the localization of QTL in the interval of 10–20 cM. Moreover, cost is involved in the propagation and evaluation of a large number of lines (Doerge 2002; Holland 2007). While several linkage analysis studies have been conducted in sorghum over the past two decades, only a limited number of genes were cloned or tagged at the gene level (Bout and Vermerris 2003; Saballos et al. 2009, 2012). Association mapping, also known as LD mapping, has emerged as an important tool to dissect the genetics of complex traits at the sequence level by exploiting the recombination events accumulated in the natural population (germplasm lines) during the course of evolution (Nordborg and Tavaré 2002; Risch and Merikangas 1996). According to Yu and Buckler (2006), association mapping has three advantages as compared to conventional linkage analysis, viz., (1) better mapping resolution, (2) reduction in research time, and (3) access to greater allele number. In addition, association mapping enables researchers to use next-generation sequencing technologies to exploit the diversity present in the natural populations, the value of which is known to crop breeders but exploited to a limited extent.

Sorghum is most suitable for association mapping of complex traits since it harbors one fourth sequence diversity that of maize (Hamblin et al. 2006), a 26-fold less population recombination than in maize (Hamblin et al. 2005), and natural homozygosity. LD is extensive enough in sorghum, which allows the simplification of a large number of SNPs into a smaller number of haplotypes resulting in reduced genotyping costs and increased statistical power (Clark 2004). Association mapping studies in sorghum using natural populations have been reported for plant growth and development (Kong et al. 2013), plant height (Murray et al. 2009; Wang et al. 2012), grain quality (de Alencar Figueiredo et al. 2010; Sukumaran et al. 2012), morphological traits (Shehzad et al. 2009), Brix (Murray et al. 2009), kernel weight and tiller number (Upadhyaya et al. 2012a), plant height and maturity (Upadhyaya et al. 2012b), endosperm quality (Hamblin et al. 2007), and agroclimatic traits (Morris et al. 2013).

Assembly of the association mapping panel comprising sorghum germplasm accessions possessing extensive genetic diversity is an important prerequisite for any association mapping study. Few association mapping panels were developed by different sorghum research groups across the world. An association mapping panel comprising 377 accessions representing all major cultivated races and important US breeding lines along with their progenitors was assembled and characterized for genetic and phenotypic diversity (Casa et al. 2008), which serves as an important genetic resource for the sorghum research community. Interested researchers can use this association panel and phenotype for their trait of interest without the need for further genotyping since the genotypic data along with appropriate statistical models are available for ready use.

A mini core comprising 242 accessions was developed from a core collection of 2,247 accessions through hierarchical cluster analysis using the phenotypic distances estimated from 11 qualitative and 10 quantitative traits and selecting about 10 % or a minimum of one accession per cluster covering a total of 21 clusters. Statistical comparisons based on homogeneity of distribution for geographical origin, biological races, qualitative traits, means, variances, phenotypic diversity indices, and phenotypic correlations indicated that the mini core collection represented the core collection (Upadhyaya et al. 2009). In addition to this, a reference set was developed comprising 374 sorghum accessions through the Generation Challenge Program as a means to enhance utilization of genetic resources in crop improvement (http://www.icrisat.org/what-we-do/crops/sorghum/Sorghum_Reference.htm).

A sorghum diversity research set (SDRS) comprising 107 sorghum accessions representing geographically diverse accessions from 27 countries in Asia and Africa was developed by Shehzad et al. (2009) through the analysis of the genetic diversity of 320 sorghum germplasm accessions with a set of 38 selected SSR markers based on three different published SSR linkage maps of sorghum. A sweet sorghum panel (SSP) was assembled by Murray et al. (2009) comprising 125 diverse accessions, which are primarily old and modern sweet sorghum cultivars along with a few grain and forage sorghums.

A core collection composed of 195 sorghum accessions originating from 39 countries and belonging to the five basic and ten intermediate races representative of the genetic diversity of core sample of 210 cultivated sorghum genotypes reported by Deu et al. (2006) was used for the association mapping of grain quality such as amylose content, protein content, lipid content, hardness, endosperm texture, and peak gelatinization temperature along with grain yield (de Alencar Figueiredo et al. 2010). In addition to the landrace collection described by Deu et al. (2006), an additional 45 inbred lines including donors for aluminum tolerance (Caniato et al. 2007) were used for the association mapping for aluminum tolerance to gain insights into the origin and evolution of aluminum tolerance and to detect functional variants (Caniato et al. 2014). Analysis of recombinant haplotypes suggested that causative polymorphisms are in introns and a transposon (MITE) insertion localized to a ~6 kb region, which is positively correlated with aluminum tolerance. However, the SNP located in the second intron of *SbMATE*, an Al-activated root citrate efflux transporter, exhibited the strongest association signal and recovered 80 % of all the aluminum-tolerant accessions in the association panel.

6.7 Mutant Resources for Analyzing Gene Function

Exploitation of natural or induced genetic variability is considered as a successful strategy in crop improvement in many food crops. Mutagenesis as a tool to create novel variation is particularly important for those crops with limited variability. Over the years, several varieties have been developed in major food crops through mutation breeding programs. Ever since H.J. Muller reported induced mutation in 1937, analysis of mutants remained an effective approach to understand the gene function (Springer 2000; Stanford et al. 2001). The rapid accumulation of genomic sequence information in the past decade has brought the reverse genetic approaches into prominence, thereby directly probing the function of specific genes by testing the in vivo consequence of disruption or overexpression of a gene on the phenotype of an organism (Tierney and Lamour 2005). This is the "reverse" of conventional approach where phenotypes are observed and then the gene responsible for that phenotype is cloned and validated.

Mutant lines are important bioresources in this regard, which can potentially accelerate the understanding of gene function through reverse genetics. Kuromori et al. (2009) while reviewing the available mutant resources for phenome analysis in plant species highlighted the importance of mutant bioresource across crop species. With the availability of various analytical platforms (particularly bioinformatics), it is feasible to discover genes involved in particular phenotypic changes. Logically, these genes need to be functionally tested in collection of mutant resources in a high-throughput manner, called phenome analysis (Alonso and Ecker 2006). Chemical mutagens, like ethyl methanesulfonate (EMS), sodium azide, and methylnitrosourea (MNU), and physical mutagens, like fast neutrons, gamma rays, and ion beam irradiation, have been used extensively to generate mutant populations since the report of the first induced mutation in 1937. However, none of these mutant populations have been systematically annotated and preserved (Sree-Ramulu 1970; Porter et al. 1978; Jenks et al. 1994). Thus, the generated resource could not be combined with genomics tools.

*T*argeting *i*nduced *l*ocal *l*esions *in* *g*enomes (TILLING) was developed as a general reverse genetic tool to derive an allelic series of induced point mutations in genes of interest (Till et al. 2004, 2006). TILLING allows rapid and low-cost discovery of induced point mutations in populations of chemically mutagenized individuals in a high-throughput manner. This has been utilized to identify mutations in genes of interest in different crops (Till et al. 2004, 2007; Talame et al. 2008; Lababidi et al. 2009) including sorghum (Xin et al. 2008, 2009; Blomstedt et al. 2012). TILLING resources have been created for various crop plants across laboratories (Barkley and Wang 2008). Xin et al. (2008) were first to report the creation of TILLING resource in sorghum through EMS mutagenesis of sorghum cultivar, BTx623. They documented the feasibility of this approach by screening the mutant population for alterations in the genes of agronomic value not associated with cyanogenesis. Recently, Blomstedt et al. (2012) developed an acyanogenic forage line, P414L, with a point mutation in *the CYP79A1 gene* of cyanogenesis biochemical pathway by combining biochemical screen and TILLING approach. In the recent years, several TILLING populations have been developed globally by different sorghum research groups, which can serve as a valuable bioresource useful in understanding gene function and high-throughput SNP discovery.

An Annotated Individually pedigreed Mutated Sorghum (AIMS) library comprising 6144 pedigreed M_4 seed pools developed through single-seed descent from individual mutagenized seeds was established (Xin et al. 2008, 2009), which contains many biologically and agronomically important mutants, such as brown midrib (*bmr*) mutants for improved biomass digestibility and ethanol yield and erect leaf (*erl*) mutants for improved capture of canopy radiation and hence biomass yield (Xin et al. 2009; Saballos et al. 2012; Sattler et al. 2012). An array of useful mutations harboring a wide range of phenotypic variation, including dwarfness, earliness, high protein digestibility, high lysine, etc., that were reported earlier in sorghum (Singh and Axtell 1973; Quinby 1975; Ejeta and Axtell 1985; Oria et al. 2000) were collected and preserved by the late Dr. Keith Schertz, a former sorghum geneticist with USDA-ARS, and this was released recently as a collection of genetic stocks through Germplasm Resources Information Network (Xin et al. 2013; www.ars-grin.gov), which is a valuable and vital resource for future genomic studies in sorghum.

6.8 Future Prospects

In the current era of crop improvement involving efficient integration of genetic information with the genomic and bioinformatics resources, focus should be on the coordination of various sorghum research groups across the globe on the sharing and utilization of resources available in the public domain. Genomics offers practical advantages for breeding cultivars by providing access to genetic variation through molecular markers and the potential to accurately measure the gene expression. With the initiation of re-sequencing projects in sorghum, whole-genome sequence information of sorghum cultivars possessing different end uses will be available in the near future resulting in the possibility of providing "genotype genomics" services that will contribute to sorghum improvement through "breeding by design." There is a need for the development of a single platform for sorghum, which can integrate the data that are scattered in different databases and also the software tools available for analysis. The main challenges facing the bioinformaticians are the development and management of databases and computational tools for data analysis in such a way that the user can define the target data and select the computational tool in order to get the output in a suitable format. The availability of different platforms for sequencing demands the development of novel algorithms and computational tools for an efficient assembly, annotation, and analysis. The application of molecular markers, comparative genomics, and annotation tools will greatly assist the identification of the genetic variation underlying an increasing number of agronomic traits and assist in the further agronomic improvement of a variety of crops.

References

Agrama HA, Widle GE, Reese JC, Campbell LR, Tuinstra MR (2002) Genetic mapping of QTLs associated with greenbug resistance and tolerance in *Sorghum bicolor*. Theor Appl Genet 104:1373–1378

Ahnert D, Lee M, Austin D, Livini C, Woodman W, Openshaw S, Smith J, Porter K, Dalon G (1996) Genetic diversity among elite sorghum inbred lines assessed with DNA markers and pedigree information. Crop Sci 36:1385–1392

Alonso JM, Ecker JR (2006) Moving forward in reverse: genetic technologies to enable genome-wide phenomic screens in *Arabidopsis*. Nat Rev Genet 7:524–536

Arun SS (2006) *In silico* EST data mining for elucidation of repeats biology and functional annotation in sorghum [*Sorghum bicolor* (L.) Moench.]. M.Sc. thesis. University of Agricultural Sciences, Dharwad

Aruna C, Bhagwat VR, Madhusudhana R, Sharma V, Hussain T, Ghorade RB, Khandalkar HG, Audilakshmi S, Seetharama N (2011) Identification and validation of genomic regions that affect shoot fly resistance in sorghum [*Sorghum bicolor* (L.) Moench]. Theor Appl Genet 122:1617–1630

Aruna C, Priya AR, Neeraja CN, Patil JV, Visarada KBRS (2012) Diversity analysis using ISSR markers for resistance to shoot pests in sorghum. Crop Prot 35:110–117

Ayana A, Bryngelsson T, Bekele E (2000) Geographic and altitudinal allozyme variation in sorghum [*Sorghum bicolor* (L.) Moench] landraces from Ethiopia and Eritrea. Hereditas 135:1–12

Barker G, Batley J, O'Sullivan H, Edwards KJ, Edwards D (2003) Redundancy based detection of sequence polymorphisms in expressed sequence tag data using autoSNP. Bioinformatics 19:421–422

Barkley NA, Wang ML (2008) Application of TILLING and EcoTILLING as reverse genetic approaches to elucidate the function of genes in plants and animals. Curr Genomics 9:212–226

Barrett T, Troup DB, Wilhite SE et al (2009) NCBI GEO: archive for high-throughput functional genomic data. Nucleic Acids Res 37:D885–D890

Batley J, Edwards D (2007) SNP applications in plants. In: Oraguzie NC, Rikkerink EHA, Gardiner SE, De

Silva HN (eds) Association mapping in plants. Springer, New York, pp 95–102

Batley J, Barker G, O'Sullivan H, Edwards KJ, Edwards D (2003) Mining for single nucleotide polymorphisms and insertions/deletions in maize expressed sequence tag data. Plant Physiol 132:84–91

Batley J, Jewell E, Edwards D (2007) Automated discovery of single nucleotide polymorphism (SNP) and simple sequence repeat (SSR) molecular genetic markers. In: Edwards D (ed) Plant bioinformatics. Humana Press, Totowa, pp 473–494

Bekele WA, Wieckhorst S, Friedt W, Snowden RJ (2013) High-throughput genomics in sorghum: from whole-genome resequencing to a SNP screening array. Plant Biotechnol J 11(9):1112–1125

Ben-Israel I, Kilian B, Nida H, Fridman E (2012) Heterotic trait locus (HTL) mapping identifies intra-locus interactions that underlie reproductive hybrid vigor in *Sorghum bicolor*. PLoS One 7:e38993

Bentsink L, Hanson J, Hanhart CJ et al (2010) Natural variation for seed dormancy in *Arabidopsis* is regulated by additive genetic and molecular pathways. Proc Natl Acad Sci U S A 107:4264–4269

Bhattramakki D, Dong J, Chhabra AK, Hart G (2000) An integrated SSR and RFLP linkage map of *Sorghum bicolor* (L.) Moench. Genome 43:988–1002

Billot C, Rivallan R, Sall MN et al (2012) A reference microsatellite kit to assess for genetic diversity of *Sorghum bicolor* (Poaceae). Am J Bot 99:e245–e250

Blomstedt CK, Gleadow RM, O'Donnell N et al (2012) A combined biochemical screen and TILLING approach identifies mutations in *Sorghum bicolor* L. Moench resulting in acyanogenic forage production. Plant Biotechnol J 10:54–66

Boivin K, Deu M, Rami JF, Trouche G, Hamon P (1999) Towards a saturated sorghum map using RFLP and AFLP markers. Theor Appl Genet 98:320–328

Bout S, Vermerris W (2003) A candidate-gene approach to clone the sorghum Brown midrib gene encoding caffeic acid O-methyltransferase. Mol Genet Genomics 269:205–214

Brachi B, Faure N, Horton M, Flahauw E, Vazquez A, Nordborg M, Bergelson J, Cuguen J, Roux F (2010) Linkage and association mapping of Arabidopsis thaliana flowering time in nature. PLoS Genet 6:1–17

Brenner S, Johnson M, Bridgham J et al (2000) Gene expression analysis by massively parallel signature sequencing (MPSS) on microbead arrays. Nat Biotechnol 18:630–634

Brown SM, Hopkins MS, Mitchell SE, Senior ML, Wang TY, Duncan RR, Gonzalez-Candelas F, Kresovich S (1996) Multiple methods for the identification of polymorphic simple sequence repeats (SSRs) in sorghum [*Sorghum bicolor* (L.) Moench]. Theor Appl Genet 93:190–198

Brown NP, Leroy C, Sander C (1998) MView: a web-compatible database search or multiple alignment viewer. Bioinformatics 14:380–381

Brown PJ, Klein PE, Bortiri E, Acharya CB, Rooney WL, Kresovich S (2006) Inheritance of inflorescence architecture in sorghum. Theor Appl Genet 113:931–942

Buchanan CD, Lim S, Salzman RA et al (2005) Sorghum bicolor's transcriptome response to dehydration, high salinity and ABA. Plant Mol Biol 58:699–720

Buckler ES, Holland JB, Bradbury PJ et al (2009) The genetic architecture of maize flowering time. Science 325:714–718

Burow GB, Franks CD, Acosta-Martinez V, Xin ZG (2009) Molecular mapping and characterization of BLMC, a locus for profuse wax (bloom) and enhanced cuticular features of Sorghum [*Sorghum bicolor* (L.) Moench.]. Theor Appl Genet 118:423–431

Calviño M, Bruggmann R, Messing J (2011) Characterization of the small RNA component of the transcriptome from grain and sweet sorghum stems. BMC Genom 12:356

Caniato FF, Guimaraes CT, Schaffert RE, Alves VMC, Kochian LV, Borem A, Klein PE, Magalhaes JV (2007) Genetic diversity for aluminum tolerance in sorghum. Theor Appl Genet 114:863–876

Caniato FF, Hamblin MT, Guimaraes CT, Zhang Z, Schaffert RE, Kochian LV, Magalhaes JV (2014) Association mapping provides insights into the origin and the fine structure of the sorghum aluminum tolerance locus, Alt_{SB}. PLoS One 9(1):e87438

Casa AM, Mitchell SE, Jensen JD, Hamblin MT, Paterson AH, Aquardo CF, Kresovich S (2006) Evidence for a selective sweep on chromosome 1 of cultivated sorghum. Crop Sci 46:S27–S40

Casa AM, Pressoir G, Brown PJ, Mitchell SE, RooneyWL TMR, Franks CD, Kresovich S (2008) Community resources and strategies for association mapping in sorghum. Crop Sci 48:30–40

Cavanagh C, Morell M, Mackay I, Powell W (2008) From mutations to MAGIC: resources for gene discovery, validation and delivery in crop plants. Curr Opin Plant Biol 11:215–221

Chagne D, Forster JW, Cogan NOI, Batley J, Edwards D (2007) Single nucleotide polymorphism discovery. In: Oraguzie NC, Rikkerink EHA, Gardiner SE, De Silva HN (eds) Association mapping in plants. Springer, New York, pp 53–76

Chantereau J, Trouche G, Rami JF, Deu M, Barro C, Grivet L (2001) RFLP mapping of QTLs for photoperiod response in tropical sorghum. Euphytica 120:183–194

Chenna R, Sugawara H, Koike T, Lopez R, Gibson TJ, Higgins DG, Thompson JD (2003) Multiple sequence alignment with the Clustal series of programs. Nucleic Acids Res 31:3497–3500

Childs KL, Miller FR, Cordonnier-Pratt MM, Pratt LH, Morgan PW, Mullet JE (1997) The sorghum photoperiod sensitive gene, Ma_3, encodes a phytochrome B. Plant Physiol 113:611–619

Clark AG (2004) The role of haplotypes in candidate gene studies. Genet Epidemiol 27:321–333

Conte MG, Gaillard S, Lanau N, Rouard M, Perin C (2008) GreenPhylDB: a database for plant comparative genomics. Nucleic Acids Res 36:D991–D998

Crasta OR, Xu WW, Rosenow DT, Mullet J, Nguyen HT (1999) Mapping of post-flowering drought resistance traits in grain sorghum: association between QTLs influencing premature senescence and maturity. Mol Gen Genet 262:579–588

de Alencar Figueiredo L, Sine B, Chantereau J, Mestres C, Fliedel G, Rami JF, Glaszmann JC, Deu M, Courtois M (2010) Variability of grain quality in sorghum: association with polymorphism in *Sh2*, *Bt2*, *SssI*, *Ae1*, *Wx* and *O2*. Theor Appl Genet 121:1171–1185

de Hoon M, Hayashizaki Y (2008) Deep cap analysis gene expression (CAGE): genome-wide identification of promoters, quantification of their expression, and network inference. Biotechniques 44:627–628

Deu M, Ratnadass A, Hamada MA, Noyer JL, Diabate M, Chantereau J (2005) Quantitative trait loci for head-bug resistance in sorghum. Afr J Biotechnol 4:247–250

Deu M, Rattunde F, Chantereau J (2006) A global view of genetic diversity in cultivated sorghums using a core collection. Genome 49:168–180

Doerge RW (2002) Mapping and analysis of quantitative trait loci in experimental populations. Nat Rev Genet 3:43–52

Dong Q, Lawrence CJ, Schlueter SD, Wilkerson MD, Kurtz S, Lushbough C, Brendel V (2005) Comparative plant genomics resources at PlantGDB. Plant Physiol 139:610–618

Du JF, Wu YJ, Fang XF, Cao JX, Zhao L, Tao SH (2010) Prediction of sorghum miRNAs and their targets with computational methods. Chinese Sci Bull 55:1263–1270

Dufour P, Deu M, Grivet L, D'Hont A, Paulet F, Bouet A, Lanaud C, Glaszmann JC, Hamon P (1997) Construction of a composite sorghum genome map and comparison with sugarcane, a related complex polyploïd. Theor Appl Genet 94:409–418

Duvick J, Fu A, Muppirala U, Sabharwal M, Wilkerson MD, Lawrence CJ, Lushbough C, Brendel V (2007) PlantGDB: a resource for comparative plant genomics. Nucleic Acids Res 36:D959–D965

Edwards D, Batley J, Cogan NOI, Forster JW, Chagne D (2007) Single nucleotide polymorphism discovery. In: Oraguzie NC, Rikkerink EHA, Gardiner SE, De Silva HN (eds) Association mapping in plants. Springer, New York, pp 53–76

Ejeta G, Axtell J (1985) Mutant gene in sorghum causing leaf "reddening" and increased protein concentration in the grain. J Hered 76:301–302

El Mannai Y, Shehzad T, Okuno K (2012) Mapping of QTLs underlying flowering time in sorghum [*Sorghum bicolor* (L.) Moench]. Breed Sci 62:151–159

Gabaldon T, Dessimoz C, Huxley-Jones J, Vilella A, Sonnhammer E, Lewis S (2009) Joining forces in the quest for orthologs. Genome Biol 10:403

Geleta N, Labuschangne MT, Chris D, Viljoen CD (2006) Genetic diversity analysis in sorghum germplasm as estimated by AFLP, SSR and morpho-agronomical markers. Biodivers Conserv 15:3251–3265

Girma Y (2009) Mining genomic resources for SNP and SNP-CAPS markers and divergence for drought tolerance in sorghum [*Sorghum bicolor* (L.) Moench]. M.Sc. (Ag) Thesis submitted to the University of Agricultural Sciences, Dharwad, India

Goff SA, Ricke D, Lan TH et al (2002) A draft sequence of the rice genome (*Oryza sativa* L. ssp *japonica*). Science 296:92–100

Goodstein DM, Shu S, Howson R et al (2012) Phytozome: a comparative platform for green plant genomics. Nucleic Acids Res 40:D1178–D1186

Hamblin MT, Mitchell SE, White GM, Gallego J, Kukatla R, Wing RA, Paterson AH, Kresovich S (2004) Comparative population genetics of the panicoid grasses: sequence polymorphism, linkage disequilibrium and selection in a diverse sample of *Sorghum bicolor*. Genetics 167:471–483

Hamblin MT, Salas Fernandez MG, Casa AM, Mitchell SE, Paterson AH, Kresovich S (2005) Equilibrium processes cannot explain high levels of short- and medium-range linkage disequilibrium in the domesticated grass *Sorghum bicolor*. Genetics 171:1247–1256

Hamblin MT, Casa AM, Sun H, Murray SC, Paterson AH, Aquardo CF, Kresovich S (2006) Challenges of detecting directional selection after a bottleneck: lessons from *Sorghum bicolor*. Genetics 173:953–964

Hamblin MT, Fernandez MGS, Tuinstra MR, Rooney WL, Kresovich S (2007) Sequence variation at candidate loci in the starch metabolism pathway in sorghum: prospects for linkage disequilibrium mapping. Crop Sci 47:S125–S134

Harbers M, Carninci P (2005) Tag-based approaches for transcriptome research and genome annotation. Nat Methods 2:495–502

Harris K, Subudhi PK, Borrell A, Jordan D, Rosenow D, Nguyen H, Klein P, Klein R, Mullet J (2007) Sorghum stay-green QTL individually reduce post-flowering drought-induced leaf senescence. J Exp Bot 58:327–338

Hart GE, Schertz KF, Peng Y, Syed NH (2001) Genetic mapping of *Sorghum bicolor* (L.) Moench QTLs that control variation in tillering and other morphological characters. Theor Appl Genet 103:1232–1242

Haussmann BIG, Mahalakshmi V, Reddy BVS, Seetharama N, Hash CT, Geiger HH (2002) QTL mapping of stay-green in two sorghum recombinant inbred populations. Theor Appl Genet 106:133–142

Haussmann BIG, Hess DE, Omanya GO, Folkertsma RT, Reddy BVS, Kayentao M, Welz HG, Geiger HH (2004) Genomic regions influencing resistance to the parasitic weed *Striga hermonthica* in two recombinant inbred populations of sorghum. Theor Appl Genet 109:1005–1016

Holland JB (2007) Genetic architecture of complex traits in plants. Curr Opin Plant Biol 10:156–161

Huang X, Paulo MJ, Boer M, Effgen S, Keizer P, Koornneef M, Eeuwijk FV (2011) Analysis of natural allelic variation in *Arabidopsis* using a multiparent recombinant inbred line population. Proc Natl Acad Sci U S A 108:4488–4493

International Barley Genome Sequencing Consortium, Mayer KF, Waugh R, Brown JW, Schulman A, Langridge P, Platzer M, Fincher GB, Muehlbauer GJ, Sato K, Close TJ, Wise RP, Stein N (2012) A physical, genetic and functional sequence assembly of the barley genome. Nature 491:711–716

Jaikishan I, Paik GR, Madhusudhana R, Elangovan M, Rajendrakumar P (2013) Development of microsatellite markers targeting $(GATA)_n$ motifs in sorghum [*Sorghum bicolor* (L.) Moench]. Mol Breed 31:223–231

Jenks MA, Joly RJ, Peters PJ, Rich PJ, Axtell JD, Ashworth EN (1994) Chemically induced cuticle mutation affecting epidermal conductance to water vapor and disease susceptibility in *Sorghum bicolor* (L.) Moench. Plant Physiol 105:1239–1245

Jordan DR, Mace ES, Henzell RG, Klein PE, Klein RR (2010) Molecular mapping and candidate gene identification of the Rf_2 gene for pollen fertility restoration in sorghum [*Sorghum bicolor* (L.) Moench]. Theor Appl Genet 120:1279–1287

Jordan DR, Klein RR, Sakrewski KG, Henzell RG, Klein PE, Mace ES (2011) Mapping and characterization of Rf_5: a new gene conditioning pollen fertility restoration in A_1 and A_2 cytoplasm in sorghum [*Sorghum bicolor* (L.) Moench]. Theor Appl Genet 123:383–396

Jordan D, Mace E, Cruikshank AW, Hunt CH, Hammer GL, Henzell RG (2012) Development and use of a sorghum nested association mapping population. Poster presented at International Plant and Animal Genome Conference, Jan., 14–18, 2012, San Diego, CA

Kebede H, Subudhi PK, Rosenow DT, Nguyen HT (2001) Quantitative trait loci influencing drought tolerance in grain sorghum (*Sorghum bicolor* L. Moench). Theor Appl Genet 103:266–276

Kebrom TH, Burson BL, Finlayson SA (2006) Phytochrome B represses Teosinte Branched1 expression and induces sorghum axillary bud outgrowth in response to light signals. Plant Physiol 140:1109–1117

Keurentjes JJ, Bentsink L, Alonso-Blanco C, Hanhart CJ, Blankestijn-De Vries H, Effgen S, Vreugdenhil D, Koornneef M (2007) Development of a near-isogenic line population of *Arabidopsis thaliana* and comparison of mapping power with a recombinant inbred line population. Genetics 175:891–905

Keurentjes JJB, Willems G, van Eeuwijk F, Nordborg M, Koornneef M (2011) A comparison of population type used for QTL mapping in *Arabidopsis thaliana*. Plant Genet Res 9:185–188

Klein RR, Rodriguez-Herrera R, Schlueter JA, Klein PE, Yu ZH, Rooney WL (2001) Identification of genomic regions that affect grain-mould incidence and other traits of agronomic importance in sorghum. Theor Appl Genet 102:307–319

Klein RR, Mullet JE, Jordan DR, Miller FR, Rooney WL, Menz MA, Franks CD, Klein PE (2008) The effect of tropical sorghum conversion and inbred development on genome diversity as revealed by high-resolution genotyping. Crop Sci 48(S1):S12–S26

Knoll J, Gunaratna N, Ejeta G (2008) QTL analysis of early-season cold tolerance in sorghum. Theor Appl Genet 116:577–587

Kong L, Dong J, Hart GE (2000) Characteristics linkage-map positions and allelic differentiation of *Sorghum bicolor* (L.) Moench DNA simple sequence repeats (SSRs). Theor Appl Genet 101:438–448

Kong W, Jin H, Franks CD, et al. (2013) Genetic analysis of recombinant inbred lines for *Sorghum bicolor* × *Sorghum propinquum*. G3 (Bethesda) 3:101–108

Kuromori T, Takahashi S, Kondou Y, Shinozaki K, Matsui M (2009) Phenome analysis in plant species using loss-of-function and gain-of-function mutants. Plant Cell Physiol 50:1215–1321

Kuzniar A, van Ham RC, Pongor S, Leunissen JA (2008) The quest for orthologs: finding the corresponding gene across genomes. Trends Genet 24:539–551

Lababidi S, Mejlhede N, Rasmussen SK, Backes G, Al-Said W, Baum M, Jahoor A (2009) Identification of barley mutants in the cultivar Lux at the Dhn loci through TILLING. Plant Breed 128:332–336

Larkin MA, Blackshields G, Brown NP et al (2007) Clustal W and Clustal X version 2.0. Bioinformatics 23:2947–2948

Li H, Hearne S, Banziger K, Li Z, Wang J (2010) Statistical properties of QTL linkage mapping in biparental genetic populations. Heredity 105:257–267

Lijavetzky D, Martinez MC, Carrari F, Hopp HE (2000) QTL analysis and mapping of pre-harvest sprouting resistance in sorghum. Euphytica 112:125–135

Lin Y, Schertz K, Paterson A (1995) Comparative analysis of QTLs affecting plant height and maturity across the Poaceae, in reference to an interspecific sorghum population. Genetics 141:391–411

Lyons E, Freeling M (2008) How to usefully compare homologous plant genes and chromosomes as DNA sequences. Plant J 53:661–673

Lyons E, Pedersen B, Kane J, Freeling M (2008) The value of non-model genomes and an example using SynMap within CoGe to dissect the hexaploidy that predates the Rosids. Trop Plant Biol 1:181–190

Mace ES, Xia L, Jordan DR, Halloran K, Parh DK, Huttner E, Wenzl P, Kilian A (2008) DArT markers: diversity analyses and mapping in *Sorghum bicolor*. BMC Genom 9:26

Mace ES, Rami JF, Bouchet S, Klein PE, Klein RR, Kilian A, Wenzl P, Xia L, Halloran K, Jordan DR (2009) A consensus genetic map of sorghum that integrates multiple component maps and high-throughput diversity array technology (DArT) markers. BMC Plant Biol 9:13

Mace ES, Singh V, Van Oosterom EJ, Hammer GL, Hunt CH, Jordan DR (2012) QTL for nodal root angle in sorghum (*Sorghum bicolor* L. Moench) co-locate with QTL for traits associated with drought adaptation. Theor Appl Genet 124:97–109

Mace ES, Tai S, Gilding EK et al (2013) Whole-genome sequencing reveals untapped genetic potential in Africa's indigenous cereal crop sorghum. Nat Commun 4:2320

Madhusudhana R, Patil JV (2013) A major QTL for plant height is linked with bloom locus in sorghum [*Sorghum bicolor* (L.) Moench]. Euphytica 191:259–268

Magalhaes JV, Liu J, Guimaraes CT et al (2007) A gene in the multidrug and toxic compound extrusion (MATE) family confers aluminum tolerance in sorghum. Nat Genet 39:1156–1161

McIntyre CL, Hermann SM, Casu RE et al (2004) Homologues of the maize rust resistance gene Rp1-D are genetically associated with a major rust resistance QTL in sorghum. Theor Appl Genet 109:875–883

McIntyre CL, Drenth J, Gonzalez N, Henzell RG, Jordan DR (2008) Molecular characterization of the waxy locus in sorghum. Genome 51:524–533

McMullen MD, Kresovich S, Villeda HS et al (2009) Genetic properties of the maize nested association mapping population. Science 325:737–740

Mochida K, Shinozaki K (2010) Genomics and bioinformatics resources for crop improvement. Plant Cell Physiol 51:497–523

Morris GP, Ramu P, Deshpande SP et al (2013) Population genomic and genome-wide association studies of agroclimatic traits in sorghum. Proc Natl Acad Sci U S A 110:453–458

Mount DM (2004) Bioinformatics: sequence and genome analysis, 2nd edn. Cold Spring Harbor Laboratory Press, Cold Spring Harbor

Muller HJ (1937) The biological effects of radiation, with especial reference to mutation. Act Sci Ind 11:477–494

Murali Mohan S, Madhusudhana R, Mathur K, Chakravarthi DVN, Rathore S, Nagaraja Reddy R, Satish K, Srinivas G, Sarada Mani N, Seetharama N (2010) Identification of quantitative trait loci associated with resistance to foliar diseases in sorghum [*Sorghum bicolor* (L.) Moench]. Euphytica 176:199–211

Murray SC, Rooney WL, Mitchell SE, Sharma A, Klein PE, Mullet JE, Kresovich S (2008) Genetic improvement of sorghum as a biofuel feedstock: II. QTL for stem and leaf structural carbohydrates. Crop Sci 48:2180–2193

Murray SC, Rooney WL, Hamblin MT, Mitchell SE, Kresovich S (2009) Sweet sorghum genetic diversity and association mapping for brix and height. Plant Gen 2:48–62

Mutegi E, Sagnard F, Semagn K, Deu M, Muraya M, Kanyenji B, de Villiers S, Kiambi D, Herselman L, Labuschagne M (2011) Genetic structure and relationships within and between cultivated and wild sorghum [*Sorghum bicolor* (L.) Moench] in Kenya as revealed by microsatellite markers. Theor Appl Genet 122:989–1004

Nagaraja Reddy R, Madhusudhana R, Murali Mohan S, Chakravarthi DVN, Seetharama N (2012) Characterization, development and mapping of unigene-derived microsatellite markers in sorghum [*Sorghum bicolor* (L.) Moench]. Mol Breed 29:543–564

Nagaraja Reddy R, Madhusudhana R, Murali Mohan S, Chakravarthi DV, Mehtre SP, Seetharama N, Patil JV (2013) Mapping QTL for grain yield and other agronomic traits in post-rainy sorghum [*Sorghum bicolor* (L.) Moench]. Theor Appl Genet 126:1921–1939

Nelson JC, Wang S, Wu Y, Li X, Antony G, White F, Yu J (2011) Single-nucleotide polymorphism discovery by high-throughput sequencing in sorghum. BMC Genom 12:352

Nordborg M, Tavaré S (2002) Linkage disequilibrium: what history has to tell us? Trends Genet 18:83–90

Oh BJ, Frederiksen RA, Magill CW (1994) Identification of molecular markers linked to head smut resistance gene (*Shs*) in sorghum by RFLP and RAPD analyses. Phytopathology 84:830–833

Oria MP, Hamaker BR, Axtell JD, Huang CP (2000) A highly digestible sorghum mutant cultivar exhibits a unique folded structure of endosperm protein bodies. Proc Natl Acad Sci U S A 97:5065–5070

Palumbi SR (1995) Nucleic acids II: the polymerase chain reaction. In: Hillis D, Moritz C (eds) Molecular systematics, 2nd edn. Sinauer Associates Inc., Sunderland, pp 205–247

Parh DK, Jordan DR, Aitken EAB, Mace ES, Junai P, McIntyre CL, Godwin ID (2008) QTL analysis of ergot resistance in sorghum. Theor Appl Genet 117:369–382

Park SJ, Huang Y, Ayoubi P (2006) Identification of expression profiles of sorghum genes in response to greenbug phloem-feeding using cDNA subtraction and microarray analysis. Planta 223:932–947

Parkinson H, Kapushesky M, Shojatalab M et al (2007) ArrayExpress – a public database of microarray experiments and gene expression profiles. Nucleic Acids Res 35:D747–D750

Paterson AH, Bowers JE, Bruggmann R et al (2009) The *Sorghum bicolor* genome and the diversification of grasses. Nature 457:551–556

Perumal R, Menz MA, Mehta PJ et al (2009) Molecular mapping of *Cg1*, a gene for resistance to anthracnose (*Colletotrichum sublineolum*) in sorghum. Euphytica 165:597–606

Phuong N, Stutzel H, Uptimoor R (2013) Quantitative trait loci associated to agronomic traits and yield components in a *Sorghum bicolor* L. Moench RIL population cultivated under pre-flowering drought and well-watered conditions. Agric Sci 4(12):781–791

Ping LX, Feng YJ, Ping GC, Acharya S (2011) Quantitative trait loci analysis of economically important traits in *Sorghum bicolor*×*S. sudanense* hybrid. Can J Plant Sci 91:81–90

Porter KS, Anxtell JD, Lechtenberg VL, Colenbrander VF (1978) Phenotype, fiber composition, and *in vitro* dry matter disappearance of chemically induced brown midrib (*bmr*) mutants of sorghum. Crop Sci 18:205–208

Proost S, Van Bel M, Sterck L, Billiau K, Van Parys T, Van de Peer Y, Vandepoele K (2009) PLAZA: a comparative genomics resource to study gene and genome evolution in plants. Plant Cell 21:718–3731

Punnuri S, Huang Y, Steets J, Yanqi W (2013) Developing new markers and QTL mapping for greenbug resistance in sorghum [*Sorghum bicolor* (L.) Moench]. Euphytica 191:191–203

Quinby JR (1975) The genetics of sorghum improvement. J Hered 66:56–62

Rajkumar FB, Kavil SP et al (2013) Molecular mapping of genomic regions harbouring QTLs for root and yield traits in sorghum [*Sorghum bicolor* (L.) Moench]. Physiol Mol Biol Plants 19(3):409–419

Rakshit S, Rakshit A, Patil JV (2012) Multiparent intercross populations in analysis of quantitative traits. J Genet 91:111–117

Rami JF, Dufour P, Trouche G, Fliedel G, Mestres C, Davrieux F, Blancard P, Hamon P (1998) Quantitative trait loci for grain quality, productivity, morphological, and agronomical traits in sorghum (*Sorghum bicolor* L. Moench). Theor Appl Genet 97:605–616

Ramu P, Kassahun B, Senthilvel S, Kumar CA, Jayashree B, Folkertsma RT, Reddy LA, Kuruvinashetti MS, Haussmann BIG, Hash CT (2009) Exploiting rice-sorghum synteny for targeted development of EST-SSRs to enrich the sorghum genetic linkage map. Theor Appl Genet 119:1193–1204

Ramu P, Billot C, Rami JF, Senthilvel S, Upadhyaya HD, Ananda Reddy L, Hash CT (2013) Assessment of genetic diversity in the sorghum reference set using EST-SSR markers. Theor Appl Genet 126:2051–2064

Reddy PS, Fakrudin B, Rajkumar, Punnuri SM, Arun SS, Kuruvinashetti MS, Das IK, Seetharama N (2008) Molecular mapping of genomic regions harboring QTLs for stalk rot resistance in sorghum. Euphytica 159:191–198

Risch N, Merikangas K (1996) The future of genetic studies of complex human diseases. Science 273:1516–1517

Ritter KB, Lynne McIntyre C, Godwin ID, Jordan DR, Chapman SC (2007) An assessment of the genetic relationship between sweet and grain sorghums, within *Sorghum bicolor* ssp. *bicolor* (L.) Moench, using AFLP markers. Euphytica 157:161–176

Ritter KB, Jordan DR, Chapman SC, Godwin ID, Mace ES, McIntyre CL (2008) Identification of QTL for sugar-related traits in sweet x grain sorghum (*Sorghum bicolor* L. Moench) recombinant inbred population. Mol Breed 22:367–384

Rouard M, Guignon V, Aluome C, Laporte MA, Droc G, Walde C, Zmasek CM, Perin C, Conte MG (2011) GreenPhylDB v2.0: comparative and functional genomics in plants. Nucleic Acids Res 39:D1095–D1102

Saballos A, Ejeta G, Sanchez E, Kang C, Vermerris W (2009) A genome-wide analysis of the cinnamyl alcohol dehydrogenase family in sorghum [*Sorghum bicolor* (L.) Moench] identifies *SbCAD2* as the brown midrib6 gene. Genetics 181:783–795

Saballos A, Sattler SE, Sanchez E, Foster TP, Xin Z, Kang C, Pedersen JF, Vermerris W (2012) Brown midrib2 (*bmr2*) encodes the major 4-coumarate: coenzyme A ligase involved in lignin biosynthesis in sorghum (*Sorghum bicolor* (L.) Moench). Plant J 70:818–830

Sanchez AC, Subudhi PK, Rosenow DT, Nguyen HT (2002) Mapping QTLs associated with drought resistance in sorghum (*Sorghum bicolor* L. Moench). Plant Mol Biol 48:713–726

Satish K, Srinivas G, Madhusudhana R, Padmaja PG, Nagaraja Reddy R, Murali Mohan S, Seetharama N (2009) Identification of quantitative trait loci for resistance to shoot a fly in sorghum [*Sorghum bicolor* (L.) Moench]. Theor Appl Genet 119:1425–1439

Sattler SE, Palmer NA, Saballos A, Greene AM, Xin Z, Sarath G, Vermerris W, Pedersen JF (2012) Identification and characterization of four missense mutations in brown midrib 12 (*bmr12*), the Caffeic O-Methyltransferase (*COMT*) of sorghum. Bioenergy Res 5:855–865

Savage D, Batley J, Erwin T, Logan E, Love CG, Lim GAC, Mongin E, Barker G, Spangenberg GC, Edwards D (2005) SNPServer: a real-time SNP discovery tool. Nucleic Acids Res 33:W493–W495

Schloss SJ, Mitchell SE, White GM, Kukatla R, Bowers JE, Paterson AH, Kresovich S (2002) Characterization of RFLP probe sequences for gene discovery and SSR development in *Sorghum bicolor* (L.) Moench. Theor Appl Genet 105:912–920

Schnable PS, Ware D, Fulton RS et al (2009) The B73 maize genome: complexity, diversity, and dynamics. Science 326:1112–1115

Shakoor N, Crasta OR, Nair R, Huang SW, Fan Z, Morris G, Kresovich S (2013) Whole genome gene expression profiling of multiple tissue types and developmental stages in *Sorghum bicolor*. Abstract: Plant and Animal Genome XXI Symposium, January 12–16, 2013, San Diego, CA, USA

Shakoor N, Nair R, Crasta OR, Morris G, Feltus A, Kresovich S (2014) A *Sorghum bicolor* expression atlas reveals dynamic genotype-specific expression profiles for vegetative tissues of grain, sweet and bioenergy sorghums. BMC Plant Biol 14:35

Shehzad T, Okuizumi H, Kawase M, Okuno K (2009) Development of SSR-based sorghum (*Sorghum bicolor* L. Moench) diversity research set of germplasm and its evaluation by morphological traits. Genet Resour Crop Evol 56:809–827

Shiringani AL, Friedt W (2011) QTL for fibre-related traits in grain×sweet sorghum as a tool for the enhancement of sorghum as a biomass crop. Theor Appl Genet 123:999–1011

Sievers F, Wilm A, Dineen D et al (2011) Fast, scalable generation of high-quality protein multiple sequence alignments using Clustal Omega. Mol Syst Biol 7:539

Singh R, Axtell JD (1973) High lysine mutant gene that improves protein quality and biological value of grain sorghum. Crop Sci 13:535–539

Singh M, Chaudhary K, Singal KR, Magill CW, Boora KS (2006) Identification and characterization of RAPD and SCAR markers linked to anthracnose resistance gene in sorghum [*Sorghum bicolor* (L.) Moench]. Euphytica 149:179–187

Singhal D, Gupta P, Sharma P, Kashyap N, Anand S, Sharma H (2011) *In-silico* single nucleotide polymorphisms (SNP) mining of *Sorghum bicolor* genome. Afr J Biotechnol 10:580–583

Smedley D, Haider S, Ballester B, Holland R, London D, Thorisson G, Kasprzyk A (2009) BioMart – biological queries made easy. BMC Genom 10:22

Springer PS (2000) Gene traps: tools for plant development and genomics. Plant Cell 12:1007–1020

Sree-Ramulu K (1970) Sensitivity and induction of mutations in sorghum. Mutat Res 10:197–206

Srinivas G, Satish K, Murali Mohan S, Nagaraja Reddy R, Madhusudhana R, Balakrishna D, Bhat BV, Howarth CJ, Seetharama N (2008) Development of genic-microsatellite markers for sorghum staygreen QTL using a comparative genomic approach with rice. Theor Appl Genet 117:283–296

Srinivas G, Satish K, Madhusudhana R, Seetharama N (2009a) Exploration and mapping of microsatellite markers from subtracted drought stress ESTs in *Sorghum bicolor* (L.) Moench. Theor Appl Genet 118:703–717

Srinivas G, Satish K, Madhusudhana R, Nagaraja Reddy R, Murali Mohan S, Seetharama N (2009b) Identification of quantitative trait loci for agronomically important traits and their association with genic-microsatellite markers in sorghum. Theor Appl Genet 118:1439–1454

Srinivasa Reddy P, Fakrudin B, Rajkumar, Punnuri SM, Arun SS, Kuruvinashetti MS, Das IK, Seetharama N (2008) Molecular mapping of genomic regions harboring QTLs for stalk rot resistance in sorghum. Euphytica 159:191–198

Stanford WL, Cohn JB, Cordes SP (2001) Gene-trap mutagenesis: past, present and beyond. Nat Rev Genet 2:756–768

Stein LD, Mungall C, Shu S et al (2002) The Generic Genome Browser, a building block for a model organism system database. Genome Res 12:1599–1610

Sukumaran S, Xiang W, Bean SR, Pedersen JF, Kresovich S, Tuinstra MR, Tesso TT, Hamblin MT, Yu J (2012) Association mapping for grain quality in a diverse sorghum collection. Plant Gen 5:126–135

Syvanen AC (2001) Accessing genetic variation genotyping single nucleotide polymorphisms. Nat Rev Genet 2:930–942

Taillon-Miller P, Gu ZJ, Li Q, Hillier L, Kwok PY (1998) Overlapping genomic sequences: a treasure trove of single nucleotide polymorphisms. Genome Res 8:748–754

Talame V, Bovina R, Sanguineti MC, Tuberosa R, Lundqvist U, Salvi S (2008) TILLMore, a resource for the discovery of chemically induced mutants in barley. Plant Biotechnol J 6:477–485

Tao Y, Manners J, Ludlow M, Henzell R (1993) DNA polymorphisms in grain sorghum (*Sorghum bicolor* (L.) Moench). Theor Appl Genet 86:679–688

Tao YZ, Jordan DR, Henzell RG, McIntyre CL (1998) Identification of genomic regions for rust resistance in sorghum. Euphytica 103:287–292

Tao YZ, Henzell RG, Jordan DR, Butler DG, Kelly AM, McIntyre CL (2000) Identification of genomic regions associated with stay green in sorghum by testing RILs in multiple environments. Theor Appl Genet 100:1225–1232

Tao YZ, Hardy A, Drenth J, Henzell RG, Franzmann BA, Jordan DR, Butler DG, McIntyre CL (2003) Identifications of two different mechanisms for sorghum midge resistance through QTL mapping. Theor Appl Genet 107:116–122

Taramino G, Tarchini R, Ferrario S, Lee M, Pe ME (1997) Characterization and mapping of simple sequence repeats (SSRs) in *Sorghum bicolor*. Theor Appl Genet 95:66–72

The Arabidopsis Genome Initiative (2000) Analysis of the genome sequence of the flowering plant *Arabidopsis thaliana*. Nature 408:796–815

Thompson JD, Gibson TJ, Plewniak F, Jeanmougin F, Higgins DG (1997) The CLUSTAL_X windows interface: flexible strategies for multiple sequence alignment aided by quality analysis tools. Nucleic Acids Res 25:4876–4882

Tierney MB, Lamour KH (2005) An introduction to reverse genetic tools for investigating gene function. Plant Health Instructor. doi:10.1094/PHI-A-2005-1025-01

Till BJ, Reynolds SH, Weil C et al (2004) Discovery of induced point mutations in maize genes by TILLING. BMC Plant Biol 4:12

Till BJ, Zerr T, Comai L, Henikoff S (2006) A protocol for TILLING and Ecotilling in plants and animals. Nat Protoc 1:2465–2477

Till BJ, Cooper J, Tai TH, Colowit P, Greene EA, Henikoff S, Comai L (2007) Discovery of chemically induced mutations in rice by TILLING. BMC Plant Biol 7:19

Tuinstra MR, Grote EM, Goldsbrough PB, Ejeta G (1997) Genetic analysis of post-flowering drought tolerance and components of grain development in sorghum. Mol Breed 3:439–448

Tuinstra MR, Ejeta G, Goldsbrough P (1998) Evaluation of near-isogenic sorghum lines contrasting for QTL markers associated with drought tolerance. Crop Sci 38:835–842

Upadhyaya HD, Pundir RPS, Dwivedi SL, Gowda CLL, Reddy VG, Singh S (2009) Developing a mini core collection of sorghum for diversified utilization of germplasm. Crop Sci 49:1769–1780

Upadhyaya H, Wang YH, Sharma S, Singh S, Hasenstein K (2012a) SSR markers linked to kernel weight and tiller number in sorghum identified by association mapping. Euphytica 187:401–410

Upadhyaya HD, Wang YH, Sharma S, Singh S (2012b) Association mapping of height and maturity across five environments using the sorghum mini core collection. Genome 55:471–479

Uptmoor R, Wenzel W, Friedt W, Donaldson G, Ayisi H, Ordon F (2003) Comparative analysis on the genetic relatedness of *Sorghum bicolor* accessions from

Southern Africa by RAPDs, AFLPs and SSRs. Theor Appl Genet 106:1316–1325

Varshney RK, Chen W, Li Y et al (2012) Draft genome sequence of pigeonpea (*Cajanus cajan*), an orphan legume crop of resource-poor farmers. Nat Biotechnol 30:83–89

Varshney RK, Song C, Saxena RK et al (2013) Draft genome sequence of chickpea (*Cicer arietinum*) provides a resource for trait improvement. Nat Biotechnol 31:240–246

Velculescu VE, Zhang L, Vogelstein B, Kinzler KW (1995) Serial analysis of gene expression. Science 270:484–487

Wang YH, Bible P, Loganantharaj R, Upadhyaya H (2012) Identification of SSR markers associated with height using pool-based genome-wide association mapping in sorghum. Mol Breed 30:281–292

Wang F, Zhao S, Han Y, Shao Y, Dong Z, Gao Y, Zhang K, Liu X, Li D, Chang J, Wang D (2013) Efficient and fine mapping of RMES1 conferring resistance to sorghum aphid *Melanaphis sacchari*. Mol Breed 31:777–784

Ware D, Jaiswal P, Ni J et al (2002) Gramene: a resource for comparative grass genomics. Nucleic Acids Res 30:103–105

Waterhouse AM, Procter JB, Martin DM, Clamp M, Barton GJ (2009) Jalview Version 2 – a multiple sequence alignment editor and analysis workbench. Bioinformatics 25:1189–1191

Wen L, Tang HV, Chen W, Chang R, Pring DR, Klein PE, Childs KL, Klein RR (2002) Development and mapping of AFLP markers linked to the sorghum fertility restorer gene rf_4. Theor Appl Genet 104:577–585

White GM, Hamblin MT, Kresovich S (2004) Molecular evolution of the phytochrome gene family in sorghum: changing rates of synonymous and replacement evolution. Mol Biol Evol 21:716–723

Winn JA, Mason RE, Robbins AL, Rooney WL, Hays DB (2009) QTL mapping of a high protein digestibility trait in *Sorghum bicolor*. Int J Plant Genomics 2009:471853

Wise MG, Schulze SR, Lin Y-R, Bowers JE, Okuizumi H, Schertz KF, Paterson AH (2002) Progress towards the positional cloning of the sorghum grain shattering gene. Plant and Animal Genome X, January 12–16, 2002, San Diego, USA

Wu Y, Huang Y (2008) Molecular mapping of QTLs for resistance to the greenbug *Schizaphis graminum* (Rondani) in *Sorghum bicolor* (Moench). Theor Appl Genet 117:117–124

Wu Y, Li X, Xiang W et al (2012) Presence of tannins in sorghum grains is conditioned by different natural alleles of Tannin1. Proc Natl Acad Sci U S A 109:10281–10286

Xin Z, Wang ML, Barkley NA, Burow G, Franks C, Pederson G, Burke J (2008) Applying genotyping (TILLING) and phenotyping analyses to elucidate gene function in a chemically induced sorghum mutant population. BMC Plant Biol 8:103

Xin Z, Wang ML, Burow G, Burke J (2009) An induced sorghum mutant population suitable for bioenergy research. Bioenergy Res 2:10–16

Xin Z, Burow G, Woodfin C, Franks CD, Klein RR, Schertz KF, Pederson GA, Burke JJ (2013) Registration of a diverse collection of sorghum genetic stocks. J Plant Reg 7:119–124

Xu W, Subudhi PK, Crasta OR, Rosenow DT, Mullet JE, Nguyen NT (2000) Molecular mapping of QTLs conferring stay-green in grain sorghum [*Sorghum bicolor* (L.) Moench]. Genome 43:461–469

Yang L, Jin G, Zhao X, Zheng Y, Xu Z, Wu W (2007) PIP: a database of potential intron polymorphism markers. Bioinformatics 23:2174–2177

Yang S, Weers BD, Morishige DT, Mullet JE (2014) *CONSTANS* is a photoperiod-regulated activator of flowering in sorghum. BMC Plant Biol 14:148

Yonemaru JI, Ando T, Mizzubayashi T, Kasuga S, Matsumoto T, Yano M (2009) Development of genome–wide simple sequence repeat markers using whole-genome shot-gun sequences of sorghum [*Sorghum bicolor* (L.) Moench]. DNA Res 16:187–193

Yu J, Buckler E (2006) Genetic association mapping and genome organization of maize. Curr Opin Biotechnol 17:155–160

Yu J, Hu S, Wang J et al (2002) A draft sequence of the rice genome (*Oryza sativa* L. ssp *indica*). Science 296:79–92

Yu J, Holland JB, McMullen MD, Buckler ES (2008) Genetic design and statistical power of nested association mapping in maize. Genetics 178:539–551

Yun-long B, Seiji Y, Maiko I, Wei CH (2006) QTLs for sugar content of stalk in sweet sorghum (*Sorghum bicolor* L. Moench). Agric Sci China 5:736–744

Zhang D, Guo H, Kim C, Lee TH, Li J, Robertson J, Wang X, Wang Z, Paterson AH (2013a) *CSGRqtl*, a comparative quantitative trait locus database for Saccharinae grasses. Plant Physiol 161:594–599

Zhang T, Zhao X, Huang L, Liu X, Zong Y, Zhu L, Yang D, Fu B (2013b) Tissue-specific transcriptomic profiling of sorghum propinquum using a rice genome array. PLoS One 8(3):e60202

Zheng LY, Guo XS, He B et al (2011) Genome-wide patterns of genetic variation in sweet and grain sorghum (*Sorghum bicolor*). Genome Biol 12:R114

Zou G, Zhai G, Feng Q et al (2012) Identification of QTLs for eight agronomically important traits using an ultra-high-density map based on SNPs generated from high-throughput sequencing in sorghum under contrasting photoperiods. J Exp Bot 63:5451–5462

Post-genome Sequencing Developments

7

K.N. Ganapathy and P. Rajendrakumar

Contents

7.1 **Introduction** 156

7.2 **Genome Organization and Comparative Genomics** 156

7.3 **Genome Annotation and Validation of Its Function** 159

7.4 **Functional Analysis and Cross Validation of Agronomically Important Genes** 162

7.5 **Detection of Single Nucleotide Polymorphism and MicroRNAs** 164
7.5.1 Single Nucleotide Polymorphism (SNP) 164
7.5.2 MicroRNA (miRNA) 165

7.6 **Whole Transcriptome Analysis** 168

7.7 **Re-sequencing of Sorghum Using NGS-Based Approaches** 169

7.8 **In Silico Mapping of Agronomically Important Genes** 171

7.9 **Epigenomic Studies** 172

7.10 **Future Prospects** 173

References 174

K.N. Ganapathy (✉)
Plant Breeding, ICAR-Indian Institute of Millets Research, Rajendranagar, Hyderabad, Telangana 500 030, India
e-mail: ganapathy@millets.res.in

P. Rajendrakumar
Biotechnology, ICAR-Indian Institute of Millets Research, Rajendranagar, Hyderabad, Telangana 500 030, India
e-mail: rajendra@millets.res.in

Abstract

The publication of the first draft of the sorghum genome assembly during 2009 opened up avenues directed towards understanding the genome organization and annotation of the genome to know the diverse classes of genes. Post-genome sequencing developments include comparative genomics for understanding the syntenic relationships with other model plants and related crop species, genome annotation through in silico approaches and validation of their functions, functional analysis of agronomically important genes, detection of SNPs and microRNAs and analyzing their role in gene function and regulation, transcriptome analysis to understand the gene networks associated with complex traits, in silico mapping of agronomically important genes, and epigenomics for understanding the non-genetic regulation of gene expression. The most effective strategy for achieving precision in the genetic improvement of sorghum is through the integration of genomic data, genetic principles, statistical knowledge, molecular biology, and plant breeding methodologies through inter-disciplinary research. This chapter focuses on the developments in genomics after the sequencing of sorghum genome, particularly the genome organization, structural and functional gene annotations, functional analysis of genes governing various biochemical pathways, and their impact in the genetic improvement of sorghum.

R. Madhusudhana et al. (eds.), *Sorghum Molecular Breeding*, DOI 10.1007/978-81-322-2422-8_7

Keywords
Comparative genomics • Annotation • Functional analysis • Transcriptomics • Epigenomics

7.1 Introduction

Plant breeding has played a major role in developing improved crop varieties resulting in a dramatic increase in grain yield and genetic enhancement of other agronomic traits. Advancement in science of genomics, especially the availability of modern sequencing tools during the past two decades provided new opportunities to meet the challenges in crop improvement. In the coming years, the science of genomics will play a major role in revolutionizing plant breeding in facilitating a better understanding of genotype–phenotype relationships of complex traits. More importantly, the crop genome sequencing provides a clear understanding about the functions of individual genes and their networks, for defining evolutionary relationships and processes, and for revealing previously unknown regulatory mechanisms that coordinate the activities of genes. Whole-genome sequence of a crop species is of wide utility for genetic conservation and subsequent utilization in plant breeding, but most of the crop genome sequencing projects are restricted to a limited number of model crops having smaller-sized genome and having economic importance. Among plant species, *Arabidopsis thaliana* was the first plant species to be sequenced (Arabidopsis Genome Initiative 2000). Subsequently genome sequencing has been completed in many agriculturally important plant species, including rice (International Rice Genome Sequencing Project 2005), sorghum (Paterson et al. 2009), maize (Schnable et al. 2009), soybean (Schmutz et al. 2010), and more recently chickpea (Varshney et al. 2013), pigeon pea (Singh et al. 2012 and Varshney et al. 2012), and wheat (The International Wheat Genome Sequencing Consortium 2014).

Compared to other cereals, smaller-sized genome, C_4 photosynthesis, drought tolerance, and capability of producing high biomass make sorghum an attractive crop for genome sequencing. Moreover, its genome makes it an ideal to be used as a tropical grass model. The sorghum genome sequencing project was initiated during 2005 by the US Department of Energy Joint Genome Institute (Community Sequencing Program) jointly with the Plant Genome Mapping Laboratory following the whole-genome shotgun sequencing strategy and subsequent validation by genetic, physical, and systemic information. The genome sequencing was completed during 2007, and the first draft of the sorghum genome assembly was published during 2009. Sorghum was the first crop genome to be sequenced exclusively using WGS sequence assemblies and thereafter assessed for integrity using high-density genetic maps and physical maps. The analysis revealed that the scaffolds of sequence assemblies from Sanger sequencing accurately span extensive repetitive DNA tracts and extend into telomeric and centromeric regions (Bevan and Uauy 2013). The wealth of large-scale sequence information generated from the sequencing of sorghum is available in the number of public databases and is freely accessible for researchers. The vast amount of genome resources resulting crop genome sequencing revolution is likely to provide a paradigm shift in the approach to plant biology and crop breeding. This chapter focuses on the developments in genomics after the sequencing of sorghum genome, particularly the genome organization, structural and functional gene annotations, functional analysis of genes governing various biochemical pathways, and their impact in the genetic improvement of sorghum.

7.2 Genome Organization and Comparative Genomics

Genome sequencing in sorghum was performed in a highly homozygous genotype BTx623 from Texas A&M University and represents the pedigree of many elite sorghum genotypes. The first draft of the sorghum sequence was published during 2009 (Paterson et al. 2009). Assembling of the genome was done by the integration of whole-genome shotgun sequence with the genetic and BAC physical map (Paterson 2013). The size of the sorghum genome (~730 Mb) is bigger than the rice (389 Mb) and smaller than that of maize (2.3 Gb) genome. One third of the sorghum genome is said to be the recombination-rich region, while the rest portion is assumed to be recombination poor. The euchromatin which represents the gene-rich regions is almost similar in sorghum (252 Mb) and rice (309 Mb).

A total of 34,496 gene models has been described in sorghum, of which approximately 27,640 belong to bonafide protein-coding genes identified using evidence-based (homology) and ab initio (based on gene content and signal detection) gene prediction methods. The rest of the gene models (5,197) were reported to contain few exons, relatively less amino acids, lack of information on ESTs and also found to be diverging from rice genomes. Chromosome-wise distribution of gene models is given in Table 7.1. Sorghum gene families were similar to *Arabidopsis*, rice, and poplar based on their number and size. Moreover, 9,503 gene families were common among *Arabidopsis*, rice, and poplar, while 15,225 genes were shared in at least one of the abovementioned species. About 25,875 high-confidence sorghum genes have orthologs in the above three species. 3,983 sorghum gene families were shared only by sorghum and rice, and 1,153 gene families were reported to be present only in sorghum. Many paralogs found in the sorghum genome possess tandem arrays, and among them the largest tandem is the 15 cytochrome P450 genes (Paterson et al. 2009).

Although cereal grass is believed to have evolved from a common ancestor, comparative genomics among the members of the grass family revealed a great deal of variation in their genome size. Transposable elements present in the intergenic space are subjected to a rapid turnover rate in grasses (Dubcovsky and Dvorak 2007). The expansion of genome size in sorghum as compared to rice is largely attributed to the presence of long terminal repeat (LTR) retrotransposons which accounts to 55 % of the sorghum genome, lower than maize (79 %) and double than rice (26 %). It was reported by Kidwell (2002) that the primary cause of the differences in the genome size is due to the accumulation of repeat elements, principally transposable elements. Among the various classes of retrotransposons, the *gypsy* and *copia*-like elements were higher in sorghum as well as in rice and lower in maize. The distribution of LTR transposons was random and they occupied the gene-poor regions. Among the sorghum DNA transposons, the CACTA-like elements (4.7 % of the genome) are predominantly found in the sorghum genome. The

Table 7.1 Chromosome-wise distribution of genes in *Sorghum bicolor* genome

Chr	Length in base pairs	Chromosome map	Gene models
1	73,840,631	Chr 1	5,136
2	77,932,606	Chr 2	3,926
3	74,441,160	Chr 3	4,102
4	68,034,345	Chr 4	3,323
5	62,352,331	Chr 5	1,755
6	62,208,784	Chr 6	2,609
7	64,342,021	Chr 7	2,015
8	55,460,251	Chr 8	1,584
9	59,635,592	Chr 9	2,328
10	60,981,646	Chr 10	2,473
-	40,203,219	88 unanchored scaffolds, concatenated with 200 N spacer	197
Total	**699,432,586**	–	**29,448**

Source: http://zoneplantgdb.gdcb.iastate.edu/SbGDB/

miniature inverted repeat elements account for 1.7 % of the genome, while the helitron elements account for 0.8 %. There is a marginal organellar DNA insertion (0.085 %) in the sorghum nuclear genome, which is 0.53 % less than that of rice.

Like any other crop species, whole-genome duplication is of wide occurrence in sorghum. Particularly, polyploidy and segmental duplication and associated gene loss are a common occurrence in the history of evolution of flowering plants. Out of the 19,929 genes exhibiting colinearity between sorghum and rice, only one copy was retained for 13,667 genes. Further, both copies were retained for 4,912 genes in rice and sorghum, while one copy each of 1,070 and 634 genes was lost in sorghum and rice, respectively, indicating that the high number of gene loss leads to divergence. It was hypothesized that occurrence of apparent segmental duplication during the whole-genome duplication actually resulted from the pan-cereal whole-genome duplication and differentiated from the remainder of the chromosome(s) owing to the concerted evolution acting independently in sorghum, rice, and perhaps other cereals (Paterson et al. 2009). Further, it is understood that sorghum has not reduplicated approximately 70 million years. This would facilitate sorghum to be used as an out-group for comparing genes or genome sequences in other crop species which have duplicated. Moreover, it would facilitate in understanding the genome evolution in closely related species such as *Saccharum*/*Miscanthus* since they are believed to have diverged 8–9 million years ago.

Similarities and differences are observed in the genome organization of sorghum and other cereals with respect to a specific class of proteins, size distribution of exons and introns, and organellar genome insertion. Among the different class of proteins, the *Pfam* domains were present in other species and over-represented in sorghum, while alpha-kafarins were found in maize but absent in rice. The nucleotide-binding site–leucine-rich repeat (NBS–LRR) class of proteins was reported to be less frequent in sorghum as compared to rice. Based on sequence comparisons, 211 NBS–LRR coding genes were observed in sorghum, 410 in rice, and 149 in *Arabidopsis*. The highest number of NBS–LRR genes (62) was observed in chromosome 5 of sorghum, while in rice it was high in chromosome 11 (106). The size distribution of exons and introns in the orthologous gene set of sorghum and rice agreed closely indicating that they are conserved. Several insertions of mitochondria and chloroplast sequence were found in the nuclear genome when these genomes were aligned. The organellar insertion in sorghum is smaller as compared to that of rice; most of them were less than 500 bp.

Comparative genome mapping in cereals has been done extensively during the last decade using genetic markers mainly using SSRs (Bennetzen 2000; Bennetzen and Ma 2003), which revealed synteny between sugarcane and sorghum. Moreover, a good macro- and micro-colinearity of gene order was observed revealing the evolutionary relationships among the cereals (Glaszmann et al. 1997; Gale and Devos 1998). The availability of the complete sorghum genome has provided opportunities for genome research in closely related crop such as sugarcane since the genome size of the latter is large (~930 Mb) with polyploidy that hinders the progress in genomics research. Both sorghum and sugarcane belong to Poaceae family and belong to sub-tribe Saccharinae and are considered to have originated from a common ancestor before 8–9 million years. The microsyntenic relationships between sugarcane and sorghum were analyzed comparing 454 pyrosequences of 20 sugarcane bacterial artificial chromosomes (BACs) with genome sequence of sorghum (Ming et al. 1998). The genic sequences of the sugarcane BACs revealed on an average of 95.2 % sequence identity with sorghum when the genome of the latter was used as a template to order sequence contigs. Within the aligned sequences, 209 genes were annotated in sugarcane and 202 in sorghum. All the 17 genes that were sugarcane specific were validated by sugarcane ESTs, and only one out of the 12 sorghum specific was validated by sorghum ESTs. This study highlighted the fact that the gene density between sugarcane BACs and corresponding sorghum sequences challenged the notion that polyploidy species might have a faster pace of gene loss due to the redundancy of multiple alleles at each locus.

Comparative sequence analysis of orthologous regions from ten diploid *Oryza* species, *Brachypodium distachyon*, sorghum, and maize was performed by Yang et al. (2012) to know the origin of an important rice gene *Ghd7* that has a major effect on several agronomic traits, including yield. Sequence analysis demonstrated the presence of high gene colinearity across the genus *Oryza* and a disruption of colinearity among non-*Oryza* species. Importantly, *Ghd7* was not present in orthologous positions except in *Oryza* species. The *Ghd7* regions had low gene densities, rich in repetitive elements, and tremendous variation in the sizes of orthologous regions. Presence of large transposable elements resulted in a high frequency of pseudogenization and events of gene movement surrounding the *Ghd7* loci. Annotation information and cytological experiments have indicated that *Ghd7* is a heterochromatic gene. Orthologs of *Ghd7* were identified in *B. distachyon*, sorghum, and maize using phylogenetic analysis, and their positions differed dramatically as a consequence of gene movements in grasses. Sequence remnants of gene movement of *Ghd7* due to illegitimate recombination in the *B. distachyon* genome were also identified.

7.3 Genome Annotation and Validation of Its Function

The initial sequence data generated from any genome sequencing projects remain highly fragmented and are of no use unless properly assembled, and genome/gene annotations are done using various computational tools and available knowledge on other/related crop species. Annotation of the sequenced genome is the most important task that adds a lot of value in the form of information of genes and their putative function. Genome annotation refers to analysis of DNA sequence of the genome/genes and assigning biological function to the sequence (Stein 2001). It is also the most difficult, time-consuming, and high-cost-intensive part of any genome sequencing project involving experimental and homology-based gene prediction methods. Genome annotation is done at different levels (Fig. 7.1).

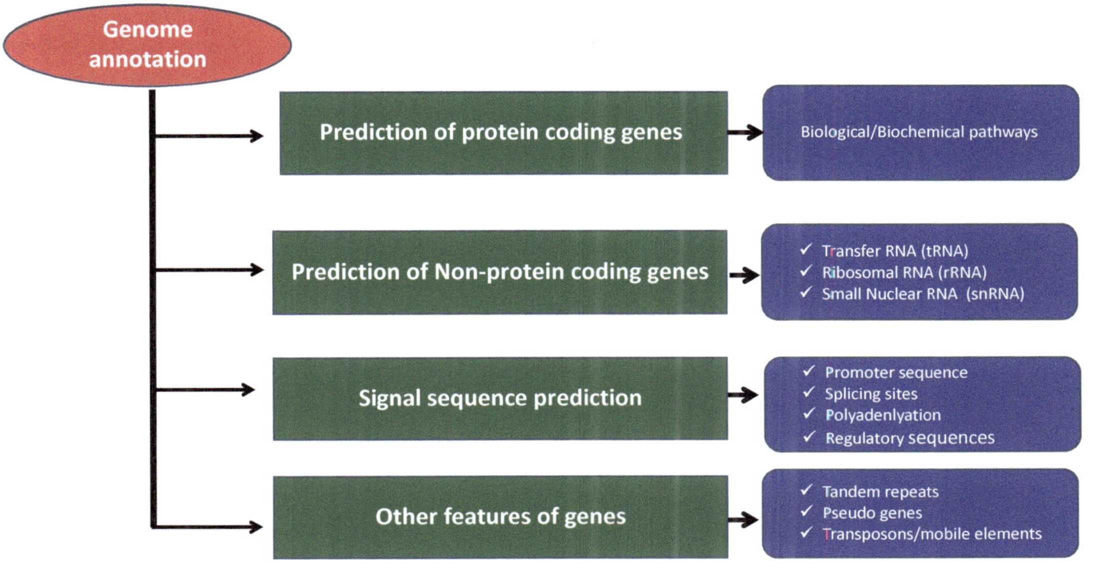

Fig. 7.1 Schematic diagram showing the stages of genome annotation

Structural Annotation: Prediction of eukaryotic genes is difficult as compared to prokaryotes since the presence of introns in the genes of eukaryotes interfere with gene prediction. Structural annotation aims at the identification of gene/genomic elements, mainly the open reading frames (ORFs), the coding regions, and the regulatory motifs (Aubourg and Rouze 2001). Several online tools are available for the genome annotation (Table 7.2).

Functional Annotation: This aims at assigning the biochemical or biological function associated with the sequence along with information on the gene regulation and interactions. Functional annotation relies on sequence similarities detected between two proteins based on their homologues indirectly indicating that they share same ancestor and therefore same biochemical function. For each predicted gene, the protein is identified from the coding region and is analyzed using blastp with the protein databases (Aubourg and Rouze 2001).

Among cereals, rice genome was the first to be sequenced, and most of the gene predictions were done using ab initio methods (without prior knowledge on gene function) rather than evidence-based methods and were further improved using cDNA and expressed sequence tags (Yuan et al. 2005). Like rice, genome sequences of maize and other crops can contribute to the annotation of genes in other cereals including sorghum. For more successful annotation of the genome, a high-quality draft sequence assembly (of at least 90 % complete) is more ideal. The most widely used statistics for describing the quality of a genome assembly are the scaffold, contig N50s, percent gaps, and coverage. An assembly with an N50 scaffold length that is gene sized is a decent target for annotation (Yandell and Ence 2012). A variety of software tools are available for genome annotations and viewing the annotations (Table 7.2). The detailed gene annotation and the functional information of genes of sorghum are available on genome website http://phytozome.jgi.doe.gov/pz/.

In sorghum, data from several sources were used to make consensus gene predictions. The transcript assemblies from The Institute for Genomic Research (TIGR) were aligned to the genome sequences of sorghum, which are repeat-masked using GenomeThreader based on splice site maize model. The EST assemblies of *Allium cepa*, *Ananas comosus*, *Avena sativa*, *Brachypodium distachyon*, *Curcuma longa*, *Hordeum vulgare*, *Oryza sativa*, *Saccharum officinarum*, *Secale cereale*, *Sorghum bicolor*, *Sorghum halepense*, *Sorghum propinquum*, *Triticum aestivum*, *Zea mays*, and *Zingiber officinale* were mapped. Optimal spliced and blastX alignments were done to predict the genes using the reference set of proteins in the SWISSPROT database, *Arabidopsis* (TAIR6), *Saccharomyces cerevisiae*, and rice (RAP2) proteomes. Optimal spliced alignment is derived from the optimal alignment of both similarity of the predicted gene product to the protein sequence and intrinsic splice site strength of the predicted intron and based on scoring. Further identification of gene models based on repeat-masked genomic sequences by ab initio methods (Fgenesh++, GeneID, GenomeScan/PASA) were also done. Jigsaw was used as a statistical tool to combine all the above supporting information. All models were scored by blastp against the UniREF90 protein database, and for each locus the best fitting model with the highest bitscore was selected. These predictions were reanalyzed through the PASA tool to predict UTRs from maize, sorghum, and sugarcane ESTs, to identify alternative splicing patterns and finally to fit all predicted models to the splice sites that resemble EST evidences of related species. The improved gene model proteins were subject to protein homology analysis using proteome databases, and the transcripts were selected based on Cscore, protein coverage, EST coverage, and its CDS overlapping with repeats. Finally, the annotation results yielded 36,338 transcript models at 34,496 loci. Out of 28,003 complete models, 6,493 candidate genes were predicted that lacked the start and/or stop codon and these were assigned as partial models.

Table 7.2 Genome annotation resources

Resources	Particulars	Availability
Structural annotation tools		
RepeatMasker	Identification of interspersed repeats and low-complexity DNA sequences	www.repeatmasker.org/
GENEMARK	Gene prediction	http://exon.gatech.edu/GeneMark/
WISE2	Comparison of a protein sequence to a genomic DNA sequence, allowing for introns and frame-shifting errors	www.ebi.ac.uk/Tools/psa/genewise/
GrailEXP	Prediction of exons, genes, promoters, polyAs, CpG islands, EST similarities, and repetitive elements	http://compbio.ornl.gov/grailexp/
GeneScan	Prediction of the location and intron–exon boundaries in a genomic sequence	http://genes.mit.edu/GENSCAN.html
yrGATE	Identification and dissemination of eukaryotic genes	www.plantgdb.org/prj/yrGATE/
PlantProm	A database of plant promoter sequences	http://linux1.softberry.com/
PlantTFDB	Plant Transcription Factor Database from about 49 species	planttfdb.cbi.edu.cn/
PLACE	A database of plant *cis*-acting regulatory DNA elements	www.dna.affrc.go.jp/PLACE/
Transfac	A database on eukaryotic transcription factors, their genomic binding sites and DNA-binding profiles	www.gene-regulation.com/pub/databases.html
PlantCare	A database of plant *cis*-acting regulatory elements and a portal to tools for in silico analysis of promoter sequences	http://bioinformatics.psb.ugent.be/webtools/plantcare/html/
Functional annotation tools		
PASA	Aligns EST and protein sequences to the genome and produces evidence-driven consensus gene models	http://pasapipeline.github.io/
MAKER	Identifies repeats, aligns ESTs and proteins to a genome, produces ab initio gene predictions, and automatically synthesizes these data into gene annotations with evidence-based quality values	http://gmod.org/wiki/MAKER
NCBI	Uses BLAST alignments together with predictions from Gnomon and GenomeScan to produce gene models	http://www.ncbi.nlm.nih.gov
Ensembl	Uses species-specific and cross-species alignments to build gene models. Also annotates non-coding RNAs	www.ensembl.org
mGene	Computational tool for the genome-wide prediction of protein-coding genes from eukaryotic DNA sequences	www.mgene.org/
SNAP	Calculates synonymous and non-synonymous substitution rates based on a set of codon-aligned nucleotide sequences	http://www.broadinstitute.org/mpg/snap/
FGENESH	Prediction of multiple genes in genomic DNA sequences	http://nhjy.hzau.edu.cn/kech/swxxx/jakj/dianzi/Bioinf6/GeneFinding/GeneFinding2.htm
Twinscan	System for predicting gene structure in eukaryotic genomic sequences	www.bioinformatics.ca/
GenomeScan	Predicting the locations and exon–intron structures of genes in genomic sequences from a variety of organisms	http://genes.mit.edu/genomescan.html

7.4 Functional Analysis and Cross Validation of Agronomically Important Genes

Functional analysis of genes/genomes is a broader term and is much related to understanding relationships between an organism's genes/genomes and its phenotype mostly based on experimental evidence. More specifically, analysis of genes refers to biochemical function (e.g., protein kinase), cellular function (signal transduction pathway), developmental function (e.g., a role in pattern formation), or adaptive function (contribution of the gene product to the fitness of the organism) (Bouchez and Hofte 1998). Moreover, it involves multiple approaches for understanding the properties and function of organism's genes and gene products. Functional genomics involve studies of natural variation in genes, RNA, and proteins over time (such as an organism's development) or space (such as its body regions), as well as studies of natural or experimental functional disruptions affecting genes, chromosomes, RNA, or proteins. Analysis of genes at a functional level provides a clear understanding of the dynamic properties of an organism at cellular and/or organism levels. This would provide a more comprehensive picture of how biological function arises from the information encoded in an organism's genome. Most of the newly identified genes show sequence similarity to the already reported genes. Although sequence homology is used for describing the function of new genes, experimental evidence is required to validate the function of the genes in most cases. Gene functions at experimental level can be obtained from spatial and temporal expression patterns by quantifying the level of mRNA and/or protein in different cell types during growth and development under specific environmental conditions. Since mRNAs are related to the expressed regions of the gene, it is possible to establish a link between a genotype and an expression phenotype. The gene knockout approach also permits the gene sequence to be linked to a phenotype for identification of gene functions (Groth et al. 2008).

One of the major breakthroughs in plant genome research is that about 54 % of higher plant genes can be assigned a function by comparing them with the sequences of genes of known function (Somerville and Somerville 1999). Merely having information that a gene encodes a kinase or transcription factor does not provide any meaning unless the pathway/processes of these genes are well understood. Availability of whole-genome sequence data will facilitate knowledge on all the genes governing a particular pathway. For instance, on the basis of sequence analysis, about 13 % of *Arabidopsis* genes are known to be involved in transcription or signal transduction (Somerville and Somerville 1999). Major approaches for studying the gene function at the RNA level include transcriptome profiling using microarray and serial analysis of gene expression (SAGE). Loss-of-gene function techniques include mutagenesis and RNA-induced gene silencing techniques (Travella et al. 2006). A cDNA library was constructed using suppression subtractive hybridization (SSH) method by Li et al. (2009a) to study the key processes governing defense mechanisms for anthracnose disease in sorghum. About 41 unique differentially expressed cDNA clones were identified, and the clones were classified into seven categories according to putative functions of their homologous sequences. The identified clones/genes were potentially involved in plant defense, signal transduction, abiotic stress, secondary metabolism, and protein synthesis and degradation. Over-expression of the genes in the tobacco plants revealed the elevated expression of the resistance genes. Functional analysis of genes involved in various biochemical pathways was accelerated after the availability of whole-genome sequence of sorghum.

High-throughput massively parallel sequencing of genes involved in sorghum-specific phytoalexin synthesis was performed by Mizuno et al. (2012) to elucidate their coordinated expression. Phytoalexins are low-molecular-weight compounds which are known to protect sorghum from fungal pathogens. The transcriptional regulation of genes of the key enzymatic steps for synthesizing sorghum-specific phytochemicals was understood. The candidate genes responsible for the missing steps of sequential reaction that causes the accumulation of phytoalexins were identified using genome-wide analysis. Moreover, the findings suggested that accumulation of 3-deoxyanthocyanidin, but not anthocyanidin, occurs upon infection with *Bipolaris sorghicola*. In another sorghum accession, DK46, anthocyanin pigment is accumulated through sequential reactions catalyzed by flavanone 3-hydroxylases, dihydroflavonol 4-reductase, and anthocyanidin reductase ANS (Liu et al. 2010) suggesting that expression of the genes encoding these proteins has changed during the history of sorghum breeding.

Sweet sorghum, a natural variant from grain sorghum, possesses high sugar content in its stalk which can be easily converted into bioethanol (Almodares and Hadi 2009). The reduced lignin content, which is the main characteristic of brown midrib (*bmr*) mutants, improves the efficiency of bioethanol conversion from biomass. Suppression subtractive hybridization (SSH) combined with cDNA microarray profiling was performed by Yan et al. (2012) to study the differential gene expression in a set of 13 *bmr* mutants that accumulate significantly less lignin than the wild type, BTx623. Among the 153 differentially expressed genes identified, 43 were upregulated and 110 downregulated in the mutants. These genes were validated by a semi-quantitative RT-PCR analysis applied to 12 of these genes. Low transcript abundance was observed for genes encoding L-phenylalanine ammonia lyase and cinnamyl alcohol dehydrogenase in the mutants than in the wild type, which is consistent with the expectation that both enzymes are associated with lignin synthesis. The genes governing lignin synthesis enzyme cinnamic acid 4-hydroxylase was found upregulated in mutants, indicating that the production of monolignol from L-PHENYLALANINE may involve more than one pathway. These differentially expressed genes could be useful for breeding sorghum with improved efficiency of bioethanol conversion from lignocellulosic biomass.

Protein expression changes in the leaf proteome were monitored by Swami et al. (2011) under salt-stressed hydroponic cultures of sorghum. The responses to salt stress were investigated after 96 h of treatment with 200 mM NaCl, and proteins with more than 1.5-fold change in expression were identified using mass spectrometry. The majority of the differentially expressed proteins were attributed to signal transduction mechanisms and inorganic ion transport and metabolism. Since the complete sorghum protein database was not established, the ESTs and partial genome sequence data were used for further identification of all affected proteins. In conclusion, this elucidated the salinity-induced proteomic alterations in sorghum leaves, and 21 such proteins were identified using 2-DE/MS. Four differentially expressed proteins could not be assigned any function using Pfam and COGnitor search and warrant detailed investigation to elucidate their exact role in salinity tolerance. Detailed functional analysis of these proteins would provide further information such as that regarding direct regulatory networks in this important crop plants. Genome-wide patterns of gene expression and evolution across reproductive tissues was studied by Davidson et al. (2012) using sequence-based approach to compare transcriptomes of *Brachypodium distachyon*, rice, and sorghum. The results revealed that only a fraction of orthologous genes showed conserved expression patterns. The high proportion of conserved orthologs was observed in genes that are upregulated in tissues like leaves, anther, pistil, and embryo, while those expressed in seeds showed diverged patterns of expression. Moreover, the genes that are highly expressed tend to be conserved at the coding sequence level. Furthermore, orthologs in syntenic genomic blocks are more likely to share correlated expression patterns compared to non-syntenic orthologs. This information could be a useful resource for comparing and validation in orphan crops where genomes have not yet been sequenced.

7.5 Detection of Single Nucleotide Polymorphism and MicroRNAs

7.5.1 Single Nucleotide Polymorphism (SNP)

Advancement in science of genome sequencing has brought down the sequencing cost, but greatly increased the speed of sequencing. In plant species, where the genome size is small, the best approach would be to sequence the whole genome for identification of SNP variation. In larger genome, where the reference sequence is not available, sequencing of the genome using reduced representation or genomic reduction approach will reduce the complexity of sequencing from the repetitive regions and would give a better representation of the sequence from the gene regions. The NGS platforms generate a large amount of sequencing data in a high-throughput manner, which can be used for SNP discovery, whose downstream applications in linkage mapping, genetic diversity assessment, association mapping, and marker-assisted selection have been demonstrated in several crop species through SNP genotyping.

Prior to the emergence of NGS, SNP discovery in complex genomes posed serious problems, mainly due to the highly repetitive nature of those genomes, and different experimental strategies were employed to avoid such regions. These strategies were unable to discover SNPs in low-copy non-coding regions and inter-genic spaces. Although SNPs in the genic regions could be detected, their frequency was generally low. The emergence of NGS technologies such as 454 Life Sciences (Roche Applied Science, Indianapolis, IN), HiSeq (Illumina, San Diego, CA), SOLiD, and Ion Torrent (Life Technologies Corporation, Carlsbad, CA) has helped in overcoming the bottlenecks associated with low-throughput and high-cost SNP discovery methods (Mardis 2008). In order to discover SNPs in a genome-wide fashion and avoid repetitive and duplicated DNA, it is very important to employ genome complexity reduction techniques coupled with NGS technologies. Earlier few genome complexity reduction techniques like high-Cot selection (Yuan et al. 2003), methylation filtering (Emberton et al. 2005), and microarray-based genomic selection (Okou et al. 2007) have been employed in SNP discovery. Due to the detection of false positives, genome complexity reduction technologies such as Complexity Reduction of Polymorphic Sequences (CRoPS) (Keygene N.V., Wageningen, the Netherlands) (van Orsouw et al. 2007) and Restriction Site Associated DNA (RAD) (Floragenics, Eugene, OR, USA) (Baird et al. 2008) were developed recently, which are computationally sound and capable of eliminating duplicated SNPs. Rapid and inexpensive discovery of SNPs within genes can be performed by the re-sequencing of the transcriptome through NGS technologies (Morozova and Marra 2008).

SNPs can be used to detect genetic variation through the process known as SNP genotyping. Different SNP genotyping assays have been developed recently, which include Illumina GoldenGate (http://www.illumina.com/), KASPar and SNPline genotyping systems (http://www.lgcgenomics.com/), iPLEX Gold technology (http://www.sequenom.com/), chip-based assays such as Affymetrix GeneChip arrays (http://www.affymetrix.com/estore/), Illumina BeadChips (http://www.illumina.com/), array-based technologies such as Infinium and GoldenGate, and genotyping by sequencing (reviewed in Kumar et al. 2012). The genotyping assay to be used is generally decided based on the number of SNPs and genotypes to be screened with some importance given to the cost of the assay as well as the level of accuracy (Kumar et al. 2012). Genotyping by sequencing (GbS) is a multiplexed approach used in the NGS platform for the construction of a sequence using reduced representation of libraries. GBS has tremendous potential in plant genomics research since it performs marker discovery and genotyping simultaneously with reduced cost. GBS has been successfully used in many crop plants such as maize, cassava, rice, and sorghum. Due to the large number of SNP variation and distribution of the SNP, it is possible to map the genome for agronomically important traits using genome-wide association mapping approaches. The GBS approach would further increase the power of sequencing with the availability of the reference genomes and advanced bioinformatics tools.

Small-scale SNP discovery based on the sequencing of the candidate genes was the order of the day prior to the availability of reference genome of sorghum and NGS technologies. DNA sequence polymorphism in 15 genes involved in the starch metabolism was analyzed by Hamblin et al. (2007) in a panel of sorghum cultivars that vary in endosperm characteristics resulting in the detection of SNPs at all loci, including a potential candidate gene for the causal mutation underlying a waxy phenotype. Genotyping of a diverse panel consisting of 125 sorghum (mostly sweet) genotypes with 322 SNPs identified three main genetic groupings of sweet sorghums using distance as well as model-based methods, which were further classified as historical and modern syrup, modern sugar/energy types, and amber types based on known origins and observed phenotypes (Murray et al. 2009). Computational approaches for SNP discovery gained prominence with the availability of a large number of sequences in public databases including expressed sequence tags (ESTs) and reference genome leading to a rapid and cost-effective SNP discovery. Following similar approach, 40,589 reliable SNPs were detected in sorghum using the online SNP and allele detection tool HaploSNPer (Singhal et al. 2011), of which 17,042, 20,500, and 3,047 were transitions, transversions, and indels, respectively. Short-read genome sequencing of eight diverse sorghum accessions using random and reduced representation libraries and alignment of eight genome equivalents (6 Gb) to the public reference genome detected 283,000 SNPs at ≥82 % confirmation probability (Nelson et al. 2011). Inbred lines, two sweet (Keller and E-Tian) and one grain (Ji2731) sorghum, were re-sequenced, and 1,057,018 SNPs, 99,948 indels of 1–10 bp in length, and 16,487 presence/absence as well as 17,111 copy number variations were identified; the majority of them resided in the genes possessing leucine-rich repeats, PPR repeats, and disease resistance (R) genes (Zheng et al. 2011). Recently, Morris et al. (2012) constructed *ApeKI*-reduced representation libraries for 971 accessions from the world germplasm collections comprising of the US sorghum association panel (SAP), the sorghum mini core collection (MCC), and the Generation Challenge Programme sorghum reference set (RS) and generated ~21 Gbp of sequence on the Illumina Genome AnalyzerIIx/HiSeq by using GBS from which ~265,000 SNPs were characterized. Analysis of SNPs could find evidence of selective sweeps around starch metabolism genes and introgressions around the known height and maturity loci. Along with genome-wide association mapping, candidate gene association mapping was employed on a diverse panel of 300 sorghum accessions by Sukumaran et al. (2012) to identify marker–trait associations for grain quality traits. The analysis of the diverse panel using 1,290 genome-wide SNPs separated the accessions into five sub-populations, viz., *durra*, *kafir*, *caudatum*, *guinea–caudatum*, and *zerazera–caudatum*, which differed in kernel hardness, acid detergent fiber, and total digestible nutrients. Candidate gene association analysis using 333 SNPs in candidate genes resulted in the identification of eight significant marker–trait associations. A SNP in starch synthase IIa (SSIIa) gene was found to be associated with kernel hardness, while SNPs in starch synthase (SSIIb) gene and pSB1120 loci were associated with starch content.

7.5.2 MicroRNA (miRNA)

MicroRNAs (miRNAs) are small RNAs (21–24 nucleotides long) that regulate gene expression in plants by getting into the posttranscriptional gene silencing pathway, leading to degradation of the target mRNA or translational repression. These miRNAs have emerged as an answer for understanding the unknown modes of gene regulation, which are highly specific to target genes. The miRNAs are encoded by a special class of genes, known as miR gene, which do not encode proteins. The transcribed RNA folds into a characteristic structure due to its partial self-complementarity and imperfect double-stranded regions from which the miRNAs originate (Bartel 2004). In plants, the excision of the mature miRNA is a multi-step reaction, which is performed by the Dicer enzyme DCL1, at least in part (Kurihara and Watanabe 2004). The miRNAs are characteristically conserved between species, which was confirmed by the fact that most of the *Arabidopsis* miRNA families have homologues in rice and other plants (Axtell and Bartel 2005). However, plant miRNAs show homology only to other plant miRNAs and animal miRNAs to animal miRNAs indicating the evolution of miRNA-dependent gene regulation mechanism separately in plants and animals (Pasquinelli et al. 2000).

In silico identification of miRNA targets complements the experimental procedures to study the diverse regulatory roles of miRNAs. Plant miRNAs generally exhibit a high degree of complementarity to their target sites, which makes it possible for the in silico identification of miRNA targets (Jones-Rhoades and Bartel 2004). Initially, the computer-aided target prediction was found to be complicated due to the small size of miRNAs (Robins et al. 2005). However, since the first discovery of miRNAs in *Caenorhabditis elegans* (Lee et al. 1993; Reinhart et al. 2000), several computational approaches and tools have been developed for a reliable prediction of miRNA targets. Numerous miRNAs were identified in various plant species through the computational approaches as well as expression analysis (Sunkar et al. 2008). Such studies will be useful for tracing the evolution of small RNAs by examining their expression in their ancestral species.

Most of the miRNA studies are aimed at the identification of miRNAs and their targets. In general, four approaches are employed for the identification of miRNAs: (1) genetic screening (Lee et al. 1993 and Wightman et al. 1993), (2) direct cloning of small RNAs (Mead and Tu 2008), (3) computational analysis (Mathews et al. 1999), and (4) expressed sequence tag (EST) analysis (Zhang et al. 2006a). Initially, the genetic screening approach (Lee et al. 1993; Wightman et al. 1993), which was similar to that of identifying other traditional genes, was used for the identification of miRNAs. This method has limitations because it is expensive, time-consuming, and dominated by chance (Bartel 2004). The second approach is an experimental approach involving the direct cloning of small RNAs after its isolation by size fractionation (Lu et al. 2005; Fu et al. 2005). This is a more efficient approach to obtain miRNAs because only small RNAs are isolated and screened. This method was further refined by Lu et al. (2005) by combining it with massively parallel signature sequencing (MPSS). In addition, high-throughput pyrosequencing was also found suitable to identify novel miRNAs in plants (Sunkar et al. 2008). The final and the more rapid approach for the identification of miRNAs and their targets is the computational approach. This approach requires DNA sequence data such as the complete genome sequence, genome survey sequences (GSSs), high-throughput genomics sequences (HTGSs), non-redundant nucleotides (NRs), and expressed sequence tags (ESTs). Zhang et al. (2006a) identified conserved miRNAs in plants using ESTs alone, which suggests the power of this approach to predict homologues or orthologs of previously known miRNAs. More importantly, the prediction of miRNAs in multiple species is possible by this approach as demonstrated by the identification of a set of miRNA and their targets using the largest data set of Triticeae ESTs (Dryanova et al. 2008). Computational programs such as MiRscan (Lim et al. 2003a, b) and MiRAlign (Wang et al. 2005) were designed and used for successful prediction of miRNA genes in *Arabidopsis* (Jones-Rhoades and Bartel 2004; Adai et al. 2005), rice (Li et al. 2005), and *Brassica* sp. (Xie et al. 2007).

Experimental identification of miRNA targets is a difficult and time-consuming process. As a consequence several computational prediction methods have been developed for target prediction for further experimental validation such as Stacking Binding Matrix (SBM), which searches candidate sequences using the information on miRNAs as well as experimentally validated target sequences (Moxon et al. 2008). New miRNA–mRNA duplexes can be found by training a miRNA target detection algorithm using the properties of known miRNA–mRNA duplexes (Sungroh and Giovanni 2006). Efficient identification of miRNA and their target transcription factors can be done using the MotifModeler informatics program (Wang et al. 2008).

Sunkar and Jagadeeswaran (2008) identified 682 miRNAs in 155 diverse plant species after searching all publicly available DNA sequences such as GSS, HTGS, ESTs, and NR. About five miRNA families, viz., miR319, miR156/157, miR169, miR165/166, and miR394, were found in 51, 45, 41, 40, and 40 diverse plant species, respectively. The miR403 homologues were detected in 16 dicots, while miR437 and miR444 homologues, as well as the variant of the miR396 family (miR396d/e), were detected only in monocots, indicating the specificity of miRNAs to the dicots and monocots.

Seventeen new miRNAs were identified in sorghum by Du et al. (2010) that were distributed evenly among 11 miRNA families using a homology-based approach as well as the miRNA secondary structure. In silico analysis of these miRNAs through online software miRU revealed that they might be involved in the regulation of 64 target genes, most of them associated with RNA processing, metabolism, cell cycle, protein degradation, stress response, and transportation. Moreover, 7 of 11 miRNA families target proteins that are associated with metabolism and stress response, suggesting their essential role in biological processes.

Recently, the small RNA component of the transcriptome was characterized by Calvino et al. (2011) from the stems of grain (BTx623) and sweet (Rio) sorghum as well as from F_2 plants derived from their cross that segregated for sugar content and flowering time. The variation in miR172 and miR395 expression was found to be correlated with flowering time, whereas that of miR169 was correlated with sugar content in stems. Importantly, genotypic differences were noticed between miR395 and miR395*, with the latter expressed as abundantly as the former in sweet sorghum but not in grain sorghum. Nine new miRNA candidates were identified, and experimental evidence was provided for previously annotated miRNAs by detecting the expression of 25 miRNA families.

Sequencing of a small RNA library and its analysis led to the identification of 113 conserved miRNA homologues belonging to 31 distinct miRNA families, of which 29 are conserved between monocots and dicots. Differential expression of several conserved and novel miRNAs was revealed by the temporal expression analysis. Some of the highly conserved miRNAs are induced in shoots and roots in response to low sulfate (miR395), during phosphate deficiency (miR399), and during copper deficiency (miR397, miR398, and miR408), which revealed the role of miRNAs in nutrient homeostasis. Approximately 125 genes that play diverse roles have been predicted as targets for conserved (100 gene targets) and novel (25 gene targets) miRNAs identified. Experimental validation of a few novel miRNAs (sbi-MIR5564a, sbi-MIR5564b, and sbi-MIR5565c) revealed abundant and ubiquitous expression of sbi-MIR5564a in all tissues analyzed, while sbi-MIR5566 and sbi-MIR5564b exhibited high expression in root tissue as compared to other tissues (Zhang et al. 2011).

A search for miR169 associated with drought tolerance and stem-sugar content in the sorghum genome led to the identification of many copies that could not be detected by the standard genome annotation methods. A new miR169 cluster was identified on chromosome 1 comprising of the previously annotated sbi-MIR169o and two new MIR169 copies (sbi-MIR169t and sbi-MIR169u). Another miR169 cluster was found on chromosome 7 consisting of sbi-MIR169l, sbi-MIR169m, and sbi-MIR169n. Two additional miR169 gene copies (miR169r andmiR169s) on chromosome 7 were identified after aligning the orthologous regions of rice and sorghum, which target different set of genes (Calvino and Messing 2013).

A bioinformatics pipeline developed by Katiyar et al. (2012) using an in-house PERL script and publicly available structure prediction tools could identify 31 new miRNAs representing ten different families from the expressed sequence tags (ESTs) and genomic survey sequence (GSS) available in the public domain. Mapping of newly identified miRNAs (31) and previously known miRNAs (148) on sorghum genome revealed that several MIR genes are arranged in clusters. About 72 potential target genes were predicted for these miRNAs; most of them are transcription factors involved in the regulation of plant growth and development. Three members of monocot species-specific MIR444 family involved in the regulation of expression of MADS-box transcription factor were identified. These newly identified miRNAs add to the growing database of miRNA and lay the foundation for further understanding of miRNA function in sorghum plant development.

7.6 Whole Transcriptome Analysis

Development of a complete transcriptome map during growth and development is essential for deciphering the sorghum genome sequence. Transcriptome profiling will help in the identification of genes and proteins encoded in the sorghum genome sequence, which subsequently allows the analysis of their function and regulation as well as their interaction in complex biological processes. Transcriptome profiling generally relies on microarray analysis (Buchanan et al. 2005; Shakoor et al. 2013) that needs genome annotation information and limited only to a fixed set of probes. Moreover, the hybridization signals cannot precisely distinguish alternative transcripts. Recently, the advances in the NGS technologies led to the emergence of a novel method known as "RNA sequencing" (RNA-seq) for mapping and quantification of transcriptomes.

Transcriptome analysis using RNA-seq in sorghum plants exposed to osmotic stress and exogenous abscisic acid (ABA) revealed transcriptional activity of 28,335 unique genes from root and shoot tissues subjected to polyethylene glycol (PEG)-induced osmotic stress or exogenous ABA (Dugas et al. 2011). A strong interplay between various metabolic (abscisic acid, salicylic acid, jasmonic acid) and plant defense pathways was noticed in response to osmotic stress and ABA. Transcriptomes of sorghum and its fungal pathogen *Bipolaris sorghicola* were analyzed simultaneously by Yazawa et al. (2013) using RNA-seq in combination with de novo transcriptome assembly since RNA-seq often relies on aligning reads to the reference genome. This study helped in the identification of genes of the fungus for its growth in sorghum such as those encoding Woronin body major protein, LysM domain-containing intracellular hyphae protein, transcriptional factors CpcA and HacA, and plant cell wall-degrading enzymes. In addition, the defense response genes of sorghum include those encoding two receptors of the simple eLRR domain protein family, transcription factors that are putative orthologs of OsWRKY45 and OsWRKY28 in rice, and a class III peroxidase associated with disease resistance in the Poaceae. Global transcriptome analysis performed in sorghum infected with *B. sorghicola* (Mizuno et al. 2012) revealed that the fungal infection activated the glyoxylate shunt in the TCA cycle and the secondary metabolic pathways of phytoalexin synthesis and sulfur-dependent detoxification. Re-sequencing of grain and sweet sorghum genotypes (Jiang et al. 2013) revealed high sequence diversity among them, but the divergence was limited at functional level. However, about 3,000 genes were differentially expressed between the grain and sweet sorghum, and the functional divergence could be due to mutations in regulatory sequences as well as DNA methylation.

The transcriptome atlas was expanded by Olson et al. (2014) by sequencing RNA from meristematic tissues, florets, and embryos, and this information along with expression levels, methylation profiles, and sequence conservation was used to predict functional gene models, and a comprehensive annotation of the sorghum transcriptome was developed. This gene annotation modified 60 % of the gene models of Sbi 1.4 version and could resolve 50 % of split gene models and also include 30 % of conserved genes that are missing from the Sbi 1.4 version. About 34,276 new potentially functional transcribed regions, including those coding for proteins, non-coding RNAs, and gene products of other classes, were identified. Several differentially expressed genes (DEGs) common between four low-N tolerant (San Chi San, China17, KS78, and high-NUE bulk) and three sensitive (CK60, BTx623, and low-NUE bulk) sorghum genotypes were identified by Gelli et al. (2014) by comparing the transcriptomes of the root tissues. The analysis revealed that the abundance of DEG transcripts associated with stress responses including oxidative stress increase in sensitive genotypes under N-stress, while an increase in the abundance of transcripts related to high-affinity nitrate transporters (NRT2.2, NRT2.3, NRT2.5, and NRT2.6) and lysine histidine transporter 1 (LHT1) was noticed in tolerant genotypes suggesting an improved efficiency in the uptake of inorganic and organic nitrogen. Moreover, increased abundance of the transcript of SEC14 cytosolic factor family protein in tolerant lines could result in increased membrane stability leading to N-stress tolerance.

A whole-transcriptome sorghum microarray chip (Sorgh-WTa520972F) was used by Shakoor et al. (2014) to identify tissue and genotype-specific expression patterns with the help of the microarray dataset generated using 78 samples involving grain (R159), sweet (Atlas and Fremont), and bioenergy (PI455230, PI152611, and AR2400) sorghums, different tissue types (shoot, root, leaf, and stem), and dissected stem tissues (pith and rind). About 19,354 genes were expressed in at least one of the 78 samples (i.e., 70.2 % of all genes on the array), and the number of transcripts expressed in the various tissues represented 56–60 % of expressed genes on the array. The greatest number of tissue-specific transcripts was expressed in the leaf and meristematic shoot tips across the genotypes PI152611, Fremont, and AR2400, whereas fewer number of tissue-specific transcripts were expressed in the seedling shoots. Tissue and genotype-specific small RNA expression was observed similar to the transcript expression. The expression of the genes SPS2 and SPS5 was consistently higher in sweet and high-biomass varieties as compared to grain varieties, highlighting the significant role played by sugar phosphate enzymes in sucrose biosynthesis.

7.7 Re-sequencing of Sorghum Using NGS-Based Approaches

The completion of reference genome sequencing of sorghum (Paterson et al. 2009) has brought the re-sequencing applications to the limelight resulting in re-sequencing sorghum genotypes through shotgun sequencing (Zheng et al. 2011; Jiang et al. 2013) and genotyping-by-sequencing (Nelson et al. 2011; Morris et al. 2012; Morishige et al. 2013) approaches. These approaches have been driven by the availability of NGS technologies that are more economical than Sanger sequencing (Sanger et al. 1977) that was developed during the 1970s. NGS technologies allow re-sequencing of large numbers of plant genomes at greater speed and lower cost as compared to traditional sequencing methods. The 454 Genome Sequencer FLX (Roche Applied Science), Illumina Genome Analyzer (Illumina), ABI SOLiD (Applied Biosystems), and Polonator G-007 (Dover) are some of the commercially known NGS technologies. Other NGS platforms include Hiseq and Miseq. Hiseq sequencing combines Illumina's reversible terminator-based sequencing by synthesis chemistry. The Hiseq 2000 platform can provide an output of 600GB in about 8–11 days (www.illumina.com). Miseq is another Illumina's PGM platform which also uses sequencing by synthesis technology. This technology can provide up to 120 MB–1.25 GB output and up to 2× 150 bp read length.

Advancements in sequencing technologies led to the advent of third-generation sequencing (TGS) approaches resulting in increased rates of sequencing, throughput, and read lengths with decreased sequencing costs and reducing the complexity of sample preparation. Ion Torrent sequencing is a TGS platform launched by Ion Torrent, a division of Life Science Technologies that involves the use of semiconductor-based high-density array of micro-reaction chambers (http://www.iontorrent.com) producing sequence reads of 100–200 bp, with up to 1 Gbp of data per run. Ion Torrent has a very smaller instrument size and is referred as personal genome machine (PGM). Single-molecule sequencing (SMS) is another TGS technology performed by Heliscope Single Molecule Sequencer, which requires no PCR amplification and the read lengths are >25 nucleotides long (Harris et al. 2008; Bowers et al. 2009). Single-molecule real-time (SMRT) sequencer performs sequencing by synthesis and produces read lengths up to 10,000 bp, enabling de novo assembly (Eid et al. 2009). The Oxford Nanopore sequencing technology employs "strand sequencing" approach in which intact DNA polymers are passed through a protein nanopore and sequenced in real time as the DNA translocates the pore. The sequencing is performed by a portable nanopore sensing technology such as GridION™ and MinION™ systems and offers 50–100 kb read length at 4 % error rate (https://www.nanoporetech.com). Third-generation sequencing techniques are predicted to be much cheaper and faster than the second-generation sequencing (SGS) or NGS technologies (Gupta 2008). Recently, plant genomes sequenced using NGS technologies are beginning to be reported. The major application of NGS technologies is given below.

QTL mapping and marker-assisted selection: SNPs are becoming the markers of choice for the molecular breeding community due to its abundance in the crop genomes and forms the basis for most molecular markers used in plant genomic research. Using NGS technologies it is now possible to sequence genes/genomes at a much faster rate, which can be subsequently used for the detection of SNPs that can be used for association mapping of the traits of interest. In the recent years, SNPs were detected through the re-sequencing of the sorghum genome using shotgun sequencing (Zheng et al. 2011) or genotype-by-sequencing methods (Nelson et al. 2011; Morris et al. 2012). Very recently, a restriction enzyme targeted genome re-sequencing method known as Digital Genotyping (DG) was developed by Morishige et al. (2013) that can be used for the genetic analysis in sorghum and other grass species having large repeat-rich genomes. The utility of DG in genetic map construction, QTL mapping, improving the assembly of the reference genome sequence and placing of super contiguous in their approximate position in the reference genome were also demonstrated. SNPs thus identified were used in the genome-wide association mapping of loci for plant height, inflorescence architecture, flowering time, maturity and anthracnose resistance (Morris et al. 2012; Thurber et al. 2013; Upadhyaya et al. 2013a, b). Very recently, a rapid method for the identification of plant QTLs by whole-genome re-sequencing of DNAs from two populations exhibiting extreme trait values for a given phenotype in a segregating progeny was reported in rice by Takagi et al. (2013). QTL for important agronomic traits, such as partial resistance to rice blast disease and seedling vigor was identified using this method. Even in sorghum, this method can be applied in population genomics studies to rapidly identify genomic regions that associated with selective sweeps between grain and sweet sorghum that could have occurred in a relatively short evolutionary period.

Transcriptome analysis and functional genomics: In addition to improved transcript coverage, the NGS technologies have brought down the sequencing cost and reduced the experimental complexity, which makes the sequencing-based transcriptome analysis more readily available and affordable to the research groups. This advancement in transcriptome sequencing is challenging the dominance of microarrays in the analysis of transcriptomes. Despite the availability of whole-genome sequences of sorghum, much of these genomic data are yet to be understood. The knowledge of transcription start sites, exon–intron structures, splice variants, polyadenylation signals, and regulatory sequences are essential for a comprehensive genome annotation. Even with rapid advances recently, complete annotation data is lacking for the most of metazoan genes (Brent 2008). Till recently, the homology-based evidence for effective annotation of protein-coding genes were provided by the Sanger-based sequencing of ESTs or FLcDNAs (Seki et al. 2002; Wortman et al. 2003; Pavy et al. 2005; Liang et al. 2008). However, the limitation in the amount of EST data generated by the Sanger sequencing method restricts its utility in the annotation of most abundantly expressed genes. For instance, about 400,000 ESTs can be generated by a single run on the 454 machine (Bainbridge et al. 2006) compared to 720 ESTs by Sanger sequencing (McCombie et al. 1992). Even though gene expression profiling through serial analysis of gene expression (SAGE) has advantages over microarray approaches (Wang 2007), it had not been widely used. However, the emergence of cost-effective NGS technologies has revived the concept behind the SAGE method. The most recent NGS technology for the mapping and quantification of transcriptomes, the RNA-seq, was used for the analysis of transcriptome expression in relation to infection of the fungus, *Bipolaris sorghicola*, in sorghum (Mizuno et al. 2012; Yazawa et al. 2013).

7.8 In Silico Mapping of Agronomically Important Genes

Genetic mapping or linkage mapping using DNA markers is being practiced for the last two decades to determine the relative position/genetic distances between markers along chromosomes. Genetic maps have been constructed using genetic mapping principles in most of the cereals including sorghum. Availability of markers very close to the identified QTL is very much essential for the fine mapping of the QTL and identification of tightly linked marker that can be used for marker-assisted selection. The availability of the reference genome of sorghum offers excellent opportunity for the integration of publicly available SSR markers and gene sequences from sorghum and other cereals on a sequence-based physical map. Linking physical map with already existing linkage map(s) provides better options for applied molecular breeding programs. Due to the integration of QTL mapping data with sequence-based physical map, in silico mapping helps in reducing the number of markers to be tested for the identification of flanking markers that exhibit polymorphism. It also helps in the selection of a set of markers representing the entire genome, which are useful for genetic diversity analysis and association mapping. If the existing linkage maps have some gaps without marker information, filling of such gaps can be achieved through in silico mapping approaches. Even though population-based conventional mapping is the most popular and reliable way to map molecular markers on the chromosomes, it is costly and time-consuming due to the development of mapping populations and large-scale genotyping. On the contrary, the in silico mapping strategy is reliable, time-saving, and economical. However, the success of in silico mapping is restricted to the availability of whole-genome sequence or genome sequences such as expressed sequence tags, sequenced BACs, or BAC-end sequences.

In sorghum, in silico mapping has been successfully used for the mapping of publicly available molecular markers and major-effect genes and also for the integration of whole-genome sequence information with a collection of QTL studies. A total of 2,113 primer pairs were designed by Li et al. (2009b) from 81,342 public genomic sequence contigs of sorghum, of which 1,710 SSR markers were found to be polymorphic in the eight sorghum genotypes tested and 1,692 of them were mapped on the ten linkage groups using in silico approaches. Further, about 202 markers conventionally mapped were also mapped in silico with 84.6 % of them mapping to the same chromosomes. With the availability of sorghum genome sequence, Mace and Jordan (2010) placed 35 major-effect genes on a consensus map. The locations of 9 out of the 35 genes were ascertained through mapping of the sequence of cloned genes or tightly linked markers for Tb_1, ma_3, Alt_{SB}, bmr_6, bmr_{12}, dw_3, *Pu*, Rf_1, and *wx*. A projection strategy based on markers common between the maps of the original study and the consensus map was used for the determination of the location of the other 26 genes. Among these, the projection of two of the major-effect genes (Sh_1 and ma_5) was based on the results of physical mapping, while the remaining 24 genes were based on the results of linkage mapping. Such integration of QTL and genome sequence information will help the sorghum breeders to be conscious of the consequences arising due to selection for major genes. A comprehensive analysis by Mace and Jordan (2011) involving the integration of whole-genome sequence information and a compendium of sorghum QTL studies published from 1995 to 2010 resulted in the projection of 771 QTL associated with 161 unique traits from 44 studies onto a sorghum consensus map. The distribution of QTL and genes was uneven across the genome and heterochromatic enrichment for QTL was noticed. This in silico projection of QTL information along with the physical map locations of sequence-based markers and predicted gene models serves as a useful resource to sorghum research groups across the globe to undertake a more detailed analysis of the traits and the formulation of an effective marker-assisted breeding approach. Physical map positions of already mapped molecular markers on one or more genetic linkage maps of sorghum were validated in silico mapping by Ramu et al. (2010). Similarly, all the published gene sequences from different cereals including sorghum were searched against the sorghum genome sequence database for their homology through BLAST and were assigned to their respective chromosome. Such integrated map will help in the identification of a set of markers representing the entire genome that are suitable to provide better resolution in diversity analyses and association mapping. It also offers new avenues for comparative mapping among the related species and development of genomic resources in closely related species, which lack them.

7.9 Epigenomic Studies

Epigenetics deals with heritable changes in gene expression that occur without DNA sequence variation. Epigenomics refers to the study of epigenetic features such as histone modifications (acetylation, methylation, phosphorylation, and ubiquitination), DNA methylation, and small RNA machinery on a large scale (Rival et al. 2010). DNA methylation by covalent modification of 5′-cytosine and posttranslational modifications of histone tail are the main epigenetic modifications that regulate gene expression (Callinan and Feinberg 2006). Regulatory RNAs (microRNAs and small interfering RNAs) account for the other means of epigenetic regulation of gene expression. The availability of complete genome sequences for rice, maize, and sorghum will help in facilitating genome-wide characterization of DNA methylation, histone modifications, and their relationships to coding as well as non-coding RNAs.

Cataloguing genome-wide DNA methylation patterns/methylation landscaping is usually performed by three approaches, viz., restriction endonuclease digestion coupled to microarray technology, bisulfite sequencing, and immunoprecipitation of 5′-methylcytosine (Callinan and Feinberg 2006). The first method involves digestion of genomic DNA by methylation-sensitive restriction enzymes such as *Msp I* and *Hpa II* (McClelland and Nelson 1988), which was initially used to identify differentially methylated sites in *Arabidopsis* by microarray analysis of small DNA fragments following digestion (Tran et al. 2005a, b). Bisulfite sequencing is the most popular method of characterizing the methylation that involves the targeted sequencing of specific genomic regions after treatment of isolated genomic DNA with bisulfate since it does not modify 5′-methylated cytosines and converts non-methylated cytosines to uracil. In the last method, methylated DNA fragments were isolated by affinity purification using proteins possessing preferential binding to methylated DNA or by immunoprecipitation using anti-^{m}C antibodies (mCIP), especially the MBD domain of the human protein MeCP2 (Cross et al. 1994). Microarray analysis or high-throughput sequencing can be used to identify methylated DNA fragments isolated by mCIP (Cokus et al. 2008; Zhang et al. 2006b; Zilberman et al. 2007). Recently, specifically designed microarrays have allowed the direct detection of methylated and non-methylated regions by employing bisulfite-treated genomic DNA for hybridization. The development of NGS technologies allows the direct detection of methylated sites, quantification of their frequency, and mapping the reads to the genomic sequence.

Profiling of histone modifications is usually done by chromatin immunoprecipitation (ChIP). Fragmented plant chromatin, either by sonication or by digestion using DNase I or micrococcal nuclease, containing certain histone modifications, is isolated by ChIP using corresponding antibodies, whose genomic locations are determined by microarray analysis or deep sequencing (ChIP-chip and ChIP-Seq, respectively) (Lippman et al. 2004; Oh et al. 2008; Elling and

Deng 2009; Zhang et al. 2009; He et al. 2010; Zhou et al. 2010). Due to the conservation of histones in eukaryotes, commercial antibodies developed for animals and fungi can also be used in plants. It is desirable to initially assess the specificities of the antibody since the success of ChIP-chip or ChIP-Seq depends on the antibodies used (Egelhofer et al. 2010). The development of new antibodies will help in the detection of histone modifications and in revealing novel interactions in future epigenomic studies.

Regulatory plant cellular RNA pools include small RNAs (sRNAs) such as small interfering RNAs (siRNAs), microRNAs (miRNAs), trans-acting small interfering RNAs (tasiRNAs), and natural antisense small interfering RNAs (nat-siRNAs), which play crucial roles in many biological processes, such as regulation of gene expression, heterochromatic silencing, and antiviral defense. These sRNAs can be isolated by two primary methods, viz., size selection from total cellular RNA (Lu et al. 2005, 2006; Zhai et al. 2008; He et al. 2010) and immunoprecipitation of sRNA-binding proteins (Mi et al. 2008; Montgomery et al. 2008; Havecker et al. 2010). The former is the most popular method that is extensively utilized to catalog sRNAs in different species, while the latter requires a protein that strongly associates with sRNAs and the ability to immunoprecipitate the protein of interest. Following ligation of sRNAs to RNA adaptors and reverse transcription, millions of sRNAs can be sequenced by the RNA-seq. Sequencing and characterization of sRNAs can be useful to infer the proteins that might be associated with the generation and processing of sRNA precursors.

7.10 Future Prospects

Availability of whole-genome sequence in sorghum and other crops has revolutionized understanding of plant genetics by unraveling the basic mechanisms in plant growth and development, cellular processes, and tolerance to various biotic and abiotic stresses. This will serve as a validation tool for comparative genomics in model cereals and also in orphan crops where genome has not yet been sequenced. With the availability of the genome sequence in model crops and understanding their syntenic relationships, the development of markers in the related crop species, in specific targeted regions, will now be a practical option. Rapid growth in sequencing enables discovery of genes and DNA markers associated with diverse agronomic traits, creating new opportunities for crop improvement. The practical applications of the sorghum genome sequencing projects are best realized only when allelic diversity patterns existing among the gene bank accessions are better understood. Information on the allelic variation patterns may contribute to functional analysis of sorghum-specific genes for genetic improvement of sorghum for agronomically important traits. Eventually integrating information on sorghum structural and functional genomics will provide an overall view of the network of genes involved in complex biological responses.

The availability of NGS and whole reference genome sequences of sorghum provides unique opportunities for exploring sequence level diversity among germplasm or traditional landraces of historic importance. The ultimate goal for re-sequencing traditional landraces is to understand the molecular basis for phenotype–genotype relationships. Diversity panels of thousands of individuals selected to sample the extent of diversity with reference genome sequences using NGS technologies will provide a platform for understanding existing genetic diversity, associating gene(s) with phenotypes, and exploiting natural genetic diversity to help develop superior genotypes using association mapping approaches. In the era of genomics, the ultimate challenge is to develop knowledge from the enormous genomic data that can be applied in crop breeding programs. Equally important is the accurate phenotyping of the trait of interest, which decides the success of the crop improvement through genomic approaches. Breeders are required to apply the genomic tools and precise phenotyping techniques to truly advance the crop improvement process and take advantage of the potential of genomics. Overall, the major gap in the genomic approaches for crop improvement is in the utilization of genomic information for development of improved crop cultivars. The most effective strategy to fulfill the gap is through inter-disciplinary research leading to the integration of knowledge of whole-genome organization, strong statistical knowledge to estimate the gene/genetic effects, good experience in molecular biology techniques, and traditional breeding methodologies, which form core components of molecular breeding.

References

Adai A, Johnson C, Mlotshwa S, Archer-Evans S, Manocha V, Vance V, Sundaresan V (2005) Computational prediction of miRNAs in *Arabidopsis thaliana*. Genome Res 15:78–91

Almodares A, Hadi MR (2009) Production of bioethanol from sweet sorghum: a review. Afr J Agric Res 4:772–780

Arabidopsis Genome Initiative (2000) Analysis of the genome sequence of the flowering plant *Arabidopsis thaliana*. Nature 408:796–815

Aubourg S, Rouze P (2001) Genome annotation. Plant Physiol Biochem 39:181–193

Axtell MJ, Bartel DP (2005) Antiquity of microRNAs and their targets in land plants. Plant Cell 17:1658–1673

Bainbridge MN, Warren RL, Hirst M et al (2006) Analysis of the prostate cancer cell line LNCaP transcriptome using a sequencing-by-synthesis approach. BMC Genom 7:246

Baird NA, Etter PD, Atwood TS, Currey MC, Shiver AL, Lewis ZA, Selker EU, Cresko WA, Johnson EA (2008) Rapid SNP discovery and genetic mapping using sequenced RAD markers. PLoS One 3:e3376

Bartel DP (2004) MicroRNAs: genomics, biogenesis, mechanism, and function. Cell 116:281–297

Bennetzen JL (2000) Comparative sequence analysis of plant nuclear genomes: microcolinearity and its many exceptions. Plant Cell 12:1021–1030

Bennetzen JL, Ma J (2003) The genetic colinearity of rice and other cereals on the basis of genomic sequence analysis. Curr Opin Plant Biol 6:128–133

Bevan MW, Uauy C (2013) Genomics reveals new landscapes for crop improvement. Genome Biol 14:206

Bouchez D, Hofte H (1998) Functional genomics in plants. Plant Physiol 118:725–732

Bowers J, Mitchell J, Beer E et al (2009) Virtual terminator nucleotides for next-generation DNA sequencing. Nat Methods 6:593–595

Brent MR (2008) Steady progress and recent breakthroughs in the accuracy of automated genome annotation. Nat Rev Genet 9:62–73

Buchanan WV, Page T, Harrison E et al (2005) Comparative transcriptome analysis reveals significant differences in gene expression and signaling pathways between developmental and dark/starvation-induced senescence in *Arabidopsis*. Plant J 42:567–585

Callinan PA, Feinberg AP (2006) The emerging science of epigenomics. Hum Mol Genet 1:R95–R101

Calvino M, Messing J (2013) Discovery of microRNA169 gene copies in genomes of flowering plants through positional information. Genome Biol Evol 5(2):402–417

Calvino M, Bruggmann R, Messing J (2011) Characterization of the small RNA component of the transcriptome from grain and sweet sorghum stems. BMC Genom 12:356

Cokus SJ, Feng S, Zhang X, Chen Z, Merriman B, Haudenschild CD, Pradhan S, Nelson SF, Pellegrini M, Jacobsen SE (2008) Shotgun bisulphite sequencing of the *Arabidopsis* genome reveals DNA methylation patterning. Nature 452:215–219

Cross SH, Charlton JA, Nan X, Bird AP (1994) Purification of CpG islands using a methylated DNA binding column. Nat Genet 6:236–244

Davidson RM, Gowda M, Moghe G, Lin H, Vaillancourt B, Shiu SH, Jiang N, Robin Buell C (2012) Comparative transcriptomics of three Poaceae species reveals patterns of gene expression evolution. Plant J 71:492–502

Dryanova A, Zakharov A, Gulick PJ (2008) Data mining for miRNAs and their targets in the Triticeae. Genome 51:433–443

Du JF, Wu YJ, Fang XF, Cao JX, Zhao L, Tao SH (2010) Prediction of sorghum miRNAs and their targets with computational methods. Chinese Sci Bull 55:1263–1270

Dubcovsky J, Dvorak J (2007) Genome plasticity a key factor in the success of polyploid wheat under domestication. Science 316:1862–1866

Dugas DV, Monaco MK, Olson A, Klein RR, Kumari S, Ware D, Klein PE (2011) Functional annotation of the transcriptome of *Sorghum bicolor* in response to osmotic stress and abscisic acid. BMC Genom 12:514

Egelhofer TA, Minoda A, Klugman S et al (2010) An assessment of histone-modification antibody quality. Nat Struct Mol Biol 18:91–93

Eid J, Fehr A, Gray J et al (2009) Real-time DNA sequencing from single polymerase molecules. Science 323:133–138

Elling AA, Deng XW (2009) Next-generation sequencing reveals complex relationships between the epigenome and transcriptome in maize. Plant Signal Behav 4:760–762

Emberton J, Ma J, Yuan Y, SanMiguel P, Bennetzen JL (2005) Gene enrichment in maize with hypomethylated partial restriction (HMPR) libraries. Genome Res 15:1441

Fu H, Tie Y, Xu C, Zhang Z, Zhu J, Shi Y, Jiang H, Sun Z, Zheng X (2005) Identification of human fetal liver miRNAs by a novel method. FEBS Lett 579:3849–3854

Gale MD, Devos KM (1998) Comparative genetics in the grasses. Proc Natl Acad Sci U S A 95:1971–1974

Gelli M, Duo Y, Konda AR, Zhang C, Holding D, Dweikat I (2014) Identification of differentially expressed genes between sorghum genotypes with contrasting nitrogen stress tolerance by genome-wide transcriptional profiling. BMC Genom 15:179

Glaszmann JC, Dufour P, Grivet L, D'Hont A, Deu M, Paulet F, Hamon P (1997) Comparative genome analysis between several tropical grasses. Euphytica 96:13–21

Groth P, Weiss B, Hans-Dieter Pohlenz, Leser U (2008) Mining phenotypes for gene function prediction. BMC Bioinf 9:136

Gupta PK (2008) Single-molecule DNA sequencing technologies for future genomics research. Trends Biotech 26:602–611

Hamblin MT, Salas Fernandez MG, Tuinstra MR, Rooney WL, Kresovich S (2007) Sequence variation at candidate loci in the starch metabolism pathway in sorghum: Prospects for linkage disequilibrium mapping. Crop Sci 47:S125–S134

Harris TD, Buzby PR, Babcock H et al (2008) Single-molecule DNA sequencing of a viral genome. Science 320:106–109

Havecker ER, Wallbridge LM, Hardcastle TJ, Bush MS, Kelly KA, Dunn RM, Schwach F, Doonan JH, Baulcombe DC (2010) The *Arabidopsis* RNA-directed DNA methylation argonautes functionally diverge based on their expression and interaction with target loci. Plant Cell 22:321–334

He G, Zhu X, Elling AA et al (2010) Global epigenetic and transcriptional trends among two rice sub-species and their reciprocal hybrids. Plant Cell 22:17–33

International Rice Genome Sequencing Project (2005) The map-based sequence of the rice genome. Nature 436:793–800

Jiang SY, Ma Z, Vanitha J, Ramachandran S (2013) Genetic variation and expression diversity between grain and sweet sorghum lines. BMC Genom 14:18

Jones-Rhoades MW, Bartel DP (2004) Computational identification of plant microRNAs and their targets, including a stress-induced miRNA. Mol Cell 14:787–799

Katiyar A, Smita S, Chinnusamy V, Pandey DM, Bansal KC (2012) Identification of miRNAs in sorghum by using bioinformatics approach. Plant Signal Behav 7:246–259

Kidwell MG (2002) Transposable elements and the evolution of genome size in eukaryotes. Genetica 115:49–63

Kumar S, Banks TW, Cloutier S (2012) SNP discovery through next-generation sequencing and its applications. Int J Plant Genomics 2012:1–15

Kurihara Y, Watanabe Y (2004) *Arabidopsis* micro-RNA biogenesis through Dicer-like 1 protein functions. Proc Natl Acad Sci U S A 101:2753–12758

Lee RC, Feinbaum RL, Ambros V (1993) The *C. elegans* heterochronic gene lin-4 encodes small RNAs with antisense complementarity to lin-14. Cell 75:843–854

Li Y, Li W, Jin YX (2005) Computational identification of novel family members of microRNA genes in *Arabidopsis thaliana* and *Oryza sativa*. Acta Biochim Biophys Sin 37:75–87

Li, L, Chu, AH, Liu, HJ, Lo C (2009a) Isolation and analysis of defense-related genes in infected sorghum seedlings. Abstract. Plant Biology symposium 2009, American Society of Plant Biologists, Honolulu, 18–22 July 2009

Li M, Yuyama N, Luo L, Hirata M, Cai H (2009b) *In silico* mapping of 1758 new SSR markers developed from public genomic sequences for sorghum. Mol Breed 24:41–47

Liang C, Liu Y, Liu L, Davis AC, Shen Y, Li Q (2008) Expressed sequence tags with cDNA termini: Previously overlooked resources for gene annotation and transcriptome exploration in *Chlamydomonas reinhardtii*. Genetics 179:83–93

Lim LP, Glasner ME, Yekta S, Burge CB, Bartel DP (2003a) Vertebrate microRNA genes. Science 299:1540

Lim LP, Lau NC, Weinstein EG, Abdelhakim A, Yekta S, Rhoades MW, Burge CB, Bartel DP (2003b) The microRNAs of *Caenorhabditis elegans*. Genes Dev 17:991–1008

Lippman Z, Gendrel AV, Black M et al (2004) Role of transposable elements in heterochromatin and epigenetic control. Nature 430:471–476

Liu H, Du Y, Chu H, Shih CH, Wong YW, Wang M, Chu IK, Tao Y, Lo C (2010) Molecular dissection of the pathogen-inducible 3-deoxyanthocyanidin biosynthesis pathway in sorghum. Plant Cell Physiol 51:1173–1185

Lu C, Tej SS, Luo S, Haudenschild CD, Meyers BC, Green PJ (2005) Elucidation of the small RNA component of the transcriptome. Science 309:1567–1569

Lu C, Kulkarni K, Souret FF et al (2006) MicroRNAs and other small RNAs enriched in the *Arabidopsis* RNA-dependent RNA polymerase-2 mutant. Genome Res 16:1276–1288

Mace ES, Jordan DR (2010) Location of major effect genes in sorghum (*Sorghum bicolor* (L.) Moench). Theor Appl Genet 121:1339–1356

Mace ES, Jordan DR (2011) Integrating sorghum whole genome sequence information with a compendium of sorghum QTL studies reveals uneven distribution of QTL and of gene-rich regions with significant implications for crop improvement. Theor Appl Genet 123:169–191

Mardis ER (2008) The impact of next-generation sequencing technology on genetics. Trends Genet 24:133–141

Mathews DH, Sabina J, Zuker M, Turner DH (1999) Expanded sequence dependence of thermodynamic parameters improves prediction of RNA secondary structure. J Mol Biol 288:911–940

McClelland M, Nelson M (1988) The effect of site-specific DNA methylation on restriction endonucleases and DNA modification methyltransferases – a review. Gene 74:291–304

McCombie WR, Adams MD, Kelley JM, FitzGerald MG, Utterback TR, Khan M, Dubnick M, Kerlavage AR, Venter JC, Fields C (1992) *Caenorhabditis elegans* expressed sequence tags identify gene families and potential disease gene homologues. Nat Genet 1:124–131

Mead EA, Tu Z (2008) Cloning, characterization and expression of microRNAs from the Asian malaria mosquito, *Anopheles stephensi*. BMC Genom 9:244

Mi S, Cai T, Hu Y et al (2008) Sorting of small RNAs into *Arabidopsis argonaute* complexes is directed by the 5′ terminal nucleotide. Cell 133:116–127

Ming R, Liu SC, Lin YR et al (1998) Detailed alignment of *Saccharum* and sorghum chromosomes: comparative organization of closely related diploid and polyploid genomes. Genetics 150:1663–1668

Mizuno H, Kawahigashi H, Kawahara Y, Kanamori H, Ogata J, Minami H, Itoh T, Matsumoto T (2012) Global transcriptome analysis reveals distinct expression among duplicated genes during sorghum-Bipolaris *sorghicola* interaction. BMC Plant Biol 12:121

Montgomery TA, Howell MD, Cuperus JT, Li D, Hansen JE, Alexander AL, Chapman EJ, Fahlgren N, Allen E, Carrington JC (2008) Specificity of ARGONAUTE7-miR390 interaction and dual functionality in TAS3 trans-acting siRNA formation. Cell 133:128–141

Morishige DT, Klein PE, Hilley JL, Sahraeian SME, Sharma A, Mullet JE (2013) Digital genotyping of sorghum – a diverse plant species with a large repeat-rich genome. BMC Genom 14:448

Morozova O, Marra MA (2008) Applications of next-generation sequencing technologies in functional genomics. Genomics 92:255–264

Morris GP, Ramu P, Deshpande SP et al (2012) Population genomic and genome-wide association studies of agroclimatic traits in sorghum. Proc Natl Acad Sci U S A 110:453–458

Moxon S, Jing R, Szittya G, Schwach F, Rusholme Pilcher RL, Moulton V, Dalmay T (2008) Deep sequencing of tomato short RNAs identifies microRNAs targeting genes involved in fruit ripening. Genome Res 18:1602–1609

Murray SC, Rooney WL, Hamblin MT, Mitchell SE, Kresovich S (2009) Sweet sorghum genetic diversity and association mapping for brix and height. Plant Gen 2:48–62

Nelson JC, Wang S, Wu Y, Li X, Antony G, White FF, Yu J (2011) Single-nucleotide polymorphism discovery by high-throughput sequencing in sorghum. BMC Genom 12:352

Oh S, Park S, van Nocker S (2008) Genic and global functions for Paf1C in chromatin modification and gene expression in *Arabidopsis*. PLoS Genet 4:e1000077

Okou DT, Steinberg KM, Middle C, Cutler DJ, Albert TJ, Zwick ME (2007) Microarray-based genomic selection for high-throughput re-sequencing. Nat Methods 4:907–909

Olson A, Klein RR, Dugas DV, Lu Z, Regulski M, Klein PE, Ware D (2014) Expanding and vetting *Sorghum bicolor* gene annotations through transcriptome and methylome sequencing. Plant Gen 7(2):1–20

Pasquinelli AE, Reinhart BJ, Slack F, Martindale MQ, Kuroda MI, Maller B, Hayward DC (2000) Conservation of the sequence and temporal expression of let-7 heterochronic regulatory RNA. Nature 408:86–89

Paterson AH (2013) The Sorghum genome sequence: a core resource for Saccharinae genomics. In: Paterson AH (ed) Genomics of saccharinae, Plant genetics and genomics: crops and models II. Springer, New York, p 107

Paterson AH, Bowers JE, Bruggmann R et al (2009) The *Sorghum bicolor* genome and the diversification of grasses. Nature 457:551–556

Pavy N, Paule C, Parsons L et al (2005) Generation, annotation, analysis and database integration of 16,500 white spruce EST clusters. BMC Genom 6:144

Ramu P, Deshpande SP, Senthilvel S, Jayashree B, Billot C, Deu M, Ananda Reddy L, Hash CT (2010) *In silico* mapping of important genes and markers available in the public domain for efficient sorghum breeding. Mol Breed 26:409–418

Reinhart BJ, Slack FJ, Basson M, Pasquinelli AE, Bettinger JC, Rougvie AE, Horvitz HR (2000) The 21-nucleotide let-7 RNA regulates developmental timing in *Caenorhabditis elegans*. Nature 403:901–906

Rival A, Beule T, Bertossi FA, Tregear J, Jaligot E (2010) Plant epigenetics: from genomes to epigenomes. Not Bot Hort Agrobot Cluj (Special Issue) 38:9–15

Robins H, Li Y, Padgett RW (2005) Incorporating structure to predict microRNA targets. Proc Natl Acad Sci U S A 102:4006–4009

Sanger F, Nicklen S, Coulson AR (1977) DNA sequencing with chain-terminating inhibitors. Proc Natl Acad Sci U S A 74:5463–5467

Schmutz J, Cannon SB, Schlueter J et al (2010) Genome sequence of the palaeopolyploid soybean. Nature 463:178–183

Schnable PS, Ware D, Fulton RS et al (2009) The B73 maize genome: complexity, diversity, and dynamics. Science 326:1112–1115

Seki M, Narusaka M, Kamiya A et al (2002) Functional annotation of a full-length *Arabidopsis* cDNA collection. Science 296:141–145

Shakoor N, Crasta OR, Nair R, Huang SW, Fan Z, Morris G, Kresovich S (2013) Whole genome gene expression profiling of multiple tissue types and developmental stages in *Sorghum bicolor*. Abstract: Plant and Animal Genome XXI symposium, San Diego, 12–16 Jan 2013

Shakoor N, Nair R, Crasta OR, Morris G, Feltus A, Kresovich S (2014) A *Sorghum bicolor* expression atlas reveals dynamic genotype-specific expression profiles for vegetative tissues of grain, sweet and bioenergy sorghums. BMC Plant Biol 14:35

Singh NK, Gupta DK, Jayaswal PK et al (2012) The first draft of the pigeonpea genome sequence. J Plant Biochem Biotechnol 21:98–112

Singhal D, Gupta P, Sharma P, Kashyap N, Anand S, Sharma H (2011) *In-silico* single nucleotide polymorphisms (SNP) mining of *Sorghum bicolor* genome. Afr J Biotechnol 10:580–583

Somerville C, Somerville S (1999) Plant functional genomics. Science 283:380–383

Stein L (2001) Genome annotation: from sequence to biology. Nat Rev Genet 2:493–503

Sukumaran S, Xiang W, Bean SR, Pedersen JF, Kresovich S, Tuinstra MR, Tesso TT, Hamblin MT, Yu J (2012) Association mapping for grain quality in a diverse sorghum collection. Plant Gen 5:126–135

Sungroh Y, Giovanni DM (2006) Computational identification of microRNAs and their targets. Birth Defect Res (Part C) 78:118–128

Sunkar R, Jagadeeswaran G (2008) *In silico* identification of conserved microRNAs in large number of diverse plant species. BMC Plant Biol 8:37

Sunkar R, Zhou X, Zheng Y, Zhang W, Zhu JK (2008) Identification of novel and candidate miRNAs in rice by high throughput sequencing. BMC Plant Biol 8:25

Swami AK, Alam SI, Sengupta N, Sarin R (2011) Differential proteomic analysis of salt stress response in *Sorghum bicolor* leaves. Environ Exp Bot 71:321–328

Takagi H, Abe A, Yoshida K et al (2013) QTL-seq: rapid mapping of quantitative trait loci in rice by whole genome resequencing of DNA from two bulked populations. Plant J 74:174–183

The International Wheat Genome Sequencing Consortium (2014) A chromosome-based draft sequence of the hexaploid bread wheat (*Triticum aestivum*) genome. Science 345(6194):1251788

Thurber CS, Ma JM, Higgins RH, Brown PJ (2013) Retrospective genomic analysis of sorghum adaptation to temperate-zone grain production. Genome Biol 14:R68

Tran RK, Henikoff JG, Zilberman D, Ditt RF, Jacobsen SE, Henikoff S (2005a) DNA methylation profiling identifies CG methylation clusters in Arabidopsis genes. Curr Biol 15:154–159

Tran RK, Zilberman D, de Bustos C et al (2005b) Chromatin and siRNA pathways cooperate to maintain DNA methylation of small transposable elements in *Arabidopsis*. Genome Biol 6:R90

Travella S, Klimm TE, Keller B (2006) RNA interference-based gene silencing as an efficient tool for functional genomics in hexaploid bread wheat. Plant Physiol 142:6–20

Upadhyaya HD, Wang YH, Gowda CLL, Sharma S (2013a) Association mapping of maturity and plant height using SNP markers with the sorghum mini core collection. Theor Appl Genet 126:2003–2015

Upadhyaya HD, Wang YH, Sharma R, Sharma S (2013b) Identification of genetic markers linked to anthracnose resistance in sorghum using association analysis. Theor Appl Genet 126:1649–1657

van Orsouw NJ, Hogers RCJ, Janssen A et al (2007) Complexity Reduction of Polymorphic Sequences (CRoPS™): a novel approach for large-scale polymorphism discovery in complex genomes. PLoS One 2:e1172

Varshney RK, Chen W, Li Y et al (2012) Draft genome sequence of pigeonpea (*Cajanus cajan*), an orphan legume crop of resource-poor farmers. Nat Biotechnol 30:83–89

Varshney RK, Song S, Saxena RK et al (2013) Draft genome sequence of chickpea (*Cicer arietinum*) provides a resource for trait improvement. Nat Biotechnol 31:240–246

Wang SM (2007) Understanding SAGE data. Trends Genet 23:42–50

Wang X, Zhang J, Li F, Gu J, He T, Zhang X, Li Y (2005) MicroRNA identification based on sequence and structure alignment. Bioinformatics 21:3610–3614

Wang G, Wang X, Wang Y, Yang JY, Li L, Nephew KP, Edenberg HJ, Zhou FC, Liu Y (2008) Identification of transcription factor and microRNA binding sites in responsible to fetal alcohol syndrome. BMC Genom 9(Suppl 1):S19

Wightman B, Ha I, Ruvkun G (1993) Posttranscriptional regulation of the heterochronic gene lin-14 by lin-4 mediates temporal pattern formation in *C. elegans*. Cell 75:855–862

Wortman JR, Haas BJ, Hannick LI, Smith RK Jr, Maiti R, Ronning CM, Chan AP, Yu C, Ayele M, Whitelaw CA, White OR, Town CD (2003) Annotation of the *Arabidopsis* genome. Plant Physiol 132:461–468

Xie FL, Huang SQ, Guo K, Zhu YY, Nie L, Yang ZM (2007) Computational identification of novel microRNAs and targets in *Brassica napus*. FEBS Lett 581:1464–1473

Yan L, Liu S, Zhao S, Kang Y, Wang D, Gu T, Xin Z, Xia G, Huang Y (2012) Identification of differentially expressed genes in sorghum (*Sorghum bicolor*) brown midrib mutants. Physiol Plant 146:375–387

Yandell M, Ence D (2012) A beginner's guide to eukaryotic genome annotation. Nat Rev Genet 13:329–341

Yang L, Liu T, Li B, Sui Y, Chen J, Shi J, Wing RA, Chen M (2012) Comparative sequence analysis of the Ghd7 orthologous regions revealed movement of Ghd7 in the grass genomes. PLoS One 7(11):e50236

Yazawa T, Kawahigashi H, Matsumoto T, Mizuno H (2013) Simultaneous transcriptome analysis of sorghum and *Bipolaris sorghicola* by using RNA-seq in combination with *de novo* transcriptome assembly. PLoS One 8:e62460

Yuan Y, SanMiguel PJ, Bennetzen JL (2003) High-Cot sequence analysis of the maize genome. Plant J 34:249–255

Yuan Q, Ouyang S, Wang A et al (2005) The institute for genomic research Osa1 rice genome annotation database. Plant Physiol 138:18–26

Zhai J, Liu J, Liu B, Li P, Meyers BC, Chen X, Cao X (2008) Small RNA-directed epigenetic natural variation in *Arabidopsis thaliana*. PLoS Genet 4:e1000056

Zhang B, Pan X, Cannon CH, Cobb GP, Anderson TA (2006a) Conservation and divergence of plant microRNA genes. Plant J 46:243–259

Zhang X, Yazaki J, Sundaresan A et al (2006b) Genome-wide high-resolution mapping and functional analysis of DNA methylation in *Arabidopsis*. Cell 126:1189–1201

Zhang X, Bernatavichute YV, Cokus S, Pellegrini M, Jacobsen SE (2009) Genome-wide analysis of mono-, di- and trimethylation of histone H3 lysine 4 in *Arabidopsis thaliana*. Genome Biol 10:R62

Zhang L, Zheng Y, Jagadeeswaran G, Li Y, Gowdu K, Sunkar R (2011) Identification and temporal expression analysis of conserved and novel microRNAs in sorghum. Genomics 98:460–468

Zheng LY, Guo XS, He B et al (2011) Genome-wide patterns of genetic variation in sweet and grain sorghum (*Sorghum bicolor*). Genome Biol 12:R114

Zhou J, Wang X, He K, Charron JB, Elling AA, Deng XW (2010) Genome-wide profiling of histone H3 lysine 9 acetylation and dimethylation in *Arabidopsis* reveals correlation between multiple histone marks and gene expression. Plant Mol Biol 72:585–595

Zilberman D, Gehring M, Tran RK, Ballinger T, Henikoff S (2007) Genome-wide analysis of *Arabidopsis thaliana* DNA methylation uncovers an interdependence between methylation and transcription. Nat Genet 39:61–69

8 Molecular Marker Development Using Bioinformatic Tools

P. Rajendrakumar

Contents

8.1 **Introduction** 180
8.2 **Bioinformatics for Molecular Marker Development** 180
8.3 **Development of Microsatellite Markers** 181
8.4 **Development of In-Del Markers** 185
8.5 **Development of Intron Polymorphism Markers** 186
8.6 **Single Nucleotide Polymorphism (SNP) and Development of Markers** 187
8.7 **Development of Molecular Markers from Organellar Genomes** 190
8.8 **Future Prospects** 191
References 191

Abstract

Availability of molecular markers is essential for various genetic and breeding applications such as assessment of genetic diversity, construction of linkage map, genetic purity testing, QTL mapping, and marker-assisted selection. Prior to in silico approaches, DNA markers were developed through experimental approaches, which were skill oriented, time consuming, and expensive. Moreover, it resulted in the development of few hundred markers involving a substantial amount of time. The advent of next-generation sequencing technologies has led to the sequencing of whole nuclear as well as organellar genomes and the transcriptome projects have resulted in the accumulation of huge amounts of expressed sequence tags (ESTs) and/or cDNA sequences. Such an ocean of DNA sequence information, including genome survey sequences (GSS), expressed sequence tags (ESTs), full-length cDNAs, and complete nuclear and organellar genome sequences, serves as a vital resource for the identification of target motifs such as simple sequence repeats (SSRs), insertions-deletions (In-Dels), and single-nucleotide polymorphisms (SNPs) through in silico approaches leading to the rapid development of DNA-based markers, which otherwise would be time consuming through conventional experimental approaches. This review discusses the development of DNA markers such as SSR, SNP, In-Del, and intron length polymorphisms using various bioinformatic tools.

Keywords
Bioinformatics • Simple sequence repeats • In-Dels • Single-nucleotide polymorphisms • Intron polymorphism

P. Rajendrakumar (✉)
Biotechnology, ICAR-Indian Institute of Millets Research, Rajendranagar, Hyderabad, Telangana 500 030, India
e-mail: rajendra@millets.res.in

R. Madhusudhana et al. (eds.), *Sorghum Molecular Breeding*, DOI 10.1007/978-81-322-2422-8_8

8.1 Introduction

Plant breeding has a long history of embracing the advances made in allied fields like plant biology, genetics, cytogenetics, quantitative genetics, molecular biology, and genomics to bring about the genetic improvement in agricultural crops. Plant breeders generally use visible phenotypes for selection, if the trait is simple and qualitatively inherited. However, for traits that are quantitatively inherited, such as grain yield and drought tolerance, selection based on the phenotype is challenging as it becomes difficult for a breeder to precisely measure such phenotypes with complex inheritance. Under such situations, indirect selection using molecular markers, popularly known as marker-assisted selection, helps a plant breeder in efficiently selecting for complex phenotypes. With the advent of molecular marker technology and the availability of the enormous number and types of molecular markers in crop plants, molecular breeding is becoming a standard practice in crop improvement programs, bringing about the desired improvement in much quicker time as compared to conventional breeding. Selection using molecular markers increases the probability of identifying superior genotypes by decreasing the number of progeny to be screened to achieve desired genetic gain, thereby enabling simultaneous improvement of negatively correlated traits (Knapp 1998). Among the several success stories, resistance to bacterial leaf blight in rice (Sundaram et al. 2008, 2009; Basavaraj et al. 2010; Rajpurohit et al. 2011), submergence tolerance in rice (Neeraja et al. 2007), and drought tolerance in maize (Ribaut and Ragot 2007; Tuberosa et al. 2007) are noteworthy.

Advancements in DNA sequencing technologies have resulted in the sequencing of complete genomes of some of the important crop species such as rice, sorghum, maize, poplar, grape, papaya, potato, *Medicago*, castor bean, and soybean. In addition to this, many transcriptome projects have led to an enormous accumulation of expressed sequence tags (ESTs) and/or cDNA sequences for nearly all economically important crop species. These sequences are mainly stored in generic databases such as GenBank at the National Center for Biotechnology Information (NCBI), the European Molecular Biology Laboratory (EMBL) nucleotide sequence database, and the DNA Database Bank of Japan (DDBJ). However, due to the accumulation of an enormous amount of sequence data of different crop species, many crop-specific databases are set up. In addition to the whole-genome nuclear as well as the organellar sequences of crop plants, the NCBI database consists of nucleotide sequences, genome survey sequences (GSS), and EST sequences. Whole genomes of four sorghum genotypes, namely, BTx623 (Paterson et al. 2009), Keller, E-Tian, and Ji2731 (Zheng et al. 2011), have been sequenced. BTx623 represents American-bred B-line, while Keller is an American-bred elite sweet sorghum line. E-Tian is a sweet sorghum line introduced to China in the early 1970s, while Ji2731 represents Chinese kaoliang grain sorghum. Very recently, Mace et al. (2013) resequenced genomes of 44 sorghum accessions representing all major races of cultivated sorghum [*Sorghum bicolor* (L.) Moench] along with its progenitors and *S. propinquum*. These lines comprised of 18 landraces, 17 improved inbreds, and 7 wild and weedy sorghums spanning the dimensions of geographic origin, end use, crop management, and taxonomic group. A total of 1,287,183 nucleotide sequences are available from *Sorghum bicolor* as of 1 July 2013, which includes nucleotides (279,058), ESTs (210,892), and GSS (797, 233). These nucleotide sequences have become a valuable target for the development of molecular markers through in silico approaches using bioinformatic tools/softwares, which is rapid and inexpensive as compared to conventional marker development approaches.

8.2 Bioinformatics for Molecular Marker Development

Bioinformatics helps to extract valuable information from the sequence data through in silico analysis; thereby, molecular breeders can use this information for crop improvement, especially in the development of molecular markers in a rapid and inexpensive way. The whole wealth of DNA sequence information, including genome survey sequences (GSS), expressed sequence tags (ESTs), full-length cDNAs, complete nuclear and organellar genome sequences, and coding sequences from genome annotation, serves as targets for the identification of DNA sequence polymorphisms such as simple sequence repeats (SSRs), insertions-deletions (In-Dels), and single-nucleotide polymorphisms (SNPs) using various bioinformatic tools, both free and commercial. These bioinformatic tools are in the form of stand-alone packages, Web-based tools, pipelines, Perl scripts, Java scripts, etc. Once DNA sequence polymorphisms are identified by these bioinformatic tools, such sequence variation will be converted into PCR-based markers by the designing suitable primers through various primer-designing softwares like GeneTool, Primer3, FastPCR, etc.

The development of DNA markers through in silico approaches requires the following components (Fig. 8.1):

1. Target sequence: This may be whole-genome nuclear/organellar sequence or short nucleotide sequences such as genome survey sequences (GSS) and expressed sequence tags (ESTs), full-length cDNAs, and complete gene and any PCR amplicon sequences.
2. Bioinformatic tools: The bioinformatic tools useful for DNA marker development include the tools for the detection of DNA sequence variations such as SSRs, SNPs, In-Dels, etc., and the tools for the designing of primers such as GeneTool, Primer3, FastPCR, etc. Designing of primers targeting the DNA variation is essential for its conversion into PCR-based markers.

8.3 Development of Microsatellite Markers

Microsatellites or simple sequence repeats (SSRs) are tandem repeats of 1–6 nucleotides, which are abundant in the eukaryotic genomes. Due to the hypervariable nature of the repeats, they are considered as suitable targets for the development of PCR-based molecular markers. Microsatellite markers have become the choice of molecular breeders due to their high level of polymorphism, locus specificity, multi-allelic and codominant nature, relative abundance, and reproducibility. Conventional methods of SSR marker development are skill oriented, time consuming, and expensive since it involves the construction of a small-insert genomic library

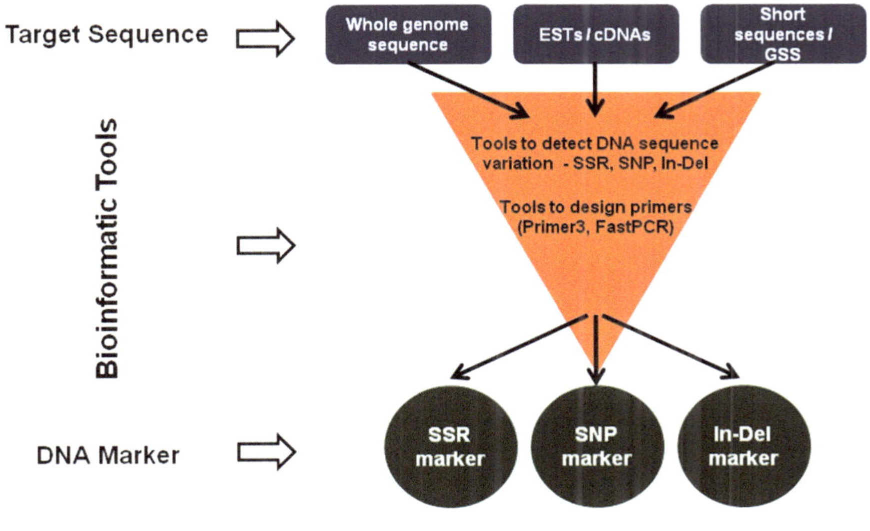

Fig. 8.1 Scheme for DNA marker development through bioinformatic approach

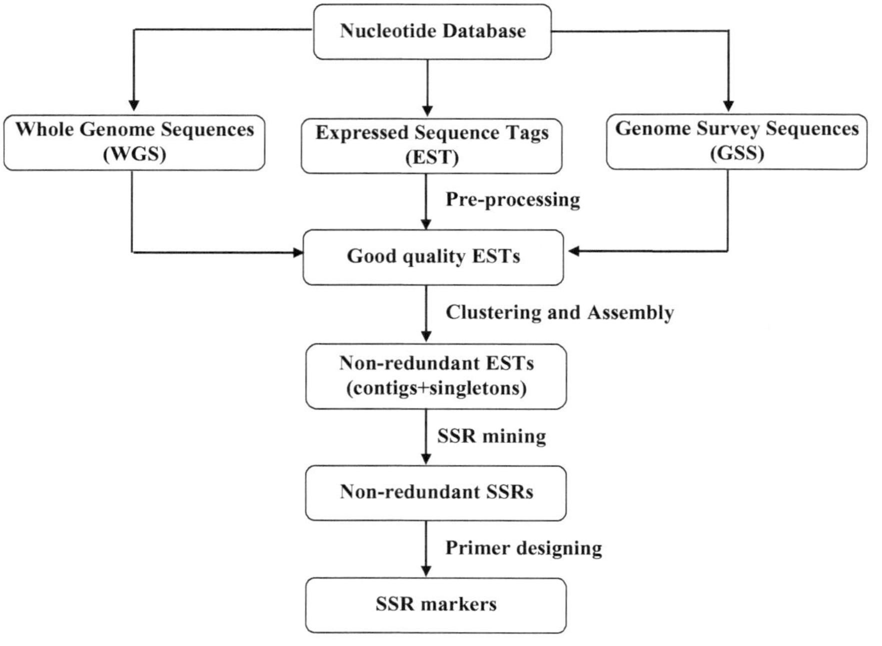

Fig. 8.2 Scheme for the development of microsatellite markers through in silico approaches

and subsequent screening of clones for the presence of SSR repeat motifs. With the availability of whole-genome sequences of sorghum as well as expressed sequence tags (ESTs) in the public databases, SSR markers can be developed rapidly and efficiently through in silico approaches (Fig. 8.2).

Microsatellite markers can be developed from two types of sequences, viz., whole-genome sequences and expressed sequence tags (ESTs). In the case of the former, it is a straightforward approach involving SSR identification and primer designing, while in the case of the latter, additional steps of preprocessing, clustering, and assembly are essential to identify non-redundant good-quality sequences in the database. ESTs are typically unedited, automatically processed, single-read sequences derived from cDNA libraries having high levels of sequence redundancy, low sequence quality, and short sequence lengths. Prior to the identification of SSRs, preprocessing of ESTs is performed to remove ambiguous sequences, thereby minimizing the chance of clustering unrelated sequences. This step helps in removing low-quality regions, contaminations (bacterial DNA, yeast DNA, etc.), vector sequences used for cloning, repeat sequences (LINEs, SINEs, LTRs, and transposons), and low-complexity sequences (poly(A) tracts, AT repeats, etc.). Dedicated tools are available for performing preprocessing of ESTs (Table 8.1). The preprocessed ESTs will be used for clustering and assembly analyses. Clustering of ESTs is done to incorporate overlapping ESTs which tag the same transcript of the same gene in a single cluster based on the similarity between any two sequences. Assembly performs the multiple alignments for each cluster and generates consensus sequences. Clustering and assembly results in identifying non-redundant sequences, viz., consensus sequences/contigs and singletons/singlets. These non-redundant sequences are used for the identification of SSRs. Clustering and assembly can be done prior to the identification of SSRs or it can follow it. Several tools are available for EST clustering and assembly (Table 8.2).

Table 8.1 Bioinformatic softwares for EST preprocessing

Software	Availability	Utility
DUST	http://www.ncbi.nlm.nih.gov/IEB/ToolBox/C_DOC/lxr/source/tools/dust.c	Removal of low-complexity regions
UniVec	http://www.ncbi.nlm.nih.gov/VecScreen/UniVec.html	Vector database
EMVec	http://www.ebi.ac.uk/Tools/sss/ncbiblast/vectors.html	Vector database
Phred/Cross_Match	http://www.phrap.com/winphrap.htm	Vector trimming and removal
Trimest	http://emboss.bioinformatics.nl/cgi-bin/emboss/trimest	Poly(A) tail trimming
Trimseq	http://emboss.bioinformatics.nl/cgi-bin/emboss/trimseq	Trimming of ambiguous ends
VecScreen	http://www.ncbi.nlm.nih.gov/VecScreen/VecScreen.html	Vector, linker, and adapter identification
Vectorstrip	http://emboss.bioinformatics.nl/cgi-bin/emboss/vectorstrip	Extraction of DNA between vector sequences
Lucy2	http://www.complex.iastate.edu/download/Lucy2/	Sequence trimming and visualization
RepeatMasker	http://www.repeatmasker.org/	Masking of repeats
Repbase	http://www.girinst.org/repbase/	Masking of repeats

Table 8.2 Bioinformatic softwares for EST clustering and assembly

Software	Availability
CAP3	http://pbil.univ-lyon1.fr/cap3.php
CLOBB	http://www.nematodes.org/bioinformatics/CLOBB/
ESSEM	http://alggen.lsi.upc.es/recerca/essem/frame-essem.html
Phrap	http://www.phrap.org/phredphrapconsed.html
miraEST	http://www.chevreux.org/projects_mira.html
ESTAP	http://staff.vbi.vt.edu/estap/
Sequencher	http://genecodes.com/
Unigene	http://www.ncbi.nlm.nih.gov/unigene/

Table 8.3 Bioinformatic softwares for microsatellite identification

Software	Reference/availability
Sputnik	Abajian (1994); http://espressosoftware.com/sputnik/index.html
Tandem Repeats Finder (TRF)	Benson (1999); http://tandem.bu.edu/trf/trf.html
SSR Identification Tool (SSRIT)	Temnykh et al. (2001); http://www.gramene.org/db/markers/ssrtool
Tandem Repeat Occurrence Locator (TROLL)	Castelo et al. (2002); http://finder.sourceforge.net/
Search for Tandem Repeats in Genomes (STRING)	Parisi et al. (2003); http://w3.uniroma1.it/valerio.parisi/STRING/
mreps	Kolpakov et al. (2003); http://bioinfo.lifl.fr/mreps/
MIcroSAtellite (MISA)	Thiel et al.(2003); http://pgrc.ipk-gatersleben.de/misa/
SSR Finder	Gao et al. (2003); http://www.fresnostate.edu/ssrfinder/
BuildSSR	Rungis et al. (2004)
Exact Tandem Repeats Analyzer (E-TRA) and Tandem Repeats Analyzer (TRA)	Karaca et al. (2005); ftp://ftp.akdeniz.edu.tr/Araclar/TRA/
Imperfect Microsatellite Extractor (IMEx)	Mudunuri and Nagarajaram (2007); http://imex.cdfd.org.in/IMEX/
SciRoKo	Kofler et al. (2007); http://kofler.or.at/bioinformatics/SciRoKo/index.html
SSR Locator	Carlos da Maia et al. (2008); http://minerva.ufpel.edu.br/~lmaia.faem/ssr_install_guide.html
WebSat	Martins et al. (2009); http://wsmartins.net/websat/
FastPCR	Kalendar et al. (2009); http://primerdigital.com/fastpcr.html
phpSSRMiner	http://bioinfo.noble.org/phpssrminer/
SSRPoly	http://appliedbioinformatics.com.au/index.php/SSRPoly

Microsatellites are classified as perfect (uninterrupted repeats), imperfect (interrupted with base substitutions), and compound (two or more repeat units) based on the arrangement of nucleotides in the repeat motifs (Weber 1990). Several bioinformatic tools/softwares/scripts are available in the public domain for the identification of microsatellites. These tools were developed by different research groups across the globe. A list of popularly used bioinformatic tools for the identification of microsatellites is given in Table 8.3. The tools available for SSR mining differ in their ability in the identification of microsatellites. For instance, the tool SSRIT identifies only the simple microsatellites, while SSR Locator, Search for Tandem Repeats in Genomes (STRING), and Tandem Repeat Occurrence Locator (TROLL) identify all the three classes of microsatellites. Some of these tools are useful only for the identification of microsatellites (SSRIT), while the other (SSR Locator, TROLL) have the option of primer designing integrated into them. Tang et al. (2008) developed PolySSR, a new pipeline to identify polymorphic SSRs which takes into account the SNPs in the flanking regions while designing PCR primers for the putatively polymorphic SSR markers thereby improving the success of the potential markers. This tool was successfully used to identify a large number of polymorphic SSRs using publicly available EST sequences of potato, tomato, rice, *Arabidopsis*, and *Brassica*.

In sorghum, microsatellite markers are developed from different sequence types such as genomic sequences, ESTs, unigenes, as well as whole-genome sequences. In general, CAP3 and SSRIT are the most popularly used bioinformatic tools for the identification of microsatellites. CAP3 and BLASTN are used for clustering and alignment, while SSRIT and MISA are used for the identification of microsatellites. During the last 5 years, about 1758 new SSR markers from genomic sequence contigs (Li et al. 2009a), 109 and 600 SSR markers from ESTs (Ramu et al. 2009; Srinivas et al. 2009), 1519 SSR markers from unigenes (Nagaraja Reddy et al. 2012), and 5599 SSR and 110 $(GATA)_n$ motif-based SSR markers from whole-genome sequences (Yonemaru et al. 2009; Jaikishan et al. 2013) were developed through in silico approaches.

8.4 Development of In-Del Markers

Even though microsatellite markers are popularly used for various genetic and plant breeding applications, there is an increasing trend in the identification of sequence length polymorphisms other than microsatellites, especially short insertions and deletions (In-Dels), and developing them into In-Del markers. This is because of the advancements in bioinformatics and the availability of huge amounts of DNA sequence data in the public domain due to next-generation sequencing technologies. Moreover, SNPs and In-Dels are becoming the preferred choice of DNA markers for molecular breeding applications due to their occurrence in high frequency, stability, amenability to high-throughput genotyping, and cost-effectiveness over other DNA markers (Henry and Edwards 2009). In-Del marker results from the insertion of transposable elements, slippage in simple sequence replication, or unequal crossover events (Britten et al. 2003). These markers can be genotyped by fragment length polymorphisms using the same experimental procedures based on size separation routinely used for SSR markers (Bhattramakki et al. 2002).

In-Dels have been recognized as an abundant source of genetic markers, next only to SNPs that are widely spread across the genome. In addition, the density of In-Del and SNP markers is more than that of SSR markers. In-Dels have been used successfully for cultivar identification and marker-assisted selection in plants (Jakse et al. 2005; Hayashi et al. 2006; Hong et al. 2008; Pacurar et al. 2012). In sorghum, Strelchenko et al. (2010a) demonstrated the utility of rice In-Del markers for the determination of genetic relationships of sorghum germplasm in Asia and Africa. Strelchenko et al. (2010b) identified the global centers of diversity of grain sorghum by using rice In-Del markers. In-Del markers have also been used in the mapping of important loci such as waxy (McIntyre et al. 2008) and tannin (Wu et al. 2012).

With the increasing number of whole-genome resequencing projects, the enormous amount of whole-genome sequence data is generated leading to the identification of DNA sequence variations such as SNPs and In-Dels in a rapid and efficient way through bioinformatic approaches. The bioinformatic tools used for the identification of In-Dels are given in Table 8.4. Through an analysis of genome-wide pattern of genetic variation in sweet and grain sorghum, Zheng et al. (2011) identified 99,948 In-Dels of 1–10 bp in length; a majority of the large-effect In-Dels

Table 8.4 Bioinformatic softwares for the identification of In-Dels

Software	Reference/availability
Pindel, SHORE, and BreakDancerMax	Ye et al. (2009), Ossowski et al. (2008), Chen et al. (2009)
Short Oligonucleotide Alignment Program (SOAP)	Li et al. (2008); http://soap.genomics.org.cn/
PolyBayes	Marth (1999); http://bioinformatics.bc.edu/~marth/PolyBayes/pages/main.html
Sequencher	http://genecodes.com/
BLASTN	Altschul et al. (1997); http://blast.ncbi.nlm.nih.gov/Blast.cgi?PAGE_TYPE=BlastSearch
VarScan	Koboldt et al. (2009); http://varscan.sourceforge.net/
Dindel	Albers et al. (2011); http://www.sanger.ac.uk/resources/software/dindel/
SAMtools	Li et al. (2009b); http://samtools.sourceforge.net/
Genome Analysis Toolkit (GATK)	McKenna et al. (2010); http://www.broadinstitute.org/gatk/
NextGENe™ Software	www.softgenetics.com/NextGENe.html
Contig Viewer	http://cgpdb.ucdavis.edu/cgpdb2/CAP3_ContigViewer_V01/
NovelSNPer	Assmus et al. (2011); http://www2.hu-berlin.de/wikizbnutztier/software/NovelSNPer/

resided in the genes with leucine-rich repeats, PPR repeats, and disease-resistance R genes with diverse biological functions, but were absent in genes that are essential for life. Similarly, whole-genome resequencing of three CMS and three restorer lines of indica rice resulted in the discovery of 160,478 insertions and 163,556 deletions across the rice genome (Gopala Krishnan et al. 2011). Very recently, Mace et al. (2013) identified about 1,982,971 In-Dels, of which 872,080 were insertions and 1,110,891 were deletions. The length of the In-Dels ranged from 1 to 66 bp, most of them were small (1–6 bp; 86 %), with a small proportion of them greater than 20 bp in length (2.5 %). Most of the In-Dels were located in inter-genic regions (83 %), while only 1.5 % of them located in coding regions.

8.5 Development of Intron Polymorphism Markers

Introns, which are the non-coding regions of genes that are transcribed but spliced out during pre-mRNA processing, are widespread and abundant in most of the genes in eukaryotic organisms. Evolutionarily, introns are less conserved in comparison to exons due to the absence of selection pressure and hence accumulate a large number of mutations. Such DNA sequence variations occur as length polymorphisms (In-Dels) and SNPs, which are being exploited in recent times for the development of genic molecular markers. Such markers are becoming more popular due to the availability of the enormous amount of expressed sequence tags (ESTs) of various plant species and also the whole-genome sequences of some economically important crop species in the public databases. These markers help in the identification of a "perfect marker" for marker-assisted selection (MAS), assessment of functional genetic diversity among germplasm lines, comparative mapping among the related species, and identification of chromosomal duplication events (Gujaria et al. 2011).

Among the two types of polymorphisms, intron length polymorphism (ILP) is easily detected by a PCR-based approach, namely, exon-primed intron-crossing PCR (EPIC-PCR) (Palumbi 1995), where primers are designed in exonic regions flanking the target introns. Very few reports are available on the development of such markers (Wei et al. 2005; Feltus et al. 2006; Chen et al. 2011; Braglia et al. 2010; Galasso et al. 2010; Poczai et al. 2010; Shang et al. 2010; Tamura et al. 2012; Liu et al. 2011; Gupta et al. 2011, 2012; Xia et al. 2012) in various crop species. Availability of whole-genome sequence of important crops with annotation information offers the opportunity for the genome-wide identification of introns leading to the development of such a marker (Table 8.5). Even though complete genome sequences are available in some crop species, genome-wide exploitation of intron polymorphism markers was reported only in rice (Wang et al. 2005) and soybean (Shu et al. 2010).

A database of potential intron polymorphism (PIP) in plants was developed by Yang et al. (2007). At present, this database has a total of 57,658 PIP markers for 59 plant species that can be readily used by the researchers. It also helps in the development of new PIP markers in any plant species if gene/EST/cDNA sequences are available. Identification of potential introns is done on the premise that it is possible to predict the exon-intron structures in homologous ESTs of other plants with the help of complete genomic sequence information of model plants such as rice and *Arabidopsis*. It is also possible to compare the PIP markers of two different crop species with that of model plant. In case, if ESTs are used, they should be subjected to clustering and assembly to identify non-redundant EST, which should be used as input for the development of PIP markers. About 4314 PIP markers of sorghum are available in this database, which are developed through in silico approach from PlantGDB-assembled unique transcripts (PUTs) available in the public database http://www.plantgdb.org/. These markers can be experimentally validated in diverse sorghum genotypes and can be used for determining phylogeny, comparative mapping, and marker-assisted selection. Recently, 37,862 potential introns were identified in sorghum using the chromosome-wise gene sequences available in public database (http://www.phytozome.net/) and primers were designed to develop potential intron polymorphism (PIP) markers. About 200 PIP markers were validated as PCR-based intron length polymorphism (ILP) markers in 24 sorghum genotypes, of which 172 gave clear and robust amplification without multiple amplicons and 48 markers were polymorphic (Jaikishan et al. 2014).

Table 8.5 Bioinformatic softwares used for the identification of potential introns

Software	Reference
Perl script involving SIM4 and BLASTN	Wang et al. (2005)
BLASTN	Wei et al. (2005)
FASTA33 and BLASTX	Keyser et al. (2009)
BLASTN	Shang et al. (2010)
Perl/Bioperl script involving SIM4	Liu et al. (2011)
SIM4 and BLASTN	Chen et al. (2011)
WebGMAP	Gupta et al. (2012)

8.6 Single Nucleotide Polymorphism (SNP) and Development of Markers

Single nucleotide polymorphisms (SNPs) are the most abundant sequence variations in the genomic DNA of genotypes of crop species. Abundance coupled with slow mutation rate within the genome (Nickerson et al. 1990) makes SNPs as one of the most commonly used genetic markers for studying complex genetic traits and genome evolution (Syvanen 2001). Moreover, SNPs in the coding region are used to directly study the genetics of expressed genes and to map functional traits. SNPs can be identified by both experimental and computational approaches. Experimental approach is laboratory oriented, time consuming, and expensive (Schlötterer 2004; Useche et al. 2001), while the computational approach makes use of the large sequence datasets present in public databases and is rapid and less expensive. The computational approaches are popularly used for the identification of SNPs because of the growing number of sequences in the public databases. A number of pipelines have been developed to detect SNPs in sequences automatically (Table 8.6), which could be categorized into two types. One group of pipeline detects SNPs using trace files or quality files, such as Phred/Phrap/PolyBayes system, while other pipelines detect SNPs using only EST redundancy in the text-based sequence files, such as autoSNP and SNiPpER.

Though computational approaches are rapid in the detection of SNPs, they do not help in

Table 8.6 Bioinformatic softwares used for the identification of SNPs

Software	Reference/availability
Phred/Phrap/Consed	Picoult-Newberg et al. (1999); http://www.phrap.org/
PolyPhred	Nickerson et al. (1997); http://droog.gs.washington.edu/polyphred/
PolyBayes	Marth et al. (1999); http://clavius.bc.edu/~marth/PolBayes/pages/main.html
autoSNP	Barker et al. (2003)
SNiPer	Huentelman et al. (2005); http://www.tgen.org/neurogenomics/data
QualitySNP	Tang et al. (2006); http://www.bioinformatics.nl/tools/snpweb
SNPserver	Savage et al. (2005); http://hornbill.cspp.latrobe.edu.au/snpdiscovery.html
HaploSNPer	Tang et al. (2008); http://www.bioinformatics.nl/tools/haplosnper/
SEAN	Huntley et al. (2006); a program written in Perl, available on request from dave.edwards@nre.vic.gov.au
Sniper	Simola and Kin (2011); http://kim.bio.upenn.edu/software/sniper.shtml
SAMtools	Li et al. (2009b); http://samtools.sourceforge.net
SNVer	Wei et al. (2011); http://snver.sourceforge.net
SOAPsnp	Li et al. (2009c); http://soap.genomics.org.cn/soapsnp.html
SNiPlay	Dereeper et al. (2011); http://sniplay.cirad.fr/cgi-bin/home.cgi
Seq-SNPing	Chang et al. (2009); http://bio.kuas.edu.tw/Seq-SNPing
SNPpipeline	Buetow et al. (1999)

distinguishing allelic and sequence variation between paralogous sequences. Moreover, they do not detect sequencing errors resulting in the frequent occurrence of false positives. This drawback is addressed by PolyBayes (Marth et al. 1999) through an enhanced paralog identification routine, but the corresponding genomic sequence and quality files in addition to the EST sequence are required. However, the utility of the PolyBayes paralog identification routine is limited since whole-genome sequences are not available for many crop species and the majority of ESTs in the public domain do not have trace or quality file.

A new haplotype-based strategy was adopted in QualitySNP (Tang et al. 2006) to detect reliable synonymous and non-synonymous SNPs from the EST sequence data available in the public domain without trace/quality files or whole-genome sequence data. SNPs have been extensively used in haplotype reconstruction, which is based on mathematical algorithms. Haplotypes represent the different alleles of a gene in a dataset. Generally, haplotype reconstruction is done by partitioning genome-wide SNPs into blocks with minimum variability and assembling them into haplotypes. The software HAP (Halperin and Eskin 2004) uses such a method of haplotype reconstruction. Haplotypes can also be reconstructed for specific genes, based on the SNPs present in the gene (Rafalski 2002). A set of SNPs discriminating all identified alleles can be used to study the association between candidate genes and phenotypes, thereby helping in the selection of individuals with specific genotypes.

SNPserver (Savage et al. 2005) is a Web-based tool for the real-time detection of SNPs related to any target sequence. This tool is based on autoSNP (Barker et al. 2003) that utilizes the frequency of occurrence of a polymorphism and co-segregation of multiple SNPs for the reliable discovery of SNPs. Even though autoSNP and QualitySNP cannot distinguish paralogs, the former detects many more false positive SNPs and requires more processing time for large datasets (Tang et al. 2006). HaploSNPer is another Web-based tool useful for the reliable detection of alleles and SNPs. This detects homologous sequences in user-specified sequence databases using a target sequence supplied by the user or on a collection of input sequences. The special feature of this tool is that it combines the QualitySNP algorithm with database search along with sequence alignment tools into an efficient pipeline.

Advancements in sequencing technologies, execution of resequencing projects, and availability of the enormous amount of ESTs along with the development of efficient computational platforms have helped in the rapid discovery of SNPs in sorghum. Sorghum researchers across the globe have utilized different types of data such as ESTs, whole-genome resequencing data, and genotyping-by-sequencing (GbS) data for the discovery of SNPs with the help of various computational tools. In sorghum, Singhal et al. (2011) aligned the ESTs from sorghum EST database with the whole sorghum genome and identified 77,094 potential and 40,589 reliable SNPs using the tool HaploSNPer (based on QualitySNP pipeline). Among the 77,094 potential SNPs, 34,398, 35,871, and 6,825 were transitions, transversions, and In-Dels, respectively. Zheng et al. (2011) identified a total of 1,057,018 SNPs, of which 83,262 were located in the coding regions using the reads of three sorghum genotypes, Keller, E-Tian, and Ji2731 and the information of physical sequence alignment and gene models available from the reference genome (BTx623). They used the SOAPsnp software, which allows the detection of heterozygosity of SNPs (Li et al. 2009c). Along with SOAP v2, Nelson et al. (2011) used another tool, NovoAlign (www.novocraft.com/products/novoalign), followed by SAMtools to align eight genome equivalents (6 Gb) to the public reference genome and identified 283,000 SNPs at ≥82 % confirmation probability. Recently, Morris et al. (2012) performed genotyping by sequencing of 971 diverse sorghum accessions comprising of the US sorghum association panel (SAP), the sorghum mini-core collection (MCC), and the Generation Challenge Program sorghum reference set (RS) and detected SNPs using the TASSEL 3.0 GBS pipeline (www.maizegenetics.net/tassel/) by mapping the sequences of these diverse lines to the sorghum reference genome (BTx623) by using BWA. They identified about 265,487 SNPs, with an average density of one SNP per 2.7 kb. It was also found that 72 % and 99 % of the 27,412 annotated genes in the reference sorghum genome were tagged by an SNP within the gene and within 10 kb, respectively. More recently, Mace et al. (2013) resequenced 44 sorghum accessions representing all major races of cultivated sorghum (*S. bicolor*), along with its progenitors, and *S. propinquum*. These lines comprised of 18 landraces, 17 improved inbreds, and seven wild and weedy sorghums. A total of 4.9 million high-quality SNPs were identified, of which majority of them (83 %) were located in inter-genic regions, with an average of 4.5 % located in coding sequences. Wild and weedy sorghum genotypes possessed higher number of SNPs as compared to the landraces and improved inbreds.

8.7 Development of Molecular Markers from Organellar Genomes

Organelle genomes, viz., chloroplast and mitochondrial genomes, have features such as conserved gene order, low recombination rates, and relatively small size, which make them the widely used tools for phylogenetic studies. Simple sequence repeats (SSRs) are also present in the organellar genomes. Phylogenetic analysis based on mononucleotide repeats and flanking nucleotide sequences from the organellar genomes (Nishikawa et al. 2005) and class I and class II SSRs in organellar genomes of rice, wheat, sorghum, and maize (Rajendrakumar et al. 2007a, 2008) were reported. Mitochondrial (http://www.ncbi.nlm.nih.gov/nuccore/115278525) and chloroplast genomes (Saski et al. 2007; http://www.ncbi.nlm.nih.gov/nuccore/118614470) of sorghum have been sequenced. Availability of organellar genome sequences in the public domain has accelerated the development of SSR markers through in silico approach, which is simple and inexpensive. The microsatellites present in the organellar genomes may be useful in the development of organellar genome-specific markers for tagging specific traits such as cytoplasmic male sterility, herbicide tolerance, etc. Only recently, perfect and imperfect SSRs of 2,161 organelle genomes (1,982 mitochondrial and 179 chloroplast genomes) were reported and maintained in a complete curated Web-oriented relational database, ChloroMitoSSRDB (Sablok et al. 2013). This database will be useful in the development of organellar SSR markers by using the information on the SSRs for the designing of primers with appropriate primer-designing softwares. SSRs from mitochondrial genomes of plant species are not covered in this database.

Conventionally, molecular phylogenetic analysis is performed using the DNA sequence variation of one (Doebley et al. 1990; Hilu and Alice 1999) or a few conserved genes (Wolfe et al. 1989; Gaut et al. 1993). Guo et al. (1996) used mitochondrial DNA variation to elucidate the evolutionary history and affinity of sorghum species through mitochondrial DNA restriction fragment analysis. However, considering more number of genes for comparison reduces inherent sampling errors and yields reliable information about the relationships. Therefore, it is imminent that analysis of genome-wide variations often provides more convincing inferences as they provide more number of datasets. A phylogenetic analysis based on organellar SSR markers by Rajendrakumar et al. (2008) revealed that sorghum was closer to maize and rice, while wheat was the farthest confirming earlier reports based on nuclear genome colinearity and chloroplast gene-based phylogenetic analysis. Apart from their use as molecular markers, the information on the number and distribution of microsatellites may help in knowing their relevance in gene function or genome evolution and also phylogenetic relationships among different crop species. Despite the availability of complete organellar genomes of many crop species, a comprehensive analysis of microsatellites has been reported only in rice, wheat, maize, and sorghum (Rajendrakumar et al. 2007a, 2008). In rice, the development of mitochondrial repeat-specific marker through in silico analysis was reported for distinguishing male sterile lines of wild-abortive

(WA) type of cytoplasm from their cognate maintainer lines (Rajendrakumar et al. 2007b). Recently, Wang et al. (2012) used in silico developed organellar SSR markers and reported that 11 chloroplast and four mitochondrial SSR markers revealed polymorphism among the six cabbage CMS types, namely, NigCMS, OguCMSR1, OguCMSR2, OguCMSR3, OguCMSHY, and PolCMS. They concluded that such organellar SSR analysis could be a feasible alternative for the characterization of different types of CMS.

8.8 Future Prospects

Bioinformatics is critical for the future of plant breeding and crop research since it can help retrieve useful information from the DNA sequence data; develop linkages between biological data such as DNA sequences, map positions, marker alleles, related genes, etc. and develop computational tools to facilitate the identification of repeat/regulatory elements and gene prediction. Development of PCR-based markers especially single-locus markers that require DNA sequence information was tedious, expensive, skill oriented, and time consuming prior to the advent of high-throughput sequencing technologies as well as bioinformatic tools for computational analysis. In the current genome sequencing era, the enormous amount of ESTs, full-length cDNAs, and genome sequences are continuously accumulated in the public domain due to the availability of next-generation sequencing technologies. These sequences can be used as targets for the identification of microsatellites, SNPs, and In-Dels by employing computational approaches and primers can be designed to convert them into PCR-based markers. Several computational tools are available in the public domain, which aid in the development of molecular markers in a rapid, economical, and time-saving manner. Even though all the resources including sequence data and computational tools are available and scattered in the public domain, there is an immense need to develop crop-specific resources integrating databases as well as computational tools for the efficient marker development with options for in silico validation of the developed markers.

References

Abajian C (1994) Sputnik. http://espressosoftware.com/sputnik/index.html

Assmus J, Schmitt AO, Bortfeldt RH, Brockmann GA (2011) NovelSNPer: a fast tool for the identification and characterization of novel SNPs and InDels. Adv Bioinforma 2011:1–11

Albers CA, Lunter G, MacArthur DG, McVean G, Ouwehand WH, Durbin R (2011) Dindel: accurate indel calls from short-read data. Genome Res 21:961–973

Altschul SF, Madden TL, Schäffer AA, Zhang J, Zhang Z, Miller W, Lipman DJ (1997) Gapped BLAST and PSI-BLAST: a new generation of protein database search programs. Nucleic Acids Res 25:3389–3402

Barker G, Batley J, O'Sullivan H, Edwards KJ, Edwards D (2003) Redundancy based detection of sequence polymorphisms in expressed sequence tag data using autoSNP. Bioinformatics 19:421–422

Basavaraj SH, Singh VK, Singh A et al (2010) Marker-assisted improvement of bacterial blight resistance in parental lines of Pusa RH10, a superfine grain aromatic rice hybrid. Mol Breed 26:293–305

Benson G (1999) Tandem repeats finder: a program to analyze DNA sequences. Nucleic Acids Res 27:573–580

Bhattramakki D, Dolan M, Hanafey M, Wineland R, Vaske D, Register JC, Tingey SV, Rafalski A (2002) Insertion–deletion polymorphisms in 30 regions of maize genes occur frequently and can be used as highly informative genetic markers. Plant Mol Biol 48:539–547

Braglia L, Manca A, Mastromauro F, Breviario D (2010) cTBP: a successful intron length polymorphism (ILP)-based genotyping method targeted to well defined experimental needs. Diversity 2:572–585

Britten RJ, Rowen L, Williams J, Cameron RA (2003) Majority of divergence between closely related DNA samples is due to indels. Proc Natl Acad Sci U S A 100:4661–4665

Buetow KH, Edmonson MN, Cassidy AB (1999) Reliable identification of large numbers of candidate SNPs from public EST data. Nat Genet 21:323–325

Carlos da Maia L, Palmieri DA, Queiroz de Souza V, Kopp MM, Félix de Carvalho FI, Costa de Oliveira A (2008) SSR locator: tool for simple sequence repeat discovery integrated with primer design and PCR simulation. Int J Syst Bioinform 4:363–374

Castelo AT, Martins W, Gao GR (2002) TROLL – tandem repeat occurrence locator. Bioinformatics 18:634–636

Chang HW, Chuang LY, Cheng YH, Ho CH, Wen CH, Yang CH (2009) Seq-SNPing: multiple-alignment tool for SNP discovery, SNP ID identification, and RFLP genotyping. OMICS J Integr Biol 13:253–260

Chen K, Wallis JW, McLellan D et al (2009) BreakDancer: an algorithm for high-resolution mapping of genomic structural variation. Nat Methods 6:677–681

Chen X, Zhang G, Wu W (2011) Investigation and utilization of intron length polymorphisms in conifers. New For 41:379–388

Dereeper A, Nicolas S, Cunff LL, Bacilieri R, Doligez A, Peros JP, Ruiz M, This P (2011) SNiPlay: a web-based tool for detection, management and analysis of SNPs. Application to grapevine diversity projects. BMC Bioinform 12:134

Doebley J, Durbin M, Golenberg EM, Clegg MT, Ma DP (1990) Evolutionary analysis of the large subunit of carboxylase (*rbcL*) nucleotide sequence among the grasses (Gramineae). Evolution 44:1097–1108

Feltus FA, Singh HP, Lohithaswa HC, Schulze SR, Silva TD, Paterson AH (2006) A comparative genomics strategy for targeted discovery of single-nucleotide polymorphisms and conserved non-coding sequences in orphan crops. Plant Physiol 140:1183–1191

Galasso I, Manca A, Braglia L, Martinelli T, Morello L, Breviario D (2010) h-TBP: an approach based on intron-length polymorphism for the rapid isolation and characterization of the multiple members of the b-tubulin gene family in *Camelina sativa* (L.) Crantz. Mol Breed 28:635–645

Gao L, Tang J, Li H, Jia J (2003) Analysis of microsatellites in major crops assessed by computational and experimental approaches. Mol Breed 12:245–261

Gaut BS, Muse SV, Clegg MT (1993) Relative rates of nucleotide substitution in the chloroplast genome. Mol Phylogenet Evol 2:89–96

Gopala Krishnan S, Waters DLE, Katiyar SK, Sadananda AR, Satyadev V, Henry R (2011) Genome-wide DNA polymorphisms in elite *indica* rice inbreds discovered by whole-genome sequencing. Plant Biotechnol J 10:623–634

Gujaria N, Kumar A, Dauthal P et al (2011) Development and use of genic molecular markers (GMMs) for construction of a transcript map of chickpea (*Cicer arietinum* L.). Theor Appl Genet 122:1577–1589

Guo JH, Skinner DZ, Liang GH (1996) Phylogenetic relationships of sorghum taxa inferred from mitochondrial DNA restriction fragment analysis. Genome 39:1027–1034

Gupta S, Kumari K, Das J, Lata C, Puranik S, Prasad M (2011) Development and utilization of novel intron length polymorphic markers in foxtail millet [*Setaria italica* (L.) P. Beauv.]. Genome 54:586–602

Gupta SK, Bansal R, Gopalakrishna T (2012) Development of intron length polymorphism markers in cowpea [*Vigna unguiculata* (L.) Walp.] and their transferability to other *Vigna* species. Mol Breed 30:1363–1370

Halperin E, Eskin E (2004) Haplotype reconstruction from genotype data using imperfect phylogeny. Bioinformatics 20:1842–1849

Hayashi K, Yoshida H, Ashikawa I (2006) Development of PCR-based allele-specific and indel marker sets for nine rice blast resistance genes. Theor Appl Genet 113:251–260

Henry RJ, Edwards K (2009) New tools for single nucleotide polymorphism (SNP) discovery and analysis accelerating plant biotechnology. Plant Biotechnol J 7:311

Hilu KW, Alice LA (1999) Evolutionary implications of *matK* indels in Poaceae. Am J Bot 86:1735–1741

Hong PC, Hong LA, Yuan DZ et al (2008) InDel and SNP markers and their applications in map-based cloning of rice genes. Rice Sci 15:251–258

Huentelman MJ, Craig DW, Shieh AD, Corneveaux JJ, Lince DH, Pearson JV, Stephan DA (2005) SNiPer: improved SNP genotype calling for Affymetrix 10K GeneChip microarray data. BMC Genomics 6:149

Huntley D, Baldo A, Johri S, Sergot M (2006) SEAN: SNP prediction and display program utilizing EST sequence clusters. Bioinformatics 22:495–496

Jaikishan I, Paik GR, Madhusudhana R, Elangovan M, Rajendrakumar P (2013) Development of microsatellite markers targeting $(GATA)_n$ motifs in sorghum [*Sorghum bicolor* (L.) Moench]. Mol Breed 31:223–231

Jaikishan I, Rajendrakumar P, Madhusudhana R, Elangovan M, Patil JV (2014) *In silico* identification of potential introns and development of PCR-based markers targeting intron length polymorphism in sorghum [*Sorghum bicolor* (L.) Moench). In: Khetmalas M, Pandey A, Gunasekaran P, Nawani N, Nagar S (eds) Abstracts, International conference on advances in biotechnology and bioinformatics ICABB 2013 and X convention of the Biotech Research Society, India organized by Dr DV Patil Vidyapeeth and The Biotech Research Society, India, pp C11

Jakse J, Martin W, McCallum J, Havey MJ (2005) Single nucleotide polymorphisms, indels and simple sequence repeats for onion cultivar identification. J Am Soc Hortic Sci 130:912–917

Kalendar R, Lee D, Schulman AH (2009) FastPCR Software for PCR primer and probe design and repeat search. G3 (Bethesda) 3:1–14

Karaca M, Bilgen M, Onus AN, Ince AG, Elmasulu SY (2005) Exact tandem repeats analyzer (E-TRA): a new program for DNA sequence mining. J Genet 84:49–54

Keyser ED, Riek JD, Bockstaele EV (2009) Discovery of species-wide EST-derived markers in *Rhododendron* by intron-flanking primer design. Mol Breed 23:171–178

Knapp S (1998) Marker-assisted selection as a strategy for increasing the probability of selecting superior genotypes. Crop Sci 38:1164–1174

Koboldt DC, Chen K, Wylie T, Larson DE, McLellan M, Mardis ER, Weinstock GM, Wilson RK, Ding L (2009) VarScan: variant detection in massively parallel sequencing of individual and pooled samples. Bioinformatics 25:2283–2285

Kofler R, Schlotterer C, Lelley T (2007) SciRoKo: a new tool for whole genome microsatellite search and investigation. Bioinformatics 23:1683–1685

Kolpakov R, Bana G, Kucherov G (2003) mreps: efficient and flexible detection of tandem repeats in DNA. Nucleic Acids Res 31:3672–3678

Li R, Li Y, Kristiansen K, Wang J (2008) SOAP: short oligonucleotide alignment program. Bioinformatics 24:713–714

Li M, Yuyama N, Luo L, Hirata M, Cai H (2009a) *In silico* mapping of 1758 new SSR markers developed from public genomic sequences for sorghum. Mol Breed 24:41–47

Li H, Handsaker B, Wysoker A, Fennell T, Ruan J, Homer N, Marth G, Abecasis G, Durbin R, 1000 Genome Project Data Processing Subgroup (2009b) The sequence alignment/map format and SAMtools. Bioinformatics 25:2078–2079

Li R, Li Y, Fang X, Wang J, Kristiansen K (2009c) SNP detection for massively parallel whole genome sequencing. Genome Res 19:1124–1132

Liu H, Lin Y, Chen G, Shen Y, Liu J, Zhang S (2011) Genome-scale identification of resistance gene analogs and the development of their intron length polymorphism markers in maize. Mol Breed 29:437–447

Mace ES, Tai S, Gilding EK et al (2013) Whole-genome sequencing reveals untapped genetic potential in Africa's indigenous cereal crop sorghum. Nat Commun 4:2320

Marth GT, Korf I, Yandell MD, Yeh RT, Gu Z, Zakeri H, Stitziel NO, Hillier L Kwok RY, Gish WR (1999) A general approach to single-nucleotide polymorphism discovery. Nat Genet 23:452–456

Martins WS, Lucas DCS, de Souza Neves KF, Bertioli DJ (2009) WebSat – a web software for microsatellite marker development. Bioinformation 3:282–283

McIntyre CL, Drenth J, Gonzalez N, Henzell RG, Jordan DR (2008) Molecular characterization of the waxy locus in sorghum. Genome 51:524–533

McKenna A, Hanna M, Banks E et al (2010) The genome analysis toolkit: a mapreduce framework for analyzing next-generation DNA sequencing data. Genome Res 20:1297–1303

Morris GP, Ramu P, Deshpande SP et al (2012) Population genomic and genome-wide association studies of agroclimatic traits in sorghum. Proc Natl Acad Sci U S A 110:453–458

Mudunuri SB, Nagarajaram HA (2007) IMEx: imperfect microsatellite extractor. Bioinformatics 23:1181–1187

Nagaraja Reddy R, Madhusudhana R, Murali Mohan S, Chakravarthi DVN, Seetharama N (2012) Characterization, development and mapping of unigene-derived microsatellite markers in sorghum [*Sorghum bicolor* (L.) Moench]. Mol Breed 29:543–564

Neeraja CN, Rodriguez RM, Pamplona A, Heuer S, Collard BCY, Septiningsih EM, Vergara G, Sanchez D, Xu K, Ismail M, Mackill DJ (2007) A marker-assisted backcross approach for developing submergence-tolerant rice cultivars. Theor Appl Genet 115:767–776

Nelson JC, Wang S, Wu Y, Li X, Antony G, White F, Yu J (2011) Single-nucleotide polymorphism discovery by high-throughput sequencing in sorghum. BMC Genomics 12:352

Nickerson DA, Kaiser R, Lappin S, Stewart J, Hood L, Landegren U (1990) Automated DNA diagnostics using an ELISA-based oligonucleotide ligation assay. Proc Natl Acad Sci U S A 87:8923–8927

Nickerson DA, Tobe VO, Taylor SL (1997) PolyPhred: automating the detection and genotyping of single nucleotide substitutions using fluorescence-based resequencing. Nucleic Acids Res 25:2745–2751

Nishikawa T, Vaughan DA, Kadowaki K (2005) Phylogenetic analysis of *Oryza* species, based on simple sequence repeats and their flanking nucleotide sequences from the mitochondrial and chloroplast genomes. Theor Appl Genet 110:696–705

Ossowski S, Schneeberger K, Clark RM, Lanz C, Warthmann N, Weigel D (2008) Sequencing of natural strains of *Arabidopsis thaliana* with short reads. Genome Res 18:2024–2033

Pacurar DL, Pacurar ML, Street N, Bussel JD, Pop TL, Gutierrez L, Bellini C (2012) A collection of INDEL markers for map-based cloning in seven *Arabidopsis* accessions. J Exp Bot 63:2491–2501

Palumbi SR (1995) Nucleic acids II: the polymerase chain reaction. In: Hillis D, Moritz C (eds) Molecular systematics, 2nd edn. Sinauer Associates Inc., Sunderland, pp 205–247

Parisi V, Fonzo VD, Pentini FA (2003) STRING: finding tandem repeats in DNA sequences. Bioinformatics 19:1733–1738

Paterson AH, Bowers JE, Bruggmann R et al (2009) The *Sorghum bicolor* genome and the diversification of grasses. Nature 457:551–556

Picoult-Newberg L, Ideker TE, Pohl MG, Taylor SL, Donaldson MA, Nickerson DA, Boyce-Jacino M (1999) Mining SNPs from EST databases. Genome Res 9:167–174

Poczai P, Cernák I, Gorji AM, Nagy S, Taller J, Polgár Z (2010) Development of intron targeting (IT) markers for potato and cross-species amplification in *Solanum nigrum* (Solanaceae). Am J Bot 97:e142–e145

Rafalski A (2002) Applications of single nucleotide polymorphisms in crop genetics. Curr Opin Plant Biol 5:94–100

Rajendrakumar P, Biswal AK, Balachandran SM, Srinivasarao K, Sundaram RM (2007a) Simple sequence repeats in organellar genomes of rice: frequency and distribution in genic and intercoding regions. Bioinformatics 23:1–4

Rajendrakumar P, Biswal AK, Balachandran SM, Ramesha MS, Viraktamath BC, Sundaram RM (2007b) A mitochondrial repeat specific marker for distinguishing wild abortive type cytoplasmic male sterile rice lines from their cognate isogenic maintainer lines. Crop Sci 47:207–211

Rajendrakumar P, Biswal AK, Balachandran SM, Sundaram RM (2008) *In silico* analysis of microsatellites in organellar genomes of major cereals for understanding their phylogenetic relationships. In Silico Biol 8:87–104

Rajpurohit D, Kumar R, Kumar M, Paul P, Awasthi A, Basha PO, Puri A, Jhang T, Singh K, Dhaliwal HS (2011) Pyramiding of two bacterial blight resistance and a semi-dwarfing gene in Type 3 Basmati using marker-assisted selection. Euphytica 178:111–126

Ramu P, Kassahun B, Senthilvel S, Ashok Kumar S, Jayashree B, Folkertsma RT, Ananda Reddy L, Kuruvinashetti MS, Haussmann BIG, Hash CT (2009) Exploiting rice-sorghum synteny for targeted development of EST-SSRs to enrich the sorghum genetic linkage map. Theor Appl Genet 119:1193–1204

Ribaut JM, Ragot M (2007) Marker-assisted selection to improve drought adaptation in maize: the backcross approach, perspectives, limitations, and alternatives. J Exp Bot 58:351–360

Rungis D, Bérubé Y, Zhang J, Ralph S, Ritland CE, Ellis BE, Douglas C, Bohlmann J, Ritland K (2004) Robust simple sequence repeat markers for spruce (*Picea* spp.) from expressed sequence tags. Theor Appl Genet 109:1283–1294

Sablok G, Mudunuri SB, Patnana S, Popova M, Fares MA, Porta NL (2013) ChloroMitoSSRDB: open source repository of perfect and imperfect repeats in organelle genomes for evolutionary genomics. DNA Res 20:127–133

Saski C, Lee SB, Fjellheim S, Guda C, Jansen RK, Luo H, Tomkins J, Rognli OA, Daniell H, Clarke JL (2007) Complete chloroplast genome sequences of *Hordeum vulgare, Sorghum bicolor* and *Agrostis stolonifera*, and comparative analyses with other grass genomes. Theor Appl Genet 115:571–590

Savage D, Batley J, Erwin T, Logan E, Love CG, Lim GA, Mongin E, Barker G, Spangenberg GC, Edwards D (2005) SNPServer: a real-time SNP discovery tool. Nucleic Acids Res 33:W493–W495

Schlötterer C (2004) The evolution of molecular markers – just a matter of fashion? Nat Rev Genet 5:63–69

Shang W, Zhou R, Jia J, Gao L (2010) RGA-ILP, a new type of functional molecular markers in bread wheat. Euphytica 172:263–273

Shu Y, Li Y, Zhu Y, Zhu Z, Lv D, Bai X, Cai H, Ji W, Guo D (2010) Genome-wide identification of intron fragment insertion mutations and their potential use as SCAR molecular markers in the soybean. Theor Appl Genet 121:1–8

Simola D, Kim J (2011) Sniper: improved SNP discovery by multiply mapping deep sequenced reads. Genome Biol 12:R55

Singhal D, Gupta P, Sharma P, Kashyap N, Anand S, Sharma H (2011) *In-silico* single nucleotide polymorphisms (SNP) mining of *Sorghum bicolor* genome. Afr J Biotechnol 10:580–583

Srinivas G, Satish K, Madhusudhana R, Seetharama N (2009) Exploration and mapping of microsatellite markers from subtracted drought stress ESTs in *Sorghum bicolor* (L.) Moench. Theor Appl Genet 118:703–717

Strelchenko P, Okuizumi H, Shehzad T, Malinovskaya E, Kawase M, Fukuoka S, Okuno K (2010a) Genetic relationships of sorghum germplasm in Asia and Africa revealed by rice cDNA-STS and indel markers. Jpn Agric Res Q 44:259–268

Strelchenko P, Romanova OI, Konarev AV, Okuno K (2010b) Identifying global centres of genetic diversity for grain sorghum with the use of rice DNA markers. Russ Agric Sci 36:168–171

Sundaram RM, Vishnupriya R, Biradar SK, Laha GS, Ashok Reddy G, Shobha Rani N, Sarma NP, Sonti RV (2008) Marker-assisted introgression of bacterial blight resistance in Samba Mahsuri, an elite indica rice variety. Euphytica 160:411–422

Sundaram RM, Vishnupriya MR, Laha GS, Rani NS, Rao PS, Balachandran SM, Reddy GA, Sarma NP, Sonti RV (2009) Introduction of bacterial blight resistance into Triguna, a high yielding, mid-early duration rice variety. Biotechnol J 4:400–407

Syvanen AC (2001) Accessing genetic variation: genotyping single nucleotide polymorphisms. Nat Rev Genet 2:930–942

Tamura K, Kiyoshi T, Yonemaru J (2012) The development of highly transferable intron-spanning markers for temperate forage grasses. Mol Breed 30:1–8

Tang J, Vosman B, Voorrips R, van der Linden CG, Leunissen J (2006) QualitySNP: a pipeline for detecting single nucleotide polymorphisms and insertions/deletions in EST data from diploid and polyploid species. BMC Bioinform 7:438

Tang J, Baldwin SJ, Jacobs JME, Linden CG, Voorrips RE, Leunissen JAM, van Eck H, Vosman B (2008) Large-scale identification of polymorphic microsatellites using an *in silico* approach. BMC Bioinform 9:374

Temnykh S, Declercke G, Lukashova A, Lipovich L, Cartinhour S, McCouch SR (2001) Computational and experimental analysis of microsatellite in rice (*O. sativa* L) frequency, length variation, transposon associations and genetic marker potential. Genome Res 11:1441–1452

Thiel T, Michalek W, Varshney RK, Graner A (2003) Exploiting EST databases for the development and characterization of gene-derived SSR-markers in barley (*Hordeum vulgare* L.). Theor Appl Genet 106:411–422

Tuberosa R, Salvi S, Giuliani S, Sanguineti MC, Bellotti M, Conti S, Landi P (2007) Genome-wide approaches to investigate and improve maize response to drought. Crop Sci 47:120–141

Useche FJ, Gao G, Harafey M, Rafalski A (2001) High-throughput identification, database storage and analysis of SNPs in EST sequences. Genome Inform 12:194–203

Wang XS, Zhao XQ, Zhu J, Wu WR (2005) Genome-wide investigation of intron length polymorphisms and their potential as molecular markers in rice (*Oryza sativa* L.). DNA Res 12:417–427

Wang Q, Zhang Y, Fang Z, Liu Y, Yang L, Zhuang M (2012) Chloroplast and mitochondrial SSR help to distinguish allo-cytoplasmic male sterile types in cabbage (*Brassica oleracea* L. var. *capitata*). Mol Breed 30:709–716

Weber JL (1990) Informativeness of human (dC-dA)n.(dGdT)n polymorphisms. Genomics 7:524–530

Wei H, Fu Y, Arora R (2005) Intron-flanking EST–PCR markers: from genetic marker development to gene structure analysis in *Rhododendron*. Theor Appl Genet 111:1347–1356

Wei Z, Wang W, Hu P, Lyon GJ, Hakonarson H (2011) SNVer: a statistical tool for variant calling in analysis of pooled or individual next-generation sequencing data. Nucleic Acids Res 39:e132

Wolfe KH, Gouy M, Yang YW, Sharp PM, Li WH (1989) Date of the monocot-dicot divergence estimated from chloroplast DNA sequence data. Proc Natl Acad Sci U S A 86:6201–6205

Wu Y, Li X, Xiang W et al (2012) Presence of tannins in sorghum grains is conditioned by different natural alleles of Tannin1. Proc Natl Acad Sci U S A 109:10281–10286

Xia S, Cheng L, Zu F, Dun X, Zhou Z, Yi B, Wen J, Ma C, Shen J, Tu J, Fu T (2012) Mapping of *BnMs4* and *BnRf* to a common microsyntenic region of *Arabidopsis thaliana* chromosome 3 using intron polymorphism markers. Theor Appl Genet 124:1193–1200

Yang L, Jin G, Zhao X, Zheng Y, Xu Z, Wu W (2007) PIP: a database of potential intron polymorphism markers. Bioinformatics 23:2174–2177

Ye K, Schulz MH, Long Q, Apweiler R, Ning Z (2009) Pindel: a pattern growth approach to detect break points of large deletions and medium sized insertions from paired-end short reads. Bioinformatics 25:2865–2871

Yonemaru JI, Ando T, Mizubayashi T, Kasuga S, Matsumoto T, Yano M (2009) Development of genome-wide simple sequence repeat markers using whole-genome shotgun sequences of sorghum [*Sorghum bicolor* (L.) Moench]. DNA Res 16:187–193

Zheng LY, Guo XS, He B et al (2011) Genome-wide patterns of genetic variation in sweet and grain sorghum (*Sorghum bicolor*). Genome Biol 12:R114

Part IV

Advances in Transgenic Research

Advances in Genetic Transformation 9

K.B.R.S. Visarada and N. Sai Kishore

Contents

9.1	**Introduction**	200
9.2	**Genetic Transformation**	200
9.2.1	Protoplast-Based Transformation	200
9.2.2	Particle Bombardment	200
9.2.3	*Agrobacterium*-Mediated Transformation	203
9.3	**Tissue Culture and Regeneration**	204
9.4	**Marker Genes**	209
9.5	**Economically Important Genes**	210
9.6	**Transformation Efficiency**	211
9.7	**Future Prospects**	212
References		212

Abstract

Transgenic sorghum has trailed behind other cereals in progress due to tissue culture limitations, lack of model genotypes, low regeneration, and lack of sustainability of regeneration through sub-cultures. Particle bombardment and *Agrobacterium*-mediated methods are frequently preferred methods for production of transgenic sorghum. Immature embryos and shoot apical meristems are the most suited as target material for genetic transformation. Transformation efficiency is improved through tailored in vitro protocols in desirable genotypes. Many agronomically important traits were introduced in sorghum genotypes to improve quality of grain and forage and to increase resistance to biotic and abiotic stresses. Despite several improvements in transgenic technology and its application for sorghum crop improvement, so far there are no reports on the release and cultivation of transgenic sorghum. Deployment of innovative genetic modification technologies that can keep away from GMO classification and biosafety concerns in sorghum can benefit the producers and consumers of sorghum.

Keywords

Sorghum • Transgenics • Tissue culture • Regeneration • Marker genes

K.B.R.S. Visarada (✉)
Genetics & Cytogenetics, ICAR-Indian Institute of Millets Research, Rajendranagar, Hyderabad, Telangana 500 030, India
e-mail: visarada@millets.res.in

N. Sai Kishore
GITAM Institute of Science, GITAM University, Visakhapatnam, Andhra Pradesh, India

R. Madhusudhana et al. (eds.), *Sorghum Molecular Breeding*, DOI 10.1007/978-81-322-2422-8_9

9.1 Introduction

Success in developing improved cultivars by genetic engineering requires an efficient gene transfer, stable integration, and predictable expression of the transgene. With the advent of genetic transformation techniques based on recombinant DNA technology, it is now possible to insert genes that confer resistance to a number of biotic stresses and to several abiotic factors into the plant genome efficiently which in turn improve the yield. Agronomically useful genes available across genera can be incorporated into the sorghum genome through gene transfer techniques along with reproducible tissue culture protocols to produce transgenic sorghum plants with enhanced yield and nutritional quality.

Though significant progress has been achieved in genetic transformation in cultivable crops, there have been very few reports on sorghum. Sorghum has trailed behind other cereals in the progress toward genetic transformation due to limitations in tissue culture, such as low regeneration frequency and accumulation of phenolic pigments. Being the most recalcitrant crop for tissue culture, regeneration, and genetic transformation, sorghum has lagged behind in the application of transgenic approach for genetic improvement as compared to other cereal crops. To date, there is no transgenic sorghum under commercial cultivation. Attempts were made in sorghum to transform with marker, selectable, and agronomically useful genes. The summary of reports on genetic transformation of sorghum is presented in Table 9.1. Despite several advancements in tissue culture techniques of sorghum, the genetic transformation is by no means either routine or easy.

9.2 Genetic Transformation

Three different methods of genetic transformation have been reported in sorghum, *viz.*, protoplast-mediated transformation, particle bombardment, and *Agrobacterium*-mediated transformation.

9.2.1 Protoplast-Based Transformation

The first report of genetic transformation of sorghum described the introduction of DNA into protoplasts by electroporation and selection of transformed cells, without achieving plant regeneration (Battraw and Hall 1991). Parameters influencing the stable transformation of sorghum protoplasts with a chimeric neomycin phosphotransferase II (*nptII*) and *β-glucuronidase* genes by electroporation were investigated. Sorghum cell suspensions can be established initially, but they do not sustain regeneration for longer periods. The cells tend to elongate and lose their regeneration potential making isolation and regeneration of protoplasts difficult. Limitations of protoplast method are that it is laborious, needs special skills, is genotype specific, and has low regeneration ability. To overcome these difficulties, leaf mesophyll tissues were used for isolation of protoplasts in dicots. Sairam et al. (1999) isolated protoplasts and regenerated plants from protoplasts isolated from leaf tissues of sorghum seedlings. However, further work on mesophyll protoplasts and genetic transformation was not followed up.

9.2.2 Particle Bombardment

The biolistic approach has been used extensively for the gene transfer in most of the monocot crops, including sorghum owing to its advantages over *Agrobacterium*-mediated approach. Hagio et al. (1991) demonstrated the first stable expression of the *hph* and *nptII* genes that conferred resistance to selectable levels of hygromycin and kanamycin, respectively. Their results extended the utility of the biolistic method as a useful DNA delivery system for the transformation of sorghum. The first transgenic sorghum plants were obtained by Casas et al. (1993, 1997) through microprojectile bombardment of immature embryos and immature inflorescence-derived calli, but with a relatively low transformation frequency

Table 9.1 Genetic transformation studies in sorghum

Target gene (s)	Transformation method	Explant	Genotype	Transformation efficiency (%)	Reference
gus, nptII	E	P	NK300	–	Battraw and Hall (1991)
gus, hph, nptII	PB	IE[a]	–	–	Hagio et al. (1991)
bar, gus	PB	IE	P898012	0.08	Casas et al. (1993)
bar, gus	PB	IF	P898012	0.33	Casas et al. (1997)
bar, chiII	PB	IE	Tx430, SRN 39	0.09	Zhu et al. (1998)
bar, gus	AT	IE	P898012, PHI391	2.1	Zhao et al. (2000)
bar, gfp	PB	IE	P898012	1.0	Able et al. (2001)
bar, chiII	PB	IE	Tx430, SRN39	–	Krishnaveni et al. (2001)
bar, gus	PB	IE	Tx430	0.18	Emani et al. (2002)
bar, gus	PB	IE	214856, 213108	1.3	Tadesse et al. (2003)
gfp, G11, tlp	AT	IE	Tx430, C401, Wetland	–	Jeoung et al. (2002)
gfp, tlp	AT	IE	Tx430, C401	2.5	Gao et al. (2005a)
*man*A	AT	IE	P8505, C401	2.88, 3.3	Gao et al. (2005b)
Bt cry1Ac	PB	SA	BTx623	1.5	Girijashankar et al. (2005)
nptII, gusplus	AT	IE	Tx430, C2-97	0.3–4.5	Howe et al. (2006)
gus, hpt	AT	IE	Sensako 85/1191	5.0	Nguyen et al. (2007)
*Sgfp, man*A	AT	IE	P898012	7.7	Gurel et al. (2009)
bar, gus	AT	IE	P898012	0.4, 0.7	Lu et al. (2009)
mtlD	PB	SM	SPV462	4.0–7.0	Maheswari et al. (2010)
hpt	PB	IE	Ramda	0.09	Raghuwanshi and Birch (2010)
bar, pmi	PB	IE	P898012	0.77	Grootboom et al. (2010)
gus, hpt	AT	IE	P898012, RTx430	1.1–7.2	Kumar et al. (2011)
nptII, gfp	PB	IE	Tx430	20.7	Liu and Godwin (2012)
Bt cry1Aa, cry1B	PB and AT	IE	CS3541, 296B, SSV84, RSSV9	0.1–0.3	Visarada et al. (2014)
moPAT, pmi, *dsRED*	AT	IE	Tx430	8–13.4	Wu et al. (2014)
gus	AT	IE	P898012	–	Urriola and Rathore (2014)

[a]Reported from a suspension culture of *Sorghum vulgare*

bar phosphinothricin acetyltransferase, *chiII* rice chitinase, *gfp* green fluorescence protein, *gus* β-glucuronidase, *hph* hygromycin phosphotransferase, *mtlD* mannitol-1-phosphate dehydrogenase, *manA/pmi* phosphomannose isomerase, *moPAT* codon-modified phosphinothricin acetyltransferase gene, *nptII* neomycin phosphotransferase II, *tlp* thaumatin-like protein

E electroporation, PB particle bombardment, AT *Agrobacterium*-mediated transformation, P protoplast, IE immature embryos, IF inflorescences, SA shoot apices, SM shoot meristem

(0.2 %), a protracted time in culture (7 months), in the genotype, PI898012, which has poor agronomic traits.

Maize genes encoding anthocyanin transcription factors (R and C1) were used by Casas et al. (1993) to optimize DNA delivery parameters in sorghum immature embryos. Frequency of transient expression of *gus* was less than 20 blue foci/embryo, which indicated that transient gene expression level in sorghum was lower than in maize due to genotype effect or interactions between genotype and acceleration pressure or other inherent characteristics of sorghum scutellar tissue. Immature embryos and immature inflorescences were used as explants by Kononowicz et al. (1995) for particle bombardment and plant regeneration. Genotype-specific differences in the response of primary explants to regeneration protocols have been found among sorghum genotypes studied and the genotypes PS98012 (immature embryos) and SRN39 (immature inflorescence) were found to be promising.

Optimization of transformation conditions, parameters for microprojectile bombardment, and strength of promoters were reported by Tadesse et al. (2003) in the genotype 214856, which is capable of producing a good-quality callus and less susceptible to phenolic pigments during selection on geneticin. Transgenic sorghum was produced by the combined use of optimized bombardment conditions, strong monocot gene promoters, and stepwise antibiotic selection. Physical and biological parameters that give the highest transient expression of the introduced *gus* reporter gene without compromising the frequency of somatic embryogenesis and regeneration capacity were studied. Variation in transient gene expression level was observed with different types of explants, and with different pressures. It was concluded that immature embryos and shoot tips were the best explants to target potential progenitor cells that are competent for embryogenesis. Evaluation of transient expression of the *gus* reporter gene under the control of a number of parameters using four different promoters revealed that the activity of all three monocot promoters (*ubi*1, *act*1, and *adh*1) was higher in sorghum than that of the CaMV 35S promoter (Tadesse et al. 2003). However, maize *ubi*1 and rice *act*1 gene promoters were preferred for stable expression of foreign genes in sorghum immature embryo and shoot tip explants. The transformation frequency ranged between 0.5 and 1.3 % for shoot tips and immature embryos, respectively. Very slow or no root formation was observed on shoots regenerated under selection on phosphinothricin or geneticin.

A low incidence of transgenic sorghum has been obtained using particle bombardment of callus explants of the genotypes, P898012 (Casas et al. 1993; Rathus et al. 2004), SRN 39 (Casas et al. 1997) and Tx430 (Zhu et al. 1998). This indicates that an efficient genotypes-independent plant regeneration system is vital for the development of transformation protocols for sorghum. Sorghum line, P898012, is reported by many because it is capable of producing a good/better embryogenic, quality callus from immature embryos and was very responsive to coconut water (Casas et al. 1993; Kaeppler and Pederson 1997; Carvalho et al. 2004). Frequently reported sorghum genotypes for successful transformation are P898012, Tx430, and C401 (Table 9.1). Immature embryos are the most preferred target tissues in sorghum. However, they are not available round the year due to the photoperiodic sensitivity of many agronomically important genotypes. Improved methods for rapid isolation and culture of immature embryos can aid in large-scale transformation experiments (Raju et al. 2007). Multiple shoot buds induced from shoot apical meristems are promising target tissues for genetic transformation, which mitigate the issues of regeneration and availability round the year. Callus cultures derived from immature inflorescences in sorghum, though have high regeneration at 30–50 days, decrease rapidly further making the recovery of transgenic plants a challenge. A highly efficient microprojectile transformation system for sorghum has been developed by Liu and Godwin (2012) using immature embryos of inbred line Tx430. After optimization of tissue culture media and parameters of microprojectile transformation, 25 independent transgenic events were obtained from 121 bombarded immature embryos with an average transformation frequency of 20.7 % in three independent experiments, the highest frequency reported so far.

Use of the cytidine analog, 5-azacytidine (azaC), in reversing the methylation-mediated transgene silencing in sorghum was demonstrated by Emani et al. (2002). It was possible to activate *gus* gene expression in T_1 seedlings and in calli derived from immature T_1 and T_2 embryos by the treatment of 5-azacytidine (azaC). The investigators suggested that methylation-based silencing is frequent in sorghum and probably responsible for several cases of transgene inactivation reported earlier in sorghum.

9.2.3 *Agrobacterium*-Mediated Transformation

Agrobacterium-mediated transformation is the most preferred method among the researchers due to its advantages in the production of transgenic plants with single or low-copy inserts. The first successful report in sorghum transformation using this method was reported by Zhao et al. (2000) using the public line P898012, in which four factors that influence sorghum transformation the most were identified, *viz.*, (1) sensitivity of immature sorghum embryos to *Agrobacterium* infection, (2) growth conditions of donor plant, (3) type of explant, and (4) cocultivation medium. Necrotic response in explants after cocultivation is a critical factor to improve the transformation efficiency. Immature embryos of sorghum proved to be very sensitive to *Agrobacterium* infection, and it was found that the level of embryo death after cocultivation was the limiting step in improving transformation efficiency. Hence, attention should be given to the number of *Agrobacterium* cells in the inoculum, selection of sorghum genotypes, and explants less sensitive to *Agrobacterium* infection. Increased percentage of embryos that formed callus (recovery of callus), reduced pigment production, and improved callus growth were observed by the addition of coconut water to the medium (Carvalho et al. 2004). Though the genotype P898012 is responsive to coconut water, its use in other genotypes may require different composition, since many genotypes did not respond to coconut water (Kaeppler and Pederson 1996). Various explants were used by Visarada et al. (2003) to explore genetic transformation of sorghum using *Agrobacterium*-mediated transformation resulting in transient *gus* expressions. However, immature embryos and calli derived from immature inflorescences were found to be ideal target tissues for obtaining high *GUS* expression. In a recent study by Wu et al. (2014), high transformation frequencies (10–33 %) were obtained through elevated copper sulfate and 6-benzylaminopurine in the resting and selection media by *Agrobacterium* infection of immature embryos using the genotype Tx430.

Optimal conditions were determined and baseline conditions in subsequent experiments were provided by Zhao et al. (2000) to achieve stable transformation of sorghum using *Agrobacterium*. It was observed that N_6 medium not only decreased the callus response from embryos (76 % with MS and 20 % with N_6) but also increased the production of phenolic pigments. The overall transformation frequency was 2.1 %. The optimization of media and other conditions for the transformation of sorghum using *Agrobacterium* was reported and a transformation frequency of 2.1 % was achieved. However, acetosyringone concentration was not optimized since a single concentration (100 μM) was used to induce *vir* operon.

Addition of amino acid L-cysteine in the plant tissue culture medium aided the recovery of putative transformants and improved the transient *gus* expression and stability of transformed explants (Sai Kishore et al. 2004). Production of transgenic sorghum plants through tailored in vitro protocols is possible in choice genotypes. However, decontamination of *Agrobacterium* by employing subtle treatments aided the recovery of transgenic plants in recalcitrant genotypes. In sorghum the resistant calli obtained after selection did not regenerate probably due to the toxic effects of decontaminating agents and selection pressure (Sai Kishore et al. 2011). Moreover, sweet sorghums secrete polyphenols into the culture medium leading to decrease in regeneration, which was overcome by frequent sub-culturing in the initial stages of post-transformation.

In planta method of *Agrobacterium*-mediated transformation helps in overcoming the hurdle of regeneration after selection. This method was demonstrated in sorghum by Elkonin et al. (2009) through the generation of transgenic plants and inheritance of the transgene to T_1 generation. Our research group developed a simplified method of *in planta* genetic transformation by a floral-dip method in sorghum, which does not involve vacuum infiltration and uses *Agrobacterium* suspension derived from solid cultures. Two independent transgenic sorghum lines were developed in sweet sorghum genotypes and the gene integration and expression were studied till T_5 generation. However, the expression of *Bt* protein in *in planta* derivatives was low. *In planta* method is simple, easy, and economic without the problem of regeneration after selection. Supplementation of agro-infiltration liquid with special components like L-cysteine and Tween-20 had a promotive effect on transformation (unpublished results).

Parameters were optimized for transformation by *Agrobacterium*-mediated and particle bombardment by Prasad Sant (2011) using the sorghum cultivars, *viz*., SA281, 296B, and Tx430. *Agrobacterium* strain LBA4404 at an inoculum density of 0.5 OD_{600nm}, heat treatment at 43 °C for 3 min., inoculation media (pH 5.2) with the surfactant Pluronic F-68 (0.02 % w/v), and a 3-day cocultivation period in dark at 22 °C was found to be optimum resulting in high frequencies of transient GFP expression in immature embryos and callus derived from immature inflorescences. The optimum conditions for particle bombardment, *viz*., use of 3–7 day-old immature embryos and 4 week-old callus from immature inflorescences, pre- and post-bombardment osmoticum treatment of 4 h and 0.6 μm gold microparticles, 1,500 kPa helium pressure, and a target distance of 15 cm, resulted in transient GFP expression for up to 14, 30, and 50 days for SA281, 296B, and Tx430, respectively. Though particle bombardment resulted in less tissue necrosis compared to *Agrobacterium*-mediated transformation, no stable transformed plants were regenerated.

Both the systems of transformation, i.e., *Agrobacterium* and particle bombardment, though successful in sorghum, have their own merits and limitations. With *Agrobacterium-based* transformation, the higher transformation efficiency achieved so far ranged from 2.1 to 4.5 % (Howe et al. 2006; Gao et al. 2005b; Zhao et al. 2000). Effective method till date remains to be the microprojectile bombardment with higher transformation efficiency of 20.7 % (Liu and Godwin 2012).

In addition to the above mentioned methods of genetic transformation, pollen-mediated method was reported in sorghum by Wang et al. (2007), in which the plasmid DNA and pollen were submerged in a 0.3 mol/l sucrose solution, subjected to ultrasonication and then used for pollination of stigmas from the male sterile line A2V4A. Confirmation of gene integration was supported by the evidence of Southern blotting.

9.3 Tissue Culture and Regeneration

A wide range of sorghum explants, either for somatic embryogenesis or organogenesis, have been tested to establish a dependable tissue culture system for regeneration, which include mature embryos, immature embryos, immature inflorescences, seedlings, leaf fragments, and anthers. However, genetic transformation achieved in sorghum so far has been established with embryogenic cultures initiated from immature embryos or immature inflorescences. Maintenance of consistent quality of the callus on medium for an extended time (2–3 months or longer) and regeneration of fertile plants are essential for successful plant transformation.

MS medium supplemented with different combinations and concentrations of plant growth hormones were used by Pola et al. (2008) to culture immature embryos. Embryogenic callus was initiated on MS medium supplemented with 2 mg l^{-1} 2,4-D and 0.5 mg l^{-1} kinetin; the addition of kinetin significantly enhanced embryogenesis. Proficiency in the development of embryogenic callus, induction of somatic embryo, and shoot regeneration was observed with 2 mg l^{-1} BAP. The regenerated shoots readily rooted on half-strength MS medium containing 1 mg l^{-1} NAA and were successfully transferred to the soil, which subsequently produced the seeds. Since the regeneration frequency of the bombarded calli derived from immature inflorescence is low, it is the least preferred explant (Kononowicz et al. 1995; Casas et al. 1997; Jogeswar et al. 2007). High-quality callus from immature inflorescence of sorghum genotypes is high, but the efficiency is still very low. It was improved to 1.01 to 3.33 % by modification of bombardment parameters (Brandao et al. 2012).

An attempt was made by Prasad Sant (2011) to establish an efficient and reproducible tissue culture and transformation system using five different sorghum cultivars, *viz.*, SA281, 296B, SC49, Wray, and Rio. Regenerable embryogenic cell lines could be established only from SA281 and 296B. In addition, embryogenic cell lines were established using florets of immature inflorescence as explants from SA281, 296B, as well as Tx430 and regenerated using sorghum callus induction media (SCIM) and Vam's wonderful regeneration media (VWRM) with the inflorescences from plants at the FL-2 stage (where the last fully opened leaf was two leaves away from the flag leaf) giving the best in vitro response. Moreover, the responses of immature inflorescences were robust in tissue culture and independent of season and growth condition. The optimum combination of plant growth regulator standardized for the micro-propagation of in vitro regenerated plantlets was 1.0 mg l^{-1} BAP and 0.5 mg l^{-1} NAA.

Apart from immature embryos and inflorescences, shoot tips from germinating seedlings are also widely used as explants in sorghum transformation (Girijashankar et al. 2005; Tadesse et al. 2003). Recently, an efficient protocol for regeneration in sorghum through somatic embryogenesis from shoot tip explants was developed by Amali et al. (2014) with the highest frequency of embryogenic callus formation (99 %), when explants were cultured on MS medium supplemented with 2.5 mg l^{-1} 2,4-D, 0.25 mg l^{-1} kinetin and 500 mg l^{-1} of casein hydrolysate. Sub-culturing of embryogenic callus on MS medium supplemented with 2.5 mg l^{-1} 2,4-D, 0.25 mg l^{-1} kinetin, 500 mg l^{-1} of casein hydrolysate, and 500 mg l^{-1} of L-proline resulted in the highest mean number of somatic embryos (33.3). The highest regeneration of plantlets (21.4 per embryogenic callus) was obtained in MS medium supplemented with 4 mg l^{-1} benzylaminopurine. A maximum number of roots (12.4) and root length (5.7 cm) were observed in half-strength MS medium supplemented with 1.0 mg l^{-1} indole-3-butyric acid and 0.8 g l^{-1} activated charcoal. The survival rates of in vitro grown plantlets transferred to greenhouse were up to 70 %, which were morphologically similar to in vivo plants.

Explants derived from meristematic tissues at the early stages of development are most amenable to tissue culture conditions (Sai Kishore et al. 2006). Highly uniform meristematic tissues are desirable for genetic transformation to minimize chimeras and somaclonal variants. Cultured immature embryos and shoot tips are the two explants of choice that have been predominantly used for sorghum genetic transformation. Two parental lines (CS3541 and 296B) that are less responsive to tissue culture were transformed by Visarada et al. (2014) by identifying the critical factors like 3d pre-cultured immature embryos as target tissues, small-size explants, treatment with L-cysteine during cocultivation, recovery of transformed embryos after antibiotic washes, and regeneration *via* multiple shoot bud induction.

The supply of immature inflorescences and embryos is not available throughout the season as sorghum flowering occurs for a few days only, thereby providing a small window for the collection of explants. To overcome this problem, Pola (2011) used leaf discs that will be available at any season for optimizing callus induction and regeneration. Efficient callus induction was obtained on MS media supplemented with 2 mg l^{-1} 2,4,5-T and 1 mg l^{-1} NAA and 0.5 mg l^{-1} zeatin, while better shoot regeneration (62.2 ± 4.6 shoots per explants) was achieved in MS medium supplemented with 2.5 mg l^{-1} thidiazuron + 1.0 mg l^{-1} BAP and 0.5 mg l^{-1} IAA. Similarly, better root induction was observed with 1.0 mg l^{-1} NAA followed by their transfer to half-strength MS medium. Various reports on tissue culture techniques and different explants with regeneration ability in sorghum are presented in Table 9.2.

Table 9.2 Summary of tissue culture studies in sorghum

Explant	Experiment	Conclusion	Reference
Immature embryos	Callus induction medium – MS medium supplemented with L-proline, L-asparagine, potassium dihydrogen phosphate, $CuSO_4$, and 2,4-D	Optimized callus induction, regeneration, and rooting medium to reduce phenolic production.	Liu and Godwin (2012)
	Regeneration medium – MS medium supplemented with BAP, IAA, and $CuSO_4$		
	Rooting medium – MS medium supplemented with NAA, IAA, IBA, and $CuSO_4$		
Germinated seeds	Shoot induction on MS medium supplemented with different concentrations of IAA, BAP, and kinetin	Developed an efficient regeneration system using two cultivars Yuantian No. 1 and M81E	Zhao et al. (2010)
Immature embryos	Regenerated callus using MS medium containing 2 mg l^{-1} 2, 4-D, 0.5 mg l^{-1} kinetin, 10 mg l^{-1} $AgNO_3$, 400 mg l^{-1} casein hydrolysate, and 200 mg l^{-1} each of L-asparagine and L-proline	Developed a protocol for long-term maintenance of callus cultures and succeeded in maintaining the embryogenic callus cultures up to 57 weeks	Pola et al. (2009)
Immature embryos	Heat treatment of immature embryos (IEs) at various temperatures for 3 min prior to *Agrobacterium* infection	Optimized a 3-min heat treatment at 43 °C prior to infection. Both heat and centrifugation increased dedifferentiation of tissue	Gurel et al. (2009)
Immature zygotic embryos	Callus initiation and regeneration potential of five sorghum genotypes on specific nutrient media	Identified the best genotype, nutrient medium combination for satisfactory regeneration	Grootboom et al. (2008)
Leaf	Shoot regeneration using different growth hormonal combinations	The highest number of somatic embryos was produced from leaf segments on MS medium supplemented with 2.0 mg l^{-1} 2,4,5-trichloro acetic acid and 1.0 mg l^{-1} zeatin	Pola et al. (2007)
Immature inflorescence	Effect of various growth regulators on somatic embryogenesis in three genotypes (SPV462, SPV839, and M35-1)	High frequency of somatic embryogenesis was obtained on MS medium supplemented with 2 mg l^{-1} 2,4-D and 0.5 mg l^{-1} kinetin	Jogeswar et al. (2007)
Immature embryos	Cold pretreatment of the immature seeds from which embryo explants were excised and by the use of activated charcoal	Developed an improved regeneration protocol suitable for transformation by limiting the production of phenolic compounds and the use of suitable culture vessels for each developmental stage in plant regeneration	Nguyen et al. (2007)

(continued)

Table 9.2 (continued)

Explant	Experiment	Conclusion	Reference
Shoot tip	Evaluated 24 sorghum genotypes and 3-day- and 5-day-old shoot tips for tissue culture response	Age of the seedlings in relation to the number of actively dividing cells that contain greater potential for in vitro response was reported for the first time using electron microscopy	Saikishore et al. (2006)
Immature embryos Immature inflorescences	A factorial experiment was conducted with a combination of 2,4-D and kinetin in 8 genotypes of *S. sudanese* and *S. bicolor*	Genotypic limitation of plant regeneration can be overcome by the addition of kinetin in callus induction media. Immature inflorescences regenerated at a higher frequency (2.71 shoot callus^{-1}) than immature embryos (1.26 shoot callus^{-1})	Gupta et al. (2006)
Shoot tip	Somatic embryogenesis and multiple shoot clumps in Indian sorghum genotypes	Somatic embryogenesis was reported by combination of 2,4-D and 6-benzyl adenine. Multiple shoot clumps were produced using BAP alone in 1-week-old shoot tips	Syamala and Devi (2003)
Shoot meristem	Multiple shoot bud induction	Multiple shoot bud induction on enlarged meristems was achieved using higher concentrations of TDZ, BAP, and NAA	Harshavardhan et al. (2002)
Immature embryo	Eleven genotypes of sorghum were experimented for their response in tissue culture	Adventitious shoot regeneration from immature embryos was achieved	Hagio (2002)
Immature embryo	Influence of phytohormones, $AgNO_3$, and maltose/sucrose as carbon source on callus induction	Callus induction with respect to regeneration was observed only by the modification of phytohormones alone. Fertile plants at high efficiency were obtained	Oldach et al. (2001)
Shoot tip	LS medium with 2,4-D and kinetin was used for induction of friable callus MS media with BAP and IAA was used for germination of somatic embryos	Somatic embryogenesis was observed in shoot tip-derived callus by the incubation of shoot tip for 6–7 weeks	Seetharama et al. (2000)
Young panicle Immature embryo	Explants were cultured on modified MS and N_6 media supplemented with Asn and Pro with different conc. of NO_3^-, NH_4^+, and PO_4^{3-}	Level of NO_3^- and NH_4+ ratio was the critical factor for the formation of embryogenic calli. Increase of the PO_4^{3-} level increased the regeneration ability	Elkonin and Pakhomova (2000)
Shoot meristem-derived callus	Regeneration of wild sorghum from suspension cultures with shoot meristem-derived callus	High frequency of somatic embryogenesis (80 %) was obtained from small clusters (300–400 wm) maintained in liquid medium with 0.25 mg l^{-1} 2,4-D	Mythili et al. (1999)

(continued)

Table 9.2 (continued)

Explant	Experiment	Conclusion	Reference
Shoot tip	Multiple shoot induction	Multiple shoot induction using BAP and 2,4-D was achieved in 18 genotypes	Zhong et al. (1998)
Immature inflorescence	For identification of high-quality callus producing genotypes, 41 diverse inbred sorghum lines were tested	Seven elite inbred genotypes and five non-elite genotypes were selected for high-quality callus based on mean rating	Kaeppler and Pederson (1997)
Seedlings	Importance of kinetin in induction of embryogenic calli from seedlings through transverse thin cell layers	Callus induction, somatic embryogenesis, and plant regeneration were reported in two cultivars from epicotyls of 15-day-old seedlings. Callus response is dependent on genotype	Gendy et al. (1996)
Immature embryo	Cefotaxime, asparagine, and praline effects on production of embryogenic callus and to enable the frequency of plant regeneration	Even though these compounds did not promote callus induction or growth of callus, they influenced the plant regeneration in sweet sorghum genotypes with high anthocyanin	Rao et al. (1995)
Immature inflorescence	Eight genotypes tested by modification of MS medium for induction and regeneration of cultures	Genotypic differences were observed in pigment production, embryogenic callus formation, shoot differentiation, and maintenance of regeneration capacity. Embryogenic calli were formed at a frequency of 8–70 % depending on the genotype	Cai and Butler (1990)
Immature inflorescence and shoot tips	Callus induction and plant regeneration	Immature inflorescence followed by shoot tips favor good callus induction and plant regeneration	Murty et al. (1990a, b)
Seedlings	Plant regeneration from 4–5-month-old callus cultures	Somaclonal variation was examined in the SC2 and SC3 generation of eight different clones and one non-tissue-cultured parental line	Bhaskaran et al. (1987)
Mature embryos	Callus and plant regeneration were induced from shoot portions of mature embryos of 5 high-tannin cultivars	Plants were regenerated on MS medium supplemented with asparagine and kinetin with a regeneration frequency of 11–48 %	Cai et al. (1987)
Anthers	Microspore-derived callus production at different temperatures	Callus was produced only from anthers incubated at 33 °C	Rose et al. (1986)
Seedlings	Sorghum plant regeneration from aluminum selection medium	Callus cultures of germinated seed	Smith et al. (1983)

9.4 Marker Genes

The most widely used selectable markers in cereal transformation are the genes encoding hygromycin phosphotransferase (*hpt*), phosphinothricin acetyltransferase (*pat* or *bar*), and neomycin phosphotransferase II (*nptII*). Use of these marker genes under the control of constitutive promoters such as CaMV 35S promoter or the *ubi*1 promoter from maize works as efficiently for selection of *Agrobacterium-transformed* cells as for biolistic-mediated transformation. The positive selectable marker *pmi* (*phosphomannose isomerase*) has been shown to be effective in the transformation of many monocots including sorghum (Gao et al. 2005a). Expression of green fluorescence protein (*gfp*) on sorghum regeneration has been reported by Able et al. (2001) using particle bombardment, in which the distance between the rupture disk and target tissue, helium inlet aperture and pressure of helium gas, and age of tungsten and spermidine were studied. The sorghum genotype Tx430 yielded the most foci/callus in transient expression, while C401 was the least amenable for stable transformation. The use of *gfp* as a reporter for optimizing the transient expression during *Agrobacterium*-mediated transformation of sorghum was demonstrated by Jeoung et al. (2002). Using two different reporter genes, the suitability of different inbred lines (Tx430, C401, CO25) was determined using different promoters, type of explants, and inbreds during the early transformation process involving both biolistic and *Agrobacterium*-mediated transformation. The results indicated that *gfp* can be used effectively as a reporter for optimizing the conditions for successful transient expression during transformation with *Agrobacterium*.

Grain sorghum was transformed by Gao et al. (2005a) with a visual reporter gene (*gfp*) and a target gene (*tlp*), encoding thaumatin-like protein, and they reported the successful use of GFP screening for efficient production of stably transformed sorghum plants without using antibiotics or herbicides as selection agents. Transformation efficiency of 2.5 % was observed, which was greater than that reported earlier by Zhao et al. (2000). A dual-marker plasmid containing the selectable marker gene, *man*A, and the reporter gene, *sgfp*, was used to transform immature sorghum embryos by employing *Agrobacterium-mediated* transformation system (Gao et al. 2005b). Both the genes were under the control of a maize *ubi*1 promoter. The phosphomannose isomerase gene *pmi* was isolated from *E. coli* and used as the selectable marker gene and mannose was used as the selection agent. Necrotic calli were rarely observed in mannose selection and it had less negative effects on plant regeneration. Gene silencing of either the *gfp* gene or the *tlp* gene in T_0 and T_1 generations was not observed by Gao et al. (2005a, b). Stable transformation experiments in sorghum immature embryos of Tx430 and C2-97 genotypes were carried out by Howe et al. (2006) using a novel strain of *A. tumefaciens* (C58) implementing *nptII* as a selectable marker, resulting in transformation frequencies in the range of 0.3–4.5 % with an average transformation frequency of ~1 %.

In most reports on genetic transformation of sorghum, it was observed that the expression of introduced *gus* gene was either very poor or totally absent. GUS enzyme activities were very low in sorghum cells compared to other *gus* gene-transformed monocot cells (Hagio et al. 1991). The majority of the *gus* transformed cells did not stain blue upon incubation with histochemical substrate X-Gluc (Battraw and Hall 1991). Moreover, GUS activity that was high in transient assays could not be detected later than 3 weeks after bombardment (Casas et al. 1993, 1997) suggesting that methylation of transgene might have occurred in sorghum cells, leading to the inhibition of the expression of the reporter gene. Jeoung et al. (2002) studied the optimization of parameters for use of *gfp* and *uid*A as visual markers and evaluated different promoters. It revealed the order of promoter strength for GUS expression as *ubi*1 > CaMV 35S > *act*1 > *adh*1. Another reporter gene, *luc*, a firefly luciferase, was used in sorghum transformation studies by Kononowicz et al. (1995).

Five different selection markers, which include *cat*, *npt II*, *hpt*, *bar*, and *man*A, representing three broad categories of selection markers (antibiotic resistance, herbicide resistance, and nutrient assimilation), were utilized in sorghum transformation. The conversion of mannose to a metabolizable six-carbon source is beneficial to plants and is an efficient and non-destructive method of screening the transformed sorghum plants under in vitro conditions. Co-bombardment was performed by Liu and Godwin (2012) with *nptII* and *gfp* genes, both under the control of the maize *ubi*1 promoter using immature embryos of sorghum resulting in high transformation efficiency.

9.5 Economically Important Genes

Transgenic technology in sorghum is extended to combat biotic and abiotic stresses and to improve the quality. Agronomically important gene, *chi ll*, encoding rice chitinase under the constitutive CaMV 35S promoter was transferred to sorghum to impart resistance to stalk rot (*Fusarium thapsinum*) by Zhu et al. (1998) and Krishnaveni et al. (2001). Transgenics for resistance to anthracnose were reported by Kosambo-Ayoo et al. (2011). Transgenic sorghum plant expressing *cry1Ac* gene under the control of the wound-inducible promoter for the protection against spotted stem borer was reported in the cultivar BTx623 through microprojectile bombardment of shoot apices after selection and regeneration through embryogenic pathway (Girijashankar et al. 2005). The study showed that the *mpiC1* promoter from maize is functional in sorghum and drives the expression of *cry1Ac* gene, but at low levels to confer partial protection against the neonate larvae of the spotted stem borer. In another study, the T_0 transgenic plants generated using *ubi1–cry1Ab* and *ubi1–cry1Ac* were found to be chimeric in nature, which may be due to the lack of the transgene *cry* in the reproductive parts (Girijashankar and Swathisree 2009). Promising *Bt* transgenic possessing *cry1Aa* gene conferring resistance to stem borer in the grain sorghum genotype M 35-1 was reported by Ratnala (2013). Transgenic sweet sorghum derivatives through backcross breeding using a promising transgenic event in the genotype M 35-1 were developed recently by Indukuri (2014), which showed promising levels of resistance in whole-plant bioassays for *Chilo partellus*. Transgenic sweet sorghum plants in the genetic background of SSV84 and RSSV9 were generated through particle bombardment and *in planta* method of *Agrobacterium*, which exhibited moderate to low levels of resistance in insect feed assays in the laboratory (Visarada et al. 2013). Transgenic sorghum expressing two different classes of *Bt* toxin proteins, Cry1Aa and Cry1B, in two elite and recalcitrant genotypes (CS3541 and 296B) was developed, which showed high levels of expression (35–500 ng/g protein) with commensurate resistance to stem borer (Visarada et al. 2014). Transgenic sorghum plants possessing *cry1C* gene showed 10–13 % leaf damage and 97–100 % larval mortality in the insect bioassays with *Chilo partellus* neonate larvae (Ignacimuthu and Premkumar 2014). Promising transgenic lines were identified in the field trials for resistance to stem borer by Balakrishna et al (2013).

High-lysine sorghum lines were generated by Tadesse et al. (2003) in the genetic background of the Ethiopian genotype 214856 through the over-expression of *dhdps-raec*1 mutated gene, which encodes the insensitive form of the *dihydrodipicolinate synthase*, the key regulatory enzyme of the lysine pathway. Marker-free transgenic sorghum plants harboring *lysyl tRNA synthetase* gene for enhanced lysine content in sorghum seed was developed by Lu et al. (2009). Contrary to the expectation, introduction of glycine-rich RNA-binding protein gene atRZ-1a from *Arabidopsis*, Bcl-2 mRNA sequence 725–1,428 representing the 3′ non-coding region of the gene from humans and rice Ca-dependent protein kinase 7 (OsCDPK7) could not improve the cold tolerance in sorghum (Mall 2010). Transgenic sorghum events expressing HMW-GS showed improvement in protein digestibility of the uncooked ground grain, and the downregulation of alpha kafirin showed the presence of distorted protein bodies in the transgenic seed (Mall 2010).

The lysine content of sorghum grain was increased by Tadesse and Jacobs (2004) through the introduction of a mutated *dhdps-rl* gene encoding a feedback-insensitive *dihydrodipicolinate synthetase* enzyme by microprojectile bombardment of immature embryos and shoot tips leading to accumulation of more amount of lysine. The *dhdps-raec1* mutated gene encodes an insensitive form of the *dihydrodipicolinate synthase*, the key regulatory enzyme of the lysine pathway. The over-expression of this gene could lead to the elevated lysine content in sorghum and improvement of nutritional quality of this crop. Lu et al. (2009) reported transgenic sorghum plants harboring a modified tRNAlys, and sorghum lys1 tRNA synthase elements for improving the lysine content in sorghum seeds. However, there was no mention on the expression of the lysine gene or amino acid content.

To impart resistance to abiotic stresses like drought, *HVA1* gene from barley was inserted into the sorghum genome through biolistic transformation by Devi et al. (2004), which accumulated barley 3 LEA protein under induced stress. Sorghum line SPV462 was transformed with the *mtlD* gene encoding for *mannitol-1-phosphate dehydrogenase* from *E. coli* to enhance tolerance to water deficit and NaCl stress (Maheswari et al. 2010). Transgenic sorghum plants maintained a 1.7–2.8-fold higher shoot and root growth, respectively, at 200 mM NaCl stress compared to untransformed control plants and demonstrated that the engineering mannitol biosynthetic pathway into sorghum can impart enhanced tolerance to water deficit and salinity. Transgenic sweet sorghum with altered lignin content was obtained by manipulating the expression of *caffeoyl-CoA-O-methyltransferase* (*CCoAOMT*) and *caffeic acid-O-methyltransferase* (*COMT*) toward production of easily degradable plant material for biofuel production (Basu et al. 2007). A young forage crop of sorghum produces HCN in the leaf tissues proving toxic to animals on grazing. Low HCN transgenic forage lines containing safe levels (<200 ppm) of HCN were developed by Pandey et al. (2010) through downregulation of dhurrin gene.

9.6 Transformation Efficiency

High-efficient transformation systems are essential for GM product development as well as gene expression studies. Much effort was applied to extend the host range of *Agrobacterium* to monocotyledonous species. *Agrobacterium* infection of seedlings/immature embryos of cereal species was employed to evaluate the competency of these explants under various conditions. Following this *Agrobacterium*-mediated transformation, protocols were developed for many important monocotyledonous crops. Transformation frequency was improved when the parameters were further optimized, by modification of medium components and optimization of coculture and resting timing periods and by the addition of *Agrobacterium* growth-inhibiting agent or bactericide such as silver nitrate. Inclusion of silver nitrate in coculture medium enhanced stable transformation in maize (Zhao et al. 2001). Silver nitrate significantly suppressed the *Agrobacterium* growth during coculture without compromising T-DNA delivery and subsequent T-DNA integration. The suppressed *Agrobacterium* growth on the target explants could facilitate plant cell recovery and resulted in increased efficiency of transformation (Cheng et al. 2003).

To date, the published reports of successful monocot transformation *via Agrobacterium* used only three different strains, i.e., LBA 4404, disarmed C58, and EHA 101 and its derivatives (EHA105, AGL0, and AGL1). Anti-necrotic compounds such as ascorbic acid, cysteine, and silver nitrate for preincubation of the explants were emphasized to be useful for efficient transformation. Super-virulent strains or superbinary vectors and acetosyringone in inoculation and coculture media were suggested to be important for efficient transformation in cereal species. Chemicals such as acetosyringone for *vir* induction are recommended in most of the cereal transformation protocols. Antibiotics such as cefotaxime, carbenicillin, and timentin have been used regularly in the *Agrobacterium*-mediated transformation of cereal crops following coculture to suppress or eliminate *Agrobacterium*. Cefotaxime worked well in *Agrobacterium-mediated* transformation of rice and maize initially, and later it was found that cefotaxime at a concentration of 250 mg l^{-1} had a detrimental effect to maize callus (Ishida et al. 1996). Zhao et al. (2000, 2001) used carbenicillin (100 mg l^{-1}) in their experiments on sorghum and maize. Hill-Ambroz and Weeks (2001) studied the expression levels of various constitutive promoters in sorghum.

Removal of myoinositol from the callus culture media in combination with a cold-shock pretreatment and the addition of L-Gln prior to and during *Agrobacterium* infection resulted in about 84 % of the treated calli being stably transformed in *Lolium* (Zhang et al. 2013). Omission of myoinositol from the callus culture media was associated with the failure of certain pathogenesis-related genes to be induced after *Agrobacterium* infection. The addition of a cold-shock and supplemental Gln appeared to have synergistic effects on infection and transformation efficiencies. Nearly 60 % of the stably transformed calli regenerated into green plantlets. It is now possible to transform even difficult monocots using tailor-made gene constructs and promoters, suitable *A. tumefaciens* strains, and a proper understanding of the entire process of efficient regeneration (Sood et al. 2011). Still, there are many challenges to reach high-efficient transformation systems in sorghum that require genotype-independent regeneration and transformation methods.

9.7 Future Prospects

Despite the advances in technology, the transgenic sorghum is not released for cultivation. Innovative genetic modifications are developed to keep away from the classification of GMOs and biosafety concerns. These modifications such as zinc finger nuclease technology, cisgenesis, RdDM, and other technologies are dependent on transgenic technology for the introduction of DNA segments. Genotype-independent and high-throughput transformation systems are the key challenges to attain highly efficient genetic transformation of sorghum. Site-directed integration of transgenes in sorghum can be a reality with the complete sequence of sorghum genome available. Apart from this, selection of effective events plays an equally important role for transgenic sorghum to see the farmer fields. Science and the arts of tissue culture, genetic transformation, and, more importantly, plant breeding have to be integrated to reach effective transgenics of sorghum. There is growing interest in including sorghum as health food to alleviate the lifestyle disorders. With the research partnerships involving public and private organizations, we can anticipate improvement of sorghum nutritional quality with enhanced levels of vitamins, minerals, and protein and also tolerance to biotic and abiotic stresses leading to the genetic improvement of sorghum in the coming years.

References

Able JA, Rathus C, Godwin ID (2001) The investigation of optimal bombardment parameters for transient and stable transgene expression in sorghum. In Vitro Cell Dev Biol Plant 37:341–348

Amali P, Kingsley SJ, Ignacimuthu S (2014) Enhanced plant regeneration involving somatic embryogenesis from shoot tip explants of *Sorghum bicolor* (L. Moench). Asian J Plant Sci Res 4(3):26–34

Balakrishna D, Vinod R, Vijaya I, Padmaja PG, Venkatesh Bhat B, Patil JV (2013) *Agrobacterium*-mediated genetic transformation of sorghum [*Sorghum bicolor* (L.) Moench] using synthetic *CRY1B* gene. In: Proceedings of the global consultation on Millets promotion for health and nutritional security, Directorate of Sorghum Research, Hyderabad, 18–20 December 2013, pp 284

Basu A, Maiti MK, Kar S, Sen SK, Pandey B (2007) Transgenic sweet sorghum with altered lignin composition and process of preparation thereof. Nagarjuna Private Limited, Hyderabad, India. Patent no. WO/2008/102241

Battraw M, Hall TC (1991) Stable transformation of *Sorghum bicolor* protoplasts with chimeric *neomycin phosphotransferase* II and *b-glucuronidase* genes. Theor Appl Genet 82:161–168

Bhaskaran S, Smith RH, Paliwal S, Schertz KF (1987) Somaclonal variation from *Sorghum bicolor* (L.) Moench. cell culture. Plant Cell Tiss Org Cult 9:189–196

Brandão RL, Carneiro NP, de Oliveira AC, Coelho GTCP, Carneiro AA (2012) Genetic transformation of immature sorghum inflorescence *via* microprojectile bombardment. In: Yelda Ozden Çiftçi (ed) Transgenic plants – advances and limitations, PhD., In Tech, China

Cai T, Butler L (1990) Plant regeneration from embryogenic callus initiated from immature inflorescences of several high-tannin sorghums. Plant Cell Tiss Org Cult 20:101–110

Cai T, Daly B, Butler L (1987) Callus induction and plant regeneration from shoot portions of mature embryos of high tannin sorghum. Plant Cell Tiss Org Cult 9 245–252

Carvalho CHS, Zehr UB, Gunaratna N, Anderson J, Kononowicz HH, Hodges TK, Axtell JD (2004) *Agrobacterium*-mediated transformation of sorghum: factors that affect transformation efficiency. Genet Mol Biol 27:259–269

Casas AM, Kononowicz AK, Zehr UB, Tomes DT, Axtell JD, Butler LG, Bressan RA, Hasegawa PM (1993) Transgenic sorghum plants *via* microprojectile bombardment. Proc Natl Acad Sci U S A 90:11212–11216

Casas AM, Kononowicz AK, Haan TG, Zhang L, Tomes DT, Bressan RA, Hasegawa PM (1997) Transgenic sorghum plants obtained after microprojectile bombardment of immature inflorescences. In Vitro Cell Dev Biol Plant 33:92–100

Cheng M, Hu T, Layton J, Liu CN, Fry JE (2003) Desiccation of plant tissues post- *Agrobacterium* infection enhances T-DNA delivery and increases stable transformation efficiency in wheat. In Vitro Cell Dev Biol Plant 39:595–604

Devi P, Zhong H, Sticklen M (2004) Production of Transgenic sorghum plants with related *HVA1* gene. In: Seetharama N, Godwin ID (eds) Sorghum tissue culture and transformation. Oxford Publishers, New Delhi, pp 75–79

Elkonin LA, Pakhomova NV (2000) Influence of nitrogen and phosphorus on induction embryogenic callus of sorghum. Plant Cell Tiss Org Cult 61:115–123

Elkonin LA, Ravin NV, Leshko EV, Volokhina IV, Chumakov MI, Skryabin AG (2009) *In planta* agrobacterial transformation of sorghum plants. Biotekhnologiya 1:23–30

Emani C, Sunilkumar G, Rathore KS (2002) Transgene silencing and reactivation in sorghum. Plant Sci 162:181–192

Gao Z, Jayaraj J, Muthukrishnan S, Claflin L, Liang GH (2005a) Efficient genetic transformation of sorghum using a visual screening marker. Genome 48:321–333

Gao Z, Xie X, Ling Y, Muthukrishnan S, Liang GH (2005b) *Agrobacterium tumefaciens* mediated sorghum transformation using a mannose selection system. Plant Biotechnol J 3:591–599

Gendy C, Sene M, Van Le B, Vidal J, Van Tran TK (1996) Somatic embryogenesis and plant regeneration in *Sorghum bicolor* (L.) Moench. Plant Cell Rep 15:900–904

Girijashankar V, Swathisree V (2009) Genetic transformation of *Sorghum bicolor*. Physiol Mol Biol Plant 15(4):287–302

Girijashankar V, Sharma HC, Sharma KK et al (2005) Development of transgenic sorghum for insect resistance against the spotted stem borer (*Chilo partellus*). Plant Cell Rep 24(9):513–522

Grootboom AW, O'Kennedy MM, Mkhonza NL, Kunert K, Chakauya E, Chikwamba RK (2008) *In vitro* culture and plant regeneration of sorghum genotypes using immature zygotic embryos as explant source. Int J Bot 4:450–455

Grootboom AW, Mkhonza NL, O'Kennedy MM, Chakauya E, Kunert K, Chikwamba RK (2010) Biolistic mediated sorghum (*Sorghum bicolor* L. Moench) transformation *via* Mannose and Bialaphos based selection systems. Int J Bot 6(2):89–94

Gupta S, Khanna VK, Singh R, Garg GK (2006) Strategies for overcoming genotypic limitations of in vitro regeneration and determination of genetic components of variability of plant regeneration traits in sorghum. Plant Cell Tiss Org Cult 86:379–388

Gurel S, Gurel E, Kaur R, Wong J, Meng L, Tan HQ, Lemaux PG (2009) Efficient reproducible *Agrobacterium*-mediated transformation of sorghum using heat treatment of immature embryos. Plant Cell Rep 28(3):429–444

Hagio T (2002) Adventitious shoot regeneration from immature embryos of sorghum. Plant Cell Tiss Org Cult 68:65–72

Hagio T, Blowers AD, Earle ED (1991) Stable transformation of sorghum cell cultures after bombardment with DNA coated microprojectiles. Plant Cell Rep 10(5):260–264

Harshavardhan D, Rani TS, Ugalanathan K, Seetharama N (2002) An improved protocol for regeneration of *Sorghum bicolor* from isolated shoot apices. Plant Biotechnol 19(3):163–171

Hill-Ambroz KL, Weeks JT (2001) Comparison of constitutive promoters for sorghum transformation. Cereal Res Commun 29:17–24

Howe A, Sato S, Dweikat I, Fromm M, Clemente T (2006) Rapid and reproducible *Agrobacterium-mediated* transformation of sorghum. Plant Cell Rep 25(8):784–791

Ignacimuthu S, Premkumar A (2014) Development of transgenic *Sorghum bicolor* (L.) Moench resistant to the *Chilo partellus* (Swinhoe) through *Agrobacterium-mediated* transformation. Mol Biol Genet Eng. doi:http://dx.doi.org/10.7243/2053-5767-2-1

Indukuri V (2014) Introgression of *cry1B* gene into sweet sorghum (SSV 84) by marker-assisted backcross breeding. Department of Botany, Andhra University, Waltair

Ishida Y, Saito H, Ohta S, Hiei Y, Komari T, Kumashiro T (1996) High efficiency transformation of maize (*Zea mays* L.) mediated by *Agrobacterium tumefaciens*. Nat Biotechnol 14(6):745–750

Jeoung JM, Krishnaveni S, Muthukrishnan S, Trick HN, Liang GH (2002) Optimization of sorghum transformation parameters using genes for green fluorescent protein and beta-glucuronidase as visual markers. Hereditas 137(1):20–28

Jogeswar G, Ranadheer D, Anjaiah V, Kavi Kishor PB (2007) High frequency somatic embryogenesis and regeneration in different genotypes of *Sorghum bicolor* (L.) Moench from immature inflorescence explants. In Vitro Cell Dev Biol Plant 43:159–166

Kaeppler HF, Pederson JF (1996) Media effects on phenotype of callus cultures initiated from photoperiod-insensitive, elite inbred sorghum lines. Maydica 41(2):83–89

Kaeppler HF, Pederson JF (1997) Evaluation of 41 elite and exotic inbred sorghum genotypes for high quality callus production. Plant Cell Tissue Organ Cult 48:71–75

Kononowicz AK, Casas AM, Tomes DT, Bressan RA, Hasegawa PM (1995) New vistas are opened for sorghum improvement by genetic transformation. Afr Crop Sci J 3(2):171–180

Kosambo-Ayoo LM, Bader M, Loerz H, Becker D (2011) Transgenic sorghum (*Sorghum bicolor* L. Moench) developed by transformation with *chitinase* and *chitosanase* genes from *Trichoderma harzianum* expresses tolerance to anthracnose. Afr J Biotechnol 10:3659–3670

Krishnaveni S, Jeoung JM, Muthukrishnan S, Liang GH (2001) Transgenic sorghum plants constitutively expressing a rice *chitinase* gene show improved resistance to stalk rot. J Genet Breed 55:151–158

Kumar V, Campbell LM, Rathore KS (2011) Rapid recovery and characterization of transformants following *Agrobacterium*-mediated T-DNA transfer to sorghum. Plant Cell Tiss Org Cult 104:137–146

Liu G, Godwin ID (2012) Highly efficient sorghum transformation. Plant Cell Rep 31(6):999–1007

Lu L, Wu X, Yin X, Morrand J, Chen X, Folk WR, Zhang ZJ (2009) Development of marker-free transgenic sorghum [*Sorghum bicolor* (L.) Moench] using standard binary vectors with bar as a selectable marker. Plant Cell Tiss Org Cult 99:97–108

Maheswari M, Varalaxmi Y, Vijayalakshmi A, Yadav SK, Sharmila P, Venkateswarlu B, Vanaja M, Saradhi PP (2010) Metabolic engineering using *mtlD* gene enhances tolerance to water deficit and salinity in sorghum. Biol Plant 54(4):647–652

Mall TK (2010) Evaluation of novel input output traits in sorghum through biotechnology. Dissertation, Agronomy, University of Nebraska-Lincoln, Lincoln, Nebraska

Murty UR, Visarada AA, Bharathi M (1990a) Developing tissue culture system for sorghum, *Sorghum bicolor* (L.) Moench. embryogenic callus induction from elite genotypes. Cereal Res Commun 18(3):257–262

Murty UR, Visarada AA, Bharathi M (1990b) Developing tissue culture system for sorghum, *Sorghum bicolor* (L.) Moench. plant regeneration from embryogenic callus. Cereal Res Commun 18(4):355–358

Mythili PK, Seetharama N, Reddy VD (1999) Plant regeneration from embryogenic cell suspension cultures of wild sorghum (*Sorghum dimidiatum* Stapf.). Plant Cell Rep 18:424–428

Nguyen TV, Thu TT, Claeys M, Angenon G (2007) *Agrobacterium*-mediated transformation of sorghum (*Sorghum bicolor* (L.) Moench) using an improved *in vitro* regeneration system. Plant Cell Tiss Org Cult 91:155–164

Oldach KH, Morgenstern A, Rother S, Girgi M, O'Kennedy MM, Lorz H (2001) Efficient *in vitro* plant regeneration from immature zygotic embryos of pearl millet [*Pennisetum glaucum* (L.) R. Br.] and *Sorghum bicolor* (L.) Moench. Plant Cell Rep 20(5):416–421

Pandey AK, Bhat BV, Balakrishna D, Seetharama N (2010) Genetic transformation of sorghum (*Sorghum bicolor* (L.) Moench.). Int J Biotech Biochem 6(1):45–53

Pola S (2011) Leaf discs as a source material for plant tissue culture studies of *Sorghum bicolor* (L.) Moench. Notulae Scientia Biologicae 3(1):70–78

Pola S, Sarada Mani N, Ramana T (2007) Enhanced shoot regeneration in tissue culture studies of *Sorghum bicolor*. J Agric Technol 3:275–286

Pola S, Sarada Mani N, Ramana T (2008) Plant tissue culture studies in *Sorghum bicolor*: immature embryo explants as the source material. Int J Plant Prod 2:1–14

Pola S, Sarada Mani N, Ramana T (2009) Long-term maintenance of callus cultures from immature embryo of *Sorghum bicolor*. World J Agric Sci 5(4):415–421

Prasad Sant RR (2011) Development of a transformation system for sorghum (*Sorghum bicolor* L.). Ph.D. thesis, Centre for Tropical Crops and Biocommodities, Queensland University of Technology, Brisbane, Australia

Raghuwanshi A, Birch RG (2010) Genetic transformation of sweet sorghum. Plant Cell Rep 29(9):997–1005

Raju RK, Nagaraju K, Annapurna A, Srinivas R, Gawali HS, Rao SV, Visarada KBRS (2007) Simple methods to generate ample explant sources for genetic transformation of sorghum. In: Poster presented at 8th agricultural science congress held during 15–17 Feb. 2007 at Tamil Nadu Agricultural University (TNAU), Coimbatore, India

Rao AM, Padma Sree K, Kavikishore PB (1995) Enhanced plant regeneration in grain and sweet sorghum by asparagine, praline and cefotaxime. Plant Cell Rep 15:72–75

Rathus C, Nguyen T, Able JA, Gray SJ, Godwin ID (2004) Optimizing sorghum transformation technology *via* somatic embryogenesis. In: Seetharama N, Godwin ID (eds) Sorghum tissue culture, transformation and genetic engineering. ICRISAT and Oxford Publishers, New Delhi

Ratnala V (2013) Genetic transformation of sorghum for stem borer resistance using *cry1Aa* gene *via* *Agrobacterium*-mediated approach. Dissertation, Department of Botany, Andhra University, Waltair, India

Rose JB, Dunwell JM, Sunderland N (1986) Anther culture of *Sorghum bicolor*. Plant Cell Tiss Org Cult 6:15–32

Sai Kishore N, Aravinda Lakshmi Y, Pashupatinath E, Ramana Kumari B, Balakrishna D, Rao SV, Seetharama N, Visarada KBRS (2004) A simple system to transgenic sorghum using shoot apical meristem. In: Proceedings of the national symposium biohorizon 2004, 12–13 March, 2004, IIT, Delhi, New Delhi, India

Sai Kishore N, Visarada KBRS, Aravinda Lakshmi Y, Pashupatinath E, Rao SV, Seetharama N (2006) *In vitro* culture methods in sorghum with shoot tip as the explant material. Plant Cell Rep 25(3):174–182

Sai Kishore N, Visarada KBRS, Rao SV, Seetharama N (2011) Progress and prospects for *Agrobacterium-mediated* genetic transformation in sorghum in comparison to other cereals. Transgenic Plant J 5(1):27–34

Sairam RV, Seetharama N, Devi PS, Verma A, Murthy UR, Potrykus I (1999) Culture and regeneration of mesophyll derived protoplasts of sorghum (*Sorghum bicolor* (L.) Moench). Plant Cell Rep 18:972–977

Seetharama N, Sairam RV, Rani TS (2000) Regeneration of sorghum from shoot tip cultures and field performance of the progeny. Plant Cell Tiss Org Cult 61:169–173

Smith RH, Bhaskaran S, Schertz K (1983) Sorghum plant regeneration from aluminum selection medium. Plant Cell Rep 2(3):129–132

Sood P, Bhattacharya A, Sood A (2011) Problems and possibilities of monocot transformation. Biol Plant 55:1–15

Syamala D, Devi P (2003) Efficient regeneration of sorghum, *Sorghum bicolor* (L.) Moench, from shoot-tip explant. Indian J Exp Biol 41(12):1482–1486

Tadesse Y, Jacobs M (2004) Nutritional quality improvement of sorghum through genetic transformation. In: Seetharama N, Godwin ID (eds) Sorghum tissue culture and transformation. Oxford Publishers, New Delhi, pp 81–84

Tadesse Y, Sagi L, Swennen R, Jacobs M (2003) Optimization of transformation conditions and production of transgenic sorghum (*Sorghum bicolor*) *via* microprojectile bombardment. Plant Cell Tiss Org Cult 75:1–18

Urriola J, Rathore KS (2014) Temporal and spatial activities of a rice glutelin promoter in transgenic sorghum. Plant Cell Tiss Org Cult 116:227–234

Visarada KBRS, Sai Kishore N, Balakrishna D, Rao SV (2003) Transient *gus* expression studies in sorghum to develop a simple protocol for *Agrobacterium*-mediated genetic transformation. J Genet Breed 57:147–154

Visarada KBRS, Rajani G, Prasad GS, Royer M (2013) Development of transgenic sweet sorghum for tolerance to stem borer. Sorghum Times 9:6

Visarada KBRS, Padmaja PG, Sai Kishore N, Pashupatinath E, Royer M, Seetharama N, Patil JV (2014) Production and evaluation of transgenic sorghum for resistance to stem borer. In Vitro Cell Dev Biol Plant 50:176–189

Wang W, Wang J, Yang C, Li Y, Liu L, Xu J (2007) Pollen-mediated transformation of *Sorghum bicolor* plants. Biotechnol Appl Biochem 48:79–83

Wu E, Lenderts B, Glassman K, Berezowska-Kaniewska M, Christensen H, Asmus T, Zhen S, Chu U, Cho MJ, Zhao ZY (2014) Optimized *Agrobacterium*-mediated sorghum transformation protocol and molecular data of transgenic sorghum plants. In Vitro Cell Dev Biol Plant 50:9–18

Zhang W, Dewey R, Boss W, Phillippy B, Qu R (2013) Enhanced *Agrobacterium*-mediated transformation efficiencies in monocot cells is associated with attenuated defense responses. Plant Mol Biol 81:273–286

Zhao ZY, Cai T, Tagliani L et al (2000) *Agrobacterium-mediated* sorghum transformation. Plant Mol Biol 44(6):789–798

Zhao ZY, Gu W, Cai T, Tagliani L, Hondred D, Bond D, Schroeder S, Rudert M, Pierce D (2001) High throughput genetic transformation mediated by *Agrobacterium tumefaciens* in maize. Mol Breed 8:323–333

Zhao L, Liu S, Song S (2010) Optimization of callus induction and plant regeneration from germinating seeds of sweet sorghum (*Sorghum bicolor* Moench). Afr J Biotechnol 9(16):2367–2374

Zhong H, Wang W, Sticklen M (1998) *In vitro* morphogenesis of *Sorghum bicolor* (L.) Moench: efficient plant regeneration from shoot apices. J Plant Physiol 153(5–6):719–726

Zhu H, Muthukrishnan S, Krishnaveni S, Wilde G, Jeoung JM, Liang GH (1998) Biolistic transformation of sorghum using a rice *chitinase* gene. J Genet Breed 52(3):243–252

Genetic Engineering for Novel Traits

10

B. Venkatesh Bhat, D. Balakrishna, P. Madhu, and J.V. Patil

Contents

10.1	**Introduction**	217
10.2	**Input Traits**	218
10.2.1	Resistance to Biotic Stresses	218
10.2.2	Herbicide Tolerance	219
10.2.3	Tolerance to Abiotic Stresses	220
10.3	**Output Traits**	221
10.3.1	Improvement in Sorghum Grain	221
10.3.2	Improvement in Sorghum Biomass and Fodder	223
10.4	**Conclusion**	224
References		224

Abstract

Accumulation of knowledge on genetic control of plant phenotypic traits in model plants and advances in plant genetic engineering have unravelled new opportunities for the genetic improvement of crops like sorghum and provided the means to manipulate them for human and environmental benefits. Trait-based breeding in sorghum needs to broadly address either those traits that enhance input efficiency or that enhance the quality and quantity of outputs. Improved resistance to biotic and abiotic stresses is the main input trait, whereas improvement of sorghum grain and biomass yields along with superior quality and amenability for efficient utilisation form the target output traits. The potential genetic manipulation approaches and candidate genes for each of these traits have been discussed in this chapter.

Keywords

Sorghum • Biotic stresses • Abiotic stresses • Grain protein • Minerals • Biomass

10.1 Introduction

Sorghum is the fifth most important cereal grain crop in the world. It feeds over 500 million people in the developing world, as an important source of energy, protein and some vitamins and

B.V. Bhat (✉)
Genetics & Cytogenetics, ICAR-Indian Institute of Millets Research, Rajendranagar, Hyderabad, Telangana 500 030, India
e-mail: bhatv@millets.res.in

D. Balakrishna
Biotechnology, ICAR-Indian Institute of Millets Research, Rajendranagar, Hyderabad, Telangana 500 030, India

P. Madhu
ICAR-Indian Institute of Millets Research, Rajendranagar, Hyderabad, Telangana 500 030, India

J.V. Patil
Plant Breeding, ICAR-Indian Institute of Millets Research, Rajendranagar, Hyderabad, Telangana 500 030, India
e-mail: hari@millets.res.in

R. Madhusudhana et al. (eds.), *Sorghum Molecular Breeding*, DOI 10.1007/978-81-322-2422-8_10

minerals. In the West, it is basically used as an animal feed and is a prospective candidate for ethanol production. Being a gluten-free cereal with more fibre, sorghum grain is of interest as food for people with celiac disease. Being a C_4 cereal and widely adapted, sorghum is becoming increasingly important, as drought and global warming are impacting the cereal production.

Novel traits, as defined by the Convention on Biological Diversity (CBD), are characteristics in an organism that have been created or introduced through a specific genetic change by genetic engineering that make it different from the unmodified organism. Novel traits of potential value in sorghum are of different types, such as those expected to be useful for efficient input utilisation and those that increase productivity, quality of output, diversified utilisation of the outputs, etc. These target traits can be broadly categorised as input traits and output traits.

10.2 Input Traits

A critical trait for any breeding programme is yield. Input traits are designed to boost or stabilise crop yield. Addressing yield directly through transgenic approaches is a considerable challenge. A more practical and obtainable goal in the short term is protection of yield through control of biotic and abiotic stresses. Examples include drought and salinity resistance, herbicide resistance, insect resistance and fungal and viral disease resistance.

10.2.1 Resistance to Biotic Stresses

10.2.1.1 Resistance to Insect Pests

Sorghum production can be severely impacted by a number of insect pests. Insect infestation not only impacts production but is known to provide an entry for secondary pathogen attack (including fungi and viruses) at the site of feeding or damage. Demonstrated molecules to develop genetically modified plants with novel insect-resistance genes include those from *Bacillus thuringiensis* (*Bt-Cry* genes), enzyme inhibitors (such as soybean trypsin inhibitors or SBTI), ribosome-inactivating proteins (RIPs), lectins, vegetative insecticidal proteins (VIPs), etc. Out of these, the *Bt* Cry proteins which are toxic to insects, but safe to other organisms, have been investigated extensively. Most of the transgenics with these *Bt* Cry proteins showed effective resistance against specific target insects and were found safe to non-target organisms including humans. The success of the *Bt* technology in maize (Armstrong et al. 1995; Barry et al. 2000), and cotton (Cattaneo et al. 2006) is a strong rationale for the evaluation of this technology in sorghum to combat specific target insects. In addition, this technology had a positive secondary effect in terms of significant reduction in accumulation of mycotoxins in plant tissues (Abbas et al. 2008; Bakan et al. 2002; Hammond et al. 2004). Tolerance against spotted stem borer was reported in sorghum transgenics carrying the *Cry1Aa* (Visarada et al. 2014), *Cry1Ac* (Girijashankar et al. 2005), *Cry1B* (Visarada et al. 2014) and *Cry1C* (Ignacimuthu and Premkumar 2014). The *Bt* technology can be sustainably utilised in conjunction with appropriate integrated pest management practices to maximise its durability over time (Kumar and Pandey 2008).

10.2.1.2 Resistance to Fungal Pathogens

Viruses and fungi that infect crop plants are a serious threat to the livelihoods of farmers in developed countries. Fungal diseases of sorghum that include foliar diseases, grain molds and downy mildew can be controlled by identifying suitable antifungal molecules that impart resistance to the infecting fungi. Transgenic plants have been produced by inserting antifungal genes to confer resistance against fungal pathogens. Genes of fungal cell wall-degrading enzymes, such as chitinase and glucanase, are frequently used to produce fungal-resistant transgenic crop plants (reviewed by Ceasar and Ignacimuthu 2012). Apart from chitinase and β-1,3-glucanase, many other antimicrobial proteins or peptides were also effective in conferring disease resistance in transgenic plants.

Three different antifungal genes were introduced in rice by various groups: the trichosanthin gene (*TCS*) by Xiaotian et al. (2000), an antifungal protein (*afp*) gene of *Aspergillus giganteus* by Coca et al. (2004) and synthetically prepared antifungal genes *Ap-CecA* and *ER-CecA* by Coca et al. (2006). Fungal-resistant finger millet (Latha et al. 2005) and pearl millet (Latha et al. 2006) were developed by inserting a synthetic antifungal protein gene *PIN* of prawn which had high homology to the cationic, antimicrobial, lytic peptide cecropin A. The other antifungal genes expressed in various plants are *hS2* gene-encoding chitinase-like protein in creeping bentgrass (Chai et al. 2002), N-terminally modified antimicrobial cationic peptide temporin-A gene in potato (Osusky et al. 2004) and mustard defensin gene in tobacco and peanut (Anuradha et al. 2008). These studies suggest that in addition to chitinase and glucanase of diverse origin, other antifungal genes can be used for developing fungi-resistant plants.

Protection of sorghum against stalk rot through the introduction of a rice *chitinase* gene was demonstrated by Zhu et al. (1998) and Krishnaveni et al. (2000). Ayoo et al. (2011) attempted to develop transgenic sorghum plants with resistance to anthracnose disease by introducing genes encoding proteins such as chitinases and chitosanases that hydrolyse fungal cell wall. They demonstrated the effect of genetic background on resistance to anthracnose in transgenics. They suggested that these transgenes could be utilised to pyramid genes for higher tolerance to anthracnose in sorghum.

10.2.1.3 Resistance to Viral Diseases

Important viral diseases of sorghum include maize stripe virus and maize dwarf mosaic viruses that cause substantial yield losses, once infected. A number of viral agents have been shown to be capable of replication in sorghum (Jensen and Giorda 2002), including members of the *Potyvirus* family such as sugarcane mosaic virus, maize dwarf mosaic virus and sorghum mosaic virus. However, only a few sorghum germplasm lines with limited resistance towards these viruses could be identified by Henzell et al. (1982). The pioneering work that demonstrated the feasibility of conferring resistance to plant viruses by the introduction of viral coat protein genes in transgenic plants (Stark and Beachy 1989) has opened up the possibility of achieving resistance against major viruses in plants. Implementing various genetic constructs that target silencing of critical gene products required for the replication of the target virus has been of interest in many a crop plants (Beachy et al. 2003; Prins 2003), including known pathogens of sorghum (Gilbert et al. 2005). Hence, such strategies offer great potential for the introduction of durable virus resistance for sorghum against viral diseases caused by maize dwarf mosaic virus and maize stripe viruses, especially in association with sorghum shoot bug (*Peregrinus maidis*) infestation.

10.2.1.4 Resistance to Parasite *Striga*

Striga or witchweed is a parasitic weed on sorghum and contains two species, *Striga hermonthica* and *S. asiatica* (Aly 2007). A study worth attention for *Striga* resistance by genetic engineering is that of targeting critical genes in the parasitic plant's life cycle by expression of RNAi constructs in the host plant which resulted in an enhanced tolerance phenotype in the *Orobanche aegyptiaca*/tomato host parasite interaction (Aly et al. 2009). However, this approach was not successful in controlling *Striga*/maize parasite interaction (Yoder and Scholes 2010). This calls for intensive research to further the understanding of the biology involved during the early stages of parasitism by *Striga*.

10.2.2 Herbicide Tolerance

Herbicide-tolerant GM crops have been developed to simplify weed control and to cut input costs. They enable farmers to use a single herbicide instead of many, reducing application costs. They also simplify crop rotation and improve farm safety, because the herbicides that are used with them degrade rapidly in the soil and are less poisonous to humans than those used on conventional crops. Herbicide tolerance has now been engineered into many crop species, including varieties of oilseed rape, maize, soybeans, sugar beet, fodder beet, cotton and rice. Genes for resistance to herbicide such as dicamba and glyphosate are good candidates for engineering resistance in sorghum. However, herbicide-resistant sorghums have already developed by private sector in the USA using mutations. The double-barrel approach of the acetolactate synthase herbicides, with both grass and broadleaf activity, also makes it a good new tool for farmers (Green and Owen 2011).

10.2.3 Tolerance to Abiotic Stresses

The ability of crops to tolerate abiotic stresses, such as drought, salinity and extreme temperatures, is likely to become increasingly important as the world population increases, competition with other land uses pushes agriculture into harsher environments, fresh water becomes scarcer and climate change predicted by some scientists increases environmental stress. The genetic basis for abiotic stress tolerance is complex but some genetic modification approaches have shown promising results. For example, plants often respond to abiotic stresses by changing their metabolism to produce sugars or similar compounds that act as osmoprotectants. One such compound is trehalose, a disaccharide similar to sucrose. Trehalose levels have been increased in GM rice by over-expressing genes encoding trehalose biosynthetic enzymes from the bacterium *E. coli* (Garg et al. 2002). This resulted in plants that showed improved performance under salt, drought and low-temperature stress conditions. Another possible solution to the problem of salt pollution, which affects millions of acres of otherwise fertile land, usually as a result of irrigation, involves the over-expression of a gene that encodes a vacuolar Na+/H+ antiport pump (Apse and Blumwald 2002). This increases the rate at which a plant can remove salt from its cytoplasm and transfer to its vacuole. Tomato plants modified in this way can tolerate salt concentrations several times higher than non-GM plants and should survive in the salt concentrations of soils that are currently considered unusable. Furthermore, the fruit does not accumulate salt and is edible. Similar technologies are being developed to address the problem of contamination of soils with heavy metals.

Several attempts have been made to verify the expression of candidate genes for imparting abiotic stress tolerance in sorghum. Transgenics expressing transcription factor-coding genes *DREB* (Dubouzet et al. 2003), *MYC, MYB* (Abe et al. 2003) and *WRKY* (Wang et al. 2007) were reported to show improved stress tolerance in crop plants, but no sorghum transgenics for these transcription factors has been reported so far. Transformation of sorghum with signalling intermediates such as calcium-dependent protein kinases did not help in improving abiotic stress tolerance, probably due to pleiotropic effects of this general signalling intermediate (Mall et al. 2011). Probably a battery of transgenes that include key transcription factors and important genes belonging to different cascades of abiotic stress response makes a good proposition to achieve usable tolerance to abiotic stresses.

10.2.3.1 Multi-genic Abiotic Stress

Addressing a plant's response to stresses that are governed in a multi-genic fashion is more challenging than single-gene traits. In order to investigate multi-genic abiotic stress response traits such as drought and heat, researchers are evaluating a coordinated expression of a suite of genes triggered by exposure to the targeted stress by the introduction of a single transcription factor (Suzuki et al. 2005; Karaba et al. 2007; Nelson et al. 2007). These transcription factor-based technologies hold great promise as a means to reduce multi-genic expressed phenotypes to a single transgene fashion (Century et al. 2008). However, the transcription factor-based strategy undoubtedly will require tight regulation, necessitating the need for tissue-specific and/or inducible promoter systems.

10.2.3.2 Nitrogen Assimilation

With respect to adaptation to low nitrogen environments, Yanagisawa et al (2004) demonstrated that expression of the maize *Dof1* transcription factor improved nitrogen assimilation in transgenic plants. However, it is feasible to directly perturb nitrogen flux in plants. Nitrogen assimilation and metabolism in plants occurs through coordinated action of a variety of enzymes acting upon a variety of substrates. Two key enzymes involved in nitrogen metabolism in plants are glutamine synthetase (GS) and glutamate synthase (GOGAT). Previous studies have shown that enhancing GS or GOGAT activities can impact nitrogen metabolism in plant species (Good et al. 2004; Cai et al. 2009). Enhancing the activity of another enzyme that impacts nitrogen flux in plants, alanine aminotransferase (Ala-AT) that catalyses the production of alanine and 2-oxoglutarate from pyruvate and glutamate has been shown to augment nitrogen-use efficiency in both rapeseed and rice (Good et al. 2007; Shrawat et al. 2008). Similar studies are required in sorghum for enhancing nitrogen assimilation. While these works aimed at enhancing nitrogen-use efficiency have gathered data sets from greenhouse or growth-chamber studies, there is very limited information on the impact of the transgenes on yield under field conditions (Brauer and Shelp 2010).

10.3 Output Traits

Output traits refer to all downstream factors such as quality enhancement and traits that produce functional foods and optimisation of food, feed and raw materials for consumers, industry and medicine. Transgenics for use in the production of functional foods are yet to reach market maturity. Current research and development includes the creation of vitamin-enriched potatoes, ‘golden rice’ containing beta-carotene (Paine et al. 2005) and apples and strawberries containing protein that acts as a prophylactic to reduce dental caries (Gillor et al. 2005). These GMOs and others like them raise hopes that public acceptance of agro-genetic engineering can be increased because their altered-use traits should mean tangible benefits for consumers. In sorghum, transgenics are being developed for improved nutritional traits in the grain and fodder, enhancing output and conversion efficiency for biofuel processing and attractiveness for industrial and food processing uses.

10.3.1 Improvement in Sorghum Grain

10.3.1.1 Grain Proteins

Sorghum grain composition, digestibility and bioavailability of nutrients are of paramount significance as sorghum is a major staple food for millions in Africa and Asia and an important livestock feed grain in developed countries. Sorghum endosperm proteins are known to have equal or lower in vitro pepsin digestibility than other cereals in raw flour and substantially lower digestibility in cooked products, reducing the bioavailability of the protein (reviewed by Duodu et al. 2003). The major sorghum proteins, prolamins, found in sorghum reside in the endosperm and are designated as kafirins. The reasons for the lower protein digestibility of cooked sorghum are multi-factorial, including the extensive polymerisation of the kafirins upon cooking and the location and organisation of the different kafirin sub-classes in the protein bodies (Duodu et al. 2003). Hence, modulation of the prolamins is a target that could be pursued in sorghum as a means to simultaneously address digestibility and nutritional quality. Oria et al. (2000) described a highly digestible, enhanced lysine sorghum mutant. The protein bodies observed within this mutant are highly folded, with a redistribution of the γ-kafirin around the body. These factors lead to increased exposure of the core α-kafirins, which translates to the increased digestibility phenotype (Duodu et al. 2003). Further understanding of the underlying biology governing protein deposition in these mutants and the influence of the various genetic modifiers will help to pinpoint candidate genes and genetic elements for genetic engineering, without negatively altering the endosperm characteristics.

Another approach to improve sorghum grain protein nutritional quality involved transformation of improved sorghum lines to suppress the synthesis of different kafirin sub-classes or backcrossing them with transgenic lines with improved protein quality (Silva et al. 2011). Co-suppression of the alpha-, gamma- and delta-kafirin sub-classes and removal of the tannin trait resulted in transgenic sorghum lines with high cooked protein digestibility. These high protein quality lines had a floury endosperm and possessed modified protein body structure, where the protein bodies were irregularly shaped with few to numerous invaginations and were less densely packed, with a dense protein matrix visible around the protein bodies. When fewer sub-classes were suppressed, i.e. gamma 1 and delta 2, the endosperm was corneous with normal protein body structure but the improvement in cooked protein digestibility appeared to be less. Apparently, co-suppression of several kafirin sub-classes is required to obtain high protein nutritional quality sorghum lines, but this seems to result in floury-type grain endosperm texture (Silva et al. 2011). Kumar et al. (2012) reported transgenic sorghum events that are downregulated in the γ- and the 29-kDa α-kafirins which showed that downregulation of γ-kafirin alone does not alter protein body formation or impact protein digestibility of cooked flour samples. They found that reduction in accumulation of a predicted 29-kDa α-kafirin altered the morphology of protein body and enhanced protein digestibility.

10.3.1.2 Lysine Content

The sorghum grain is poor in lysine content, limiting its value as food and feed. Grootboom (2010) employed an RNAi co-suppression strategy that resulted in 45.23 and 77.55 % increase in whole-seed and endosperm lysine, respectively. The co-suppression RNAi constructs targeted the endosperm-specific suppression of three lysine-poor storage proteins, namely, δ-kaf-2, γ- kaf-1 and −2 and an enzyme that catalyses seed lysine degradation, lysine keto-glutarate reductase (LKR). The transgenic co-suppression of the target kafirins resulted in the endosperm structural change from a hard, corneous endosperm to a soft, floury endosperm, consistent with γ-zein suppression in the opaque-2 maize mutant (Grootboom 2010).

10.3.1.3 Wheat Quality for Bread Making

The ability to make bread and a range of other processed foods (pasta, noodles, cakes, biscuits, etc.) from wheat flour, but not from other cereal flours, is determined by the unique properties of the grain storage proteins. These are deposited in discrete protein bodies in the cells of the starchy endosperm but coalesce to form a continuous matrix, or network, in the cell during the later stages of grain maturation. When flour is mixed with water to form dough, the gluten proteins in the individual flour particles come together to form a continuous network in the dough. The co-expression of wheat *1Ax1* sub-unit with puroindoline by transformation increased the dough strength in wheat cultivars (Li et al. 2012). Hence, it is proposed to transfer the wheat glutenin gene *1Ax1* into sorghum to make it amenable to bread making suitable for the bakery industry.

10.3.1.4 Vitamin A Content

Enhancing vitamin A production in plants may be accomplished by addition or enhancement of genes involved in carotenoid biosynthesis. For example, β-carotene synthesis reached as high as 37 μg/g in golden rice 2 with *phytoene synthase (PSY)* gene from maize and *carotene desaturase (CRT-I)* from *Erwinia uredovora* (Paine et al. 2005). Transgenic sorghum lines with the same genes as in golden rice (but controlled by sorghum kafirin promoters) in addition to sorghum low-phytic acid 1 (LPA-1) gene, help to improve zinc and iron bioavailability from the sorghum grains. The results of in vitro digestion studies in these sorghum transgenics revealed enhanced total and bioaccessible provitamin A carotenoid levels in sorghum (Tristan et al. 2013).

10.3.1.5 Content and Bioavailability of Minerals

Plant seeds are potentially important sources of minerals for nutrition of humans and livestock, but a high proportion of the minerals present is unavailable as they are in the form of mixed salts of phytic acid (myoinositol-1,2,3,4,5,6-hexakisphosphate). Thus, phytate accounts for over 70 % of the total phosphorus as well as substantial amounts of Mg^{2+}, K^{+}, Fe^{3+}, Zn^{2+}, Ca^{2+} and Cu^{2+}. Phytates act as storage reserves in the seed and are degraded during germination. For example, phytin granules are abundant in the embryo and aleurone of cereal grains. However, animals cannot digest phytate and consequently it is excreted. The low availability of calcium, iron and zinc in cereals and other plant foods can also contribute to nutritional deficiency in humans, particularly women and children in developing countries. Genetic engineering can be used to digest the phytin and increase the mineral availability in seeds, by expression of genes encoding phytase. Genes from *Aspergillus* species have been used for this as they express extracellular phytase enzymes and have been produced commercially. The expression of phytase has been reported in crops such as soybean (Denbow et al. 1998), oilseed rape (Zhang et al. 2000) and wheat (Brinch-Pedersen et al. 2003), using the *phyA* gene from *Aspergillus niger*. Feeding studies with transgenic soybean showed a 50 % reduction in phosphate excretion by broiler chickens (Denbow et al. 1998), while feeding transgenic canola to piglets and broilers showed similar positive effects on growth to those achieved by supplementation with exogenous phytase enzyme (Zhang et al. 2000). Drakakaki et al. (2005) also expressed the phytase from *Aspergillus niger* in maize using the rice glutelin promoter and a CaCo-2 cell model to show increased availability and uptake of iron. A similar approach in sorghum can enhance increased availability and uptake of these minerals.

An alternative or complementary approach is to increase the amounts of other iron-binding compounds in the seed. Ferritin is an iron-binding protein which provides a storage reserve of iron in plants, bacteria and animals (Theil 1987). The expression of ferritin genes from soybean (Goto et al. 1998, 1999) and *Phaseolus* (Lucca et al. 2001) in developing seeds of rice has been shown to result in two- to threefold increase in the iron content of the grain, demonstrating the feasibility of using genetic engineering to increase iron availability. A similar increase in iron accumulation occurred when the soybean ferritin was expressed in rice grain, but this was associated with decreased iron in the leaves (Qu et al. 2005). Hence, accumulation may ultimately be limited by iron uptake and transport. Finally, Drakakaki et al. (2005) showed that combined expression of soybean ferritin and fungal phytase in rice resulted in a 20–70 % increase in the iron content of maize seeds. Thus, ferritin makes the case for a potential transgene for sorghum genetic engineering.

10.3.2 Improvement in Sorghum Biomass and Fodder

Obvious target traits for improvement of sorghum as a bioenergy crop include enhancing biomass and its conversion efficiency, removal of factors that reduce the quality of biomass and improvement for harvesting, processing, storage and transportation of biomass. Many of these traits such as low lignin, more sugars, higher cellulose content, etc. are available in the primary gene pool itself and the genes for low lignin (brown midrib trait) has been transferred to several sorghum lines with good biomass potential. In the short to medium term, no promising transgene of interest for biomass improvement is on the horizon.

Transgene-based improvement of sorghum fodder is similar to enhancing biomass, but additional traits such as fodder quality, ability to regenerate after cutting (ratoon-like), palatability and silage-making ability are important. Many of these traits are multi-genic in nature. Some of the anti-quality traits are also present in sorghum fodder. One of the major anti-quality factors of sorghum fodder is the cyanogenic (HCN-producing) glycoside dhurrin in stem and leaves that is toxic to the feeding livestock. A study aimed at obtaining sorghum plants with reduced expression levels of the *CYP79A1* gene using antisense approach was undertaken (Bhat et al. 2009). Transgenics with *CYP79A1* antisense gene were developed and stable T_3 plants exhibited HCN as low as 7 ppm with an overall mean of 70 ppm compared to mean HCN of 221 ppm in the control (Pandey 2010). The study effectively demonstrated that the antisense strategy was effective in producing sorghum plants with low cyanogenic potential.

10.4 Conclusion

The output of sorghum is fast approaching a plateau after successful exploitation of genetic diversity and heterotic potential in the primary gene pool, especially in terms of grain and biomass yield. Genetic engineering is beneficially applied to improve those aspects of sorghum production and processing and utilisation where no effective donors are available in the primary gene pool and management practices are of no avail. The constant evolution of technologies such as genetic engineering has enabled introgression of alien genes as well as regulating gene expression for deriving beneficial outputs. As more candidate genes and regulatory elements are unravelled by the advances in metabolomics and genomics, possibilities of applying genetic engineering for beneficial production and utilisation of sorghum would increase in the future. Well-documented instances of genetic engineering adding to productivity, quality and utilisation of plants are available in major crops and model plant species. Sorghum genetic engineering is still in its beginnings, and release and regulatory mechanisms for GM crops in major sorghum-growing countries are yet to be available and streamlined. The present status thus calls for adding to the repertoire of useful traits for sorghum genetic engineering by selection of appropriate candidate genes and their verification and testing their efficacy in containment conditions. Simultaneously, efforts to achieve required expression levels of endogenous genes either by over-expression or by downregulation need to be attempted by genetic engineering as these may call for precision experimentation and thorough testing to ensure optimum expression and absence of undesirable phenotypes/effects in diverse environments.

References

Abbas HK, Accinelli C, Zablotowicz RM, Abel CA, Bruns HA, Dong Y, Shier WT (2008) Dynamics of mycotoxin and *Aspergillus flavus* levels in aging *Bt* and non-*Bt* corn residues under Mississippi no-till conditions. J Agric Food Chem 56:7578–7585

Abe H, Urao T, Ito T, Seki M, Shinozaki K, Yamaguchi-Shinozaki K (2003) *Arabidopsis AtMYC2* (*bHLH*) and *AtMYB2* (*MYB*) function as transcriptional activators in abscisic acid signaling. Plant Cell 15:63–78

Aly R (2007) Conventional and biotechnological approaches for control of parasitic weeds. In Vitro Cell Dev Biol-Plant 43:304–317

Aly R, Cholakh H, Joel DM, Leibman D, Steinitz B, Zelcer A, Naglis A, Yarden O, Gal- On A (2009) Gene silencing of mannose 6-phosphate reductase in the parasitic weed *Orobanche aegyptiaca* through the production of homologous dsRNA sequences in the host plant. Plant Biotechnol J 7:487–498

Anuradha TS, Divya K, Jami SK, Kirti PB (2008) Transgenic tobacco and peanut plants expressing a mustard defensin show resistance to fungal pathogens. Plant Cell Rep 27:1777–1786

Apse MP, Blumwald E (2002) Engineering salt tolerance in plants. Curr Opin Biotechnol 13:146–150

Armstrong CL, Parker GB, Pershing JC et al (1995) Field evaluation of European corn borer control in progeny of 173 transgenic corn events expressing an insecticidal protein from *Bacillus thuringiensis*. Crop Sci 35:550–557

Ayoo LMK, Bader M, Loerz H, Becker D (2011) Transgenic sorghum (*Sorghum bicolor L. Moench*) developed by transformation with chitinase and chitosanase genes from *Trichoderma harzianum* expresses tolerance to anthracnose. Afr J Biotechnol 10(19):3659–3670

Bakan B, Melcion D, Richard-Molard D, Cahagnier B (2002) Fungal growth and *Fusarium* mycotoxin content in isogenic traditional maize and genetically modified maize grown in France and Spain. J Agric Food Chem 50:728–731

Barry BD, Darrah LL, Huckla DL, Antonio AQ, Smith GS, O'Day MH (2000) Performance of transgenic corn hybrids in Missouri for insect control and yield. J Econ Entomol 93:993–999

Beachy RN, Fraley RT, Rogers SG (2003) Protection of plants against viral infection. Monsanto Technology LLC and Washington University, United States patent no. 6,608, 241, 19 Aug 2003

Bhat BV, Pandey, AK, Ramya S, Pushpa K, Balakrishna D, N Seetharama (2009) *CYP79A1* antisense gene for reducing HCN toxicity in forage sorghum. In: National symposium on "Emerging Trends in Forage Research and Livestock Production" organized by Society for Forage Research during February 16–17, 2009 at Jaisalmer, Rajasthan, India

Brauer EK, Shelp BJ (2010) Nitrogen use efficiency: re-consideration of the bioengineering approach. Botany 88:103–109

Brinch-Pedersen H, Hatzack F, Sorensen LD, Holm PB (2003) Concerted action of endogenous and heterologous phytase on phytic acid degradation in seeds of transgenic wheat (*Triticum aestivum* L.). Transgenic Res 12:649–659

Cai H, Zhou Y, Xiao J, Li X, Zhang Q, Lian X (2009) Over-expressed glutamine synthetase gene modifies nitrogen metabolism and abiotic stress responses in rice. Plant Cell Rep 28:527–537

Cattaneo MG, Yafuso C, Schmidt CW et al (2006) Farm-scale evaluation of the impacts of transgenic cotton on biodiversity, pesticide use, and yield. Proc Natl Acad Sci U S A 103:7571–7576

Ceasar SA, Ignacimuthu S (2012) Genetic engineering of crop plants for fungal resistance: role of antifungal genes. Biotechnol Lett 34(6):995–1002

Century K, Reuber TL, Ratcliffe OJ (2008) Regulating the regulators: the future prospects for transcription-factor-based agricultural biotechnology products. Plant Physiol 147:20–29

Chai B, Maqbool SB, Hajela RK, Green D, Vargas JM, Warkentin D, Sabzikar R, Sticklen MB (2002) Cloning of a chitinase-like cDNA *(hs2)*, its transfer to creeping bentgrass (*Agrostis palustris* Huds.) and development of brown patch (*Rhizoctonia solani*) disease resistant transgenic lines. Plant Sci 163:183–193

Coca M, Bortolotti C, Rufat M, Peñas G, Eritja R, Tharreau D, Martinez del Pozo A, Messeguer J, Segundo SB (2004) Transgenic rice plants expressing the antifungal AFP protein from *Aspergillus giganteus* show enhanced resistance to the rice blast fungus *Magnaporthe grisea*. Plant Mol Biol 54:245–259

Coca M, Peñas G, Gómez J, Campo S, Bortolotti C, Messeguer J, San Segundo B (2006) Enhanced resistance to the rice blast fungus *Magnaporthe grisea* conferred by expression of a *cecropin A* gene in transgenic rice. Planta 223:392–406

Denbow DM, Grabau EA, Lacy GH, Kornegay ET, Russell DR, Umbeck PF (1998) Soybeans transformed with a fungal phytase gene improve phosphorus availability for broilers. Poult Sci 77:878–881

Drakakaki G, Marcel S, Glahn RP, Lund EK, Pariagh S, Fischer R, Christou P, Stoger E (2005) Endosperm-specific co-expression of recombinant soybean ferritin and *Aspergillus* phytase in maize results in significant increases in the levels of bioavailable iron. Plant Mol Biol 59:869

Dubouzet JG, Sakuma Y, Ito Y, Kasuga M, Dubouzet EG, Miura S, Seki M, Shinozaki K, Yamaguchi- Shinozaki K (2003) *OsDREB* genes in rice, *Oryza sativa* L., encode transcription activators that function in drought-, high-salt- and cold-responsive gene expression. Plant J 33:751–763

Duodu KG, Taylor JRN, Belton PS, Hamaker BR (2003) Factors affecting sorghum protein digestibility. J Cereal Sci 38:117–131

Garg AK, Kim J-K, Owens TG, Ranwala AP, Choi YD, Kochian LV, Wu RJ (2002) Trehalose accumulation in rice plants confers high tolerance levels to different abiotic stresses. Proc Natl Acad Sci U S A 99:15898–15903

Gilbert RA, Gallo-Meagher M, Comstock JC, Miller JD, Jain M, Abouzid A (2005) Agronomic evaluation of sugarcane lines transformed for resistance to sugarcane mosaic virus strain E. Crop Sci 45:2060–2067

Gillor O, Nigro LM, Riley MA (2005) Genetically engineered bacteriocins and their potential as the next generation of antimicrobials. Curr Pharm Des 11(8):1067–1075

Girijashankar V, Sharma HC, Sharma KK et al (2005) Development of transgenic sorghum for insect resistance against the spotted stem borer (*Chilo partellus*). Plant Cell Rep 24:513–522

Good AG, Shrawat AK, Muench DG (2004) Can less yield more? Is reducing nutrient input into the environment compatible with maintaining crop production? Trends Plant Sci 9:597–605

Good AG, Johnson SJ, De Pauw M, Carroll RT, Savidov N, Vidmar J, Lu Z, Taylor GJ, Stroeher V (2007) Engineering nitrogen use efficiency with alanine aminotransferase. Can J Bot 85:252–262

Goto F, Yoshihara T, Saiki H (1998) Iron accumulation in tobacco plants expressing the soybean ferritin gene. Transgenic Res 7:173–180

Goto F, Yoshihara T, Shigemoto N, Toki S, Takaiwa F (1999) Iron fortification of rice seed by the soybean ferritin gene. Nat Biotechnol 17:282–286

Green JM, Owen MDK (2011) Herbicide-resistant crops: utilities and limitations for herbicide-resistant weed management. J Agric Food Chem 59(11):5819–5829

Grootboom AW (2010) Effect of RNAi down-regulation of three lysine-deficient kafirins on the seed lysine content of sorghum [*Sorghum bicolor* (L.) Moench]. Unpublished Ph.D. dissertation, Faculty of Natural and Agricultural Sciences, University of Pretoria, South Africa

Hammond BG, Campbell KW, Pilcher CD, Degooyer TA, Robinson AE, McMillen BL, Spangler SM, Riordan SG, Rice LG, Richard JL (2004) Lower fumonisin mycotoxin levels in the grain of *Bt* corn grown in the United States in 2000–2002. J Agric Food Chem 52:1390–1397

Henzell RG, Persley DM, Greber RS, Fletcher DS, Van Slobbe L (1982) Development of grain sorghum lines with resistance to sugarcane mosaic and other sorghum diseases. Plant Dis 6:900–901

Ignacimuthu S, Premkumar A (2014) Development of transgenic *Sorghum bicolor* (L.) Moench resistant to the *Chilo partellus* (Swinhoe) through *Agrobacterium-mediated* transformation. Mol Biol Genet Eng 2:1. http://dx.doi.org/10.7243/2053-5767-2-1

Jensen SG, Giorda LM (2002) Virus diseases of sorghum and millet in the Americas and Australia. In: Leslie JF (ed) Sorghum and millets diseases. Iowa State Press, Ames, pp 403–410

Karaba A, Dixit S, Greco R, Aharoni A, Trijatmiko KR, Marsch-Martinez N, Nataraja K, Udayakumar M, Pereira A (2007) Improvement of water use efficiency in rice by expression of *HARDY*, an *Arabidopsis* drought and salt gene. Proc Natl Acad Sci U S A 104:15270–15275

Krishnaveni S, Jeoung JM, Muthukrishnan S, Liang GH (2000) Transgenic sorghum plants constitutively expressing a rice chitinase gene show improved resistance to stalk rot. J Genet Breed 55:151–158

Kumar S, Pandey KC (2008) *Bacillus thuringiensis* (*Bt*) transgenic crop: an environment friendly insect-pest management strategy. J Environ Biol 29:641–653

Kumar T, Dweikat I, Sato S, Ge Z, Nersesian N, Chen H, Elthon T, Bean S, Loerger BP, Tilley M, Clemente T (2012) Modulation of kernel storage proteins in grain sorghum (*Sorghum bicolor* (L.) Moench). Plant Biotechnol J 10(5):533–544

Latha MA, Dasvantha Reddy V, Madhavi Latha A, Venkateswara Rao K (2005) Production of transgenic plants resistant to leaf blast disease in finger millet (*Eleusine coracana* (L.) Gaertn.). Plant Sci 169:657–667

Latha MA, Rao KV, Reddy TP, Reddy VD (2006) Development of transgenic pearl millet (*Pennisetum glaucum* (L.) R. Br.) plants resistant to downy mildew. Plant Cell Rep 25:927–935

Li Y, Wang Q, Li X et al (2012) Co-expression of the high molecular weight glutenin sub-unit 1Ax1 and puroindoline improves dough mixing properties in durum wheat (*Triticum turgidum* L. ssp. *durum*). PLoS One 7(11):e50057

Lucca P, Hurrell R, Potrykus I (2001) Genetic engineering approaches to improve the bioavailability of the level of iron in rice grains. Theor Appl Genet 102:392–397

Mall TK, Dweikat I, Sato SJ, Neresian N, Xu K, Ge Z, Wang D, Elthon T, Clemente T (2011) Expression of the rice *CDPK-7* in sorghum: molecular and phenotypic analyses. Plant Mol Biol 75(4–5):467–479

Nelson DE, Repetti PP, Adams TR et al (2007) Plant nuclear factor Y (NF-Y) B sub-units confer drought tolerance and lead to improved corn yields on water-limited acres. Proc Natl Acad Sci U S A 104:16450–16455

Oria MP, Hamaker BR, Axtell JD, Huang CP (2000) A highly digestible sorghum mutant cultivar exhibits a unique folded structure of endosperm protein bodies. Proc Natl Acad Sci U S A 97:5065–5070

Osusky M, Osuska L, Hancock RE, Kay W, Misra S (2004) Transgenic potatoes expressing a novel cationic peptide are resistant to late blight and pink rot. Transgenic Res 13:181–190

Paine JA, Shipton CA, Chaggar S et al (2005) Improving the nutritional value of Golden Rice through increased pro-vitamin A content. Nat Biotechnol 23(4):482–487

Pandey AK (2010) Developing low-HCN producing forage sorghum (*Sorghum bicolor* [L.]) using anti-sense approach. Ph.D. thesis. Jawaharlal Nehru Technological University, Hyderabad, India

Prins M (2003) Broad virus resistance in transgenic plants. Trends Biotechnol 21:373–375

Qu LQ, Yoshihara T, Ooyama A, Goto F, Takaiwa F (2005) Iron accumulation does not parallel the high expression level of ferritin in transgenic rice seeds. Planta 222:225–233

Shrawat AK, Carroll RT, DePauw M, Taylor GJ, Good AG (2008) Genetic engineering of improved nitrogen use efficiency in rice by the tissue-specific expression of alanine aminotransferase. Plant Biotechnol J 6:722–732

Silva LS, Jung R, Zhao Z, Glassman K, Taylor J, Taylor JRN (2011) Effect of suppressing the synthesis of different kafirin sub-classes on grain endosperm texture, protein body structure and protein nutritional quality in improved sorghum lines. J Cereal Sci 54:160–167

Stark DM, Beachy RN (1989) Protection against potyvirus infection in transgenic plants: evidence for broad spectrum resistance. Nat Biotechnol 7:1257–1262

Suzuki N, Rizhsky L, Liang H, Shuman J, Shulaev V, Mittler R (2005) Enhanced tolerance to environmental stress in transgenic plants expressing the transcriptional coactivator multiprotein bridging factor 1c1[w]. Plant Physiol 139:1313–1322

Theil EC (1987) Ferritin: structure, gene regulation and cellular function in animals, plants, and microorganisms. Annu Rev Biochem 56:289–315

Tristan EL, De Moura FF, Zhao ZY, Albertsen MC, Che P, GlassmanK FMG (2013) Bioaccessibility of carotenoids from transgenic provitamin A biofortified sorghum. J Agric Food Chem 61(24):5764–5771

Visarada KBRS, Padmaja PG, Saikishore N, Pashupatinath E, Royer M, Seetharama N, Patil JV (2014) Production and evaluation of transgenic sorghum for resistance to stem borer. In Vitro Cell Dev Biol-Plant 50:176–189

Wang H, Hao J, Chen X, Hao Z, Wang X, Lou Y, Peng Y, Guo Z (2007) Over-expression of rice *WRKY89* enhances ultraviolet B tolerance and disease resistance in rice plants. Plant Mol Biol 65:799–815

Xiaotian M, Lijiang W, Chengcai AN, Huayi Y, Zhangliang C (2000) Resistance to rice blast *(Pyricularia oryzae)* caused by the expression of trichosanthin gene in transgenic rice plants transferred through *Agrobacterium* method. Chin Sci Bull 45:1774–1778

Yanagisawa S, Akiyama A, Kisaka H, Uchimiya H, Miwa T (2004) Metabolic engineering with Dof1 transcription factor in plants: improved nitrogen assimilation and growth under low-nitrogen conditions. Proc Natl Acad Sci U S A 101:7833–7838

Yoder JI, Scholes JD (2010) Host plant resistance to parasitic weeds: recent progress and bottlenecks. Curr Opin Plant Biol 13(4):478–484

Zhang ZB, Kornegay ET, Radcliffe JS, Wilson JH, Veit HP (2000) Comparison of phytase from genetically engineered *Aspergillus* and canola in weanling pig diets. J Anim Sci 78:2868–2878

Zhu H, Muthukrishnan S, Krishnaveni S, Wilde G, Jeoung JM, Liang GH (1998) Biolistic transformation of sorghum using a rice chitinase gene. J Genet Breed 52:243–252